T0392931

Suspension Plasma Spray Coating of Advanced Ceramics

Suspension Plasma Spray Coating of Advanced Ceramics presents the significance of suspension plasma spray coating of ceramics for thermal barrier applications. It covers suspension formation and optimization in different oxide and non-oxide mixtures and ceramic matrix composites (CMC) of submicron- and nanosized powders.

Enabling readers to understand the importance of thermally inert and insulating ceramic coatings on metals and alloys, the book explains how to improve their utilization in applications, such as in turbine blades or diesel engines, gas turbines, and coating methods. This book also discusses advanced topics on nanomaterials coatings in monolithic or composite forms as thermal barriers through organic- and nonorganic-based suspensions using high-energy plasma spray methods.

Features:

- Presents significant thermal barrier properties using high-energy plasma spray methods
- Explores advanced surface modification techniques
- Covers monolithic, composite, and solid solution ceramics coating
- Discusses high-precision coating methods

The book will be useful for professional engineers working in surface modification and for researchers studying materials in science and engineering, corrosion, and abrasion.

Suspension Plasma Spray Coating of Advanced Ceramics

Thermal Barrier Applications

Navid Hosseinabadi
Hossein Ali Dehghanian

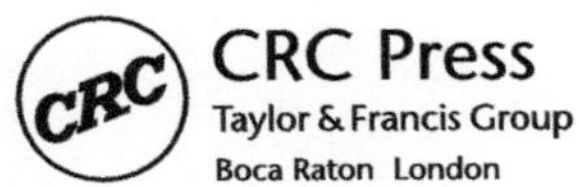

CRC Press is an imprint of the
Taylor & Francis Group, an informa business

First edition published 2022
by CRC Press
6000 Broken Sound Parkway NW, Suite 300, Boca Raton, FL 33487–2742

and by CRC Press
4 Park Square, Milton Park, Abingdon, Oxon, OX14 4RN

CRC Press is an imprint of Taylor & Francis Group, LLC

Library of Congress Cataloging-in-Publication Data
Names: Hosseinabadi, Navid, author. | Dehghanian, Hossein Ali, author.
Title: Suspension plasma spray coating of advanced ceramics : thermal barrier applications / Navid Hosseinabadi and Hossein Ali Dehghanian.
Description: First edition. | Boca Raton, FL : CRC Press, 2023. | Includes bibliographical references and index.
Summary: "Suspension Plasma Spray Coating of Advanced Ceramics presents the significance of suspension plasma spray coating of ceramics for thermal barrier applications. It covers suspension formation and optimization in different oxide and non-oxide mixtures and ceramic matrix composites (CMC) of sub-micron and nanosized powders. The book will be useful for professional engineers working in surface modification and researchers studying materials science. This book discusses advanced topics on nanomaterials coatings in monolithic or composite forms as thermal barriers through organic and non-organic based suspensions using high energy plasma spray methods"— Provided by publisher.
Identifiers: LCCN 2022010390 (print) | LCCN 2022010391 (ebook) | ISBN 9781032257853 (hbk) | ISBN 9781032257877 (pbk) | ISBN 9781003285014 (ebk)
Subjects: LCSH: Plasma spraying. | Ceramic materials—Protection. | Glazing (Ceramics) | Thermal barrier coatings.
Classification: LCC TS695.15 .H67 2023 (print) | LCC TS695.15 (ebook) | DDC 621.044—dc23/eng/20220601
LC record available at https://lccn.loc.gov/2022010390
LC ebook record available at https://lccn.loc.gov/2022010391

ISBN: 978-1-032-25785-3 (hbk)
ISBN: 978-1-032-25787-7 (pbk)
ISBN: 978-1-003-28501-4 (ebk)

DOI: 10.1201/9781003285014

Typeset in Times LT Std
by Apex CoVantage, LLC

Dedication

Navid Hosseinabadi would like to dedicate this book to
His wife, son, and parents.

Hossein Ali Dehghanian would like to dedicate this book to
Sara, Mani, Maria, and his parents.

Contents

General List of Symbols

a	particle radius [m]
a_i	particle radius of species/size i [m]
A	Hamaker constant [J]
c	mass concentration [kg m^{-3}]
D	rate-of-strain tensor [s^{-1}]
D_f	fractal dimension [-]
D_{ij}	components of the rate-of-strain tensor [s^{-1}]
D	diffusivity tensor [m^2 s^{-1}]
D_0	Stokes-Einstein-Sutherland diffusivity, Eq. (1.5) [m^2 s^{-1}]
D_{ij}	components of the diffusivity tensor [m^2 s^{-1}]
D_r	rotational diffusivity [s^{-1}]
$D_{r,0}$	limiting rotational diffusivity for zero volume fraction [s^{-1}]
D^s	self-diffusivity tensor [m^2 s^{-1}]
$D^{s\ ij}$	components of the self-diffusivity tensor [m^2 s^{-1}]
D^{ss}	short-time self-diffusion coefficient [m^2 s^{-1}]
E	elasticity modulus [Pa]
e	electronic charge [C]
F	force [N]
g	gravity or acceleration constant [m s^{-2}]
$g(r)$	radial distribution function [-]
G	modulus [N m^{-2}]
G'	storage modulus [N m^{-2}]
G''	loss modulus [N m^{-2}]
G_{pl}	plateau modulus [N m^{-2}]
h	surface-to-surface distance between particles [m]
I	unit tensor [-]
k	coefficient in the power law model [N s^n m^{-1}]
k'	coefficient [s]
k_B	Boltzmann's constant [J K^{-1}]
k_H	Huggins coefficient [-]
L	length [m]
m	power law index in Cross model [-]
n	number density [m^{-3}]
N	number of particles [-]
N_A	Avogadro's number [mol^{-1}]
N_i	ith normal stress difference [Pa]
P	pressure [Pa]
P_y	compressive yield stress [Pa]
q	scattering vector [nm^{-1}]
R	radius [m]
R_g	radius of gyration [m]
r	distance from center of particle [m]

S	entropy [J K^{-1}]
t	time [s]
T	temperature [K]
U	relative velocity between particles [m s^{-1}]
v	local speed [m s^{-1}]
V	volume [m^3]
v	velocity vector [m]
v_i	velocity component in the i direction; i = x, y, or z [m s^{-1}]
W	stability ratio [-]
W^{shear}	stability ratio for shear-induced cluster formation [-]
x	Cartesian coordinate, in simple shear flow, the flow direction [m]
y	Cartesian coordinate, in simple shear flow, the velocity gradient direction [m]
z	Cartesian coordinate, in simple shear flow, the vorticity direction [m]
II_i	second invariant of tensor i

SUBSCRIPTS

eff	effective
el	elastic contribution
ext	extensional
floc	floc
g	glass
gel	gel
lin	linearity limit
m	suspending medium/mean value
M	Maxwell
max	maximum value
p	particle
pl	plastic
r	relative
y	yield condition
0	limiting value in the zero-shear limit
∞	limiting value at high shear rate or frequency

SUPERSCRIPTS

B	Brownian, with yield stress Bingham
C	Casson
d	dispersion
g	gravity
h	hydrodynamic
hcY	hard-core Yukawa (potential)
hs	hard sphere
H	Herschel-Bulkley

I	interparticle contribution
m	power law index in Cross model
n	power law index for shear stress
s	surface
*	complex

DIMENSIONLESS NUMBERS

Bo	Boussinesq number,
De	Deborah number (ratio of characteristic material time to characteristic process time)
Ha	Hartmann number
Mn	Mason number
Mn_{mag}	magnetic Mason number
Pe_D	Péclet number for microrheology,
Pe_i	Péclet number for the ions,
Pe	Péclet number for microrheology,
Re	Reynolds number
Re_p	particle Reynolds number
St	Stokes number
Wi	Weissenberg number

BASIC PHYSICAL CONSTANTS AND VALUES

Constant		Value
e	Elementary charge	$1.602\ 176\ 487 \times 10^{-19}$ C
g	Standard acceleration of gravity	$9.806\ 65$ m s^{-2}
k_B	Boltzmann's constant	$1.380\ 650\ 4 \times 10^{-23}$ J K^{-1}
m_u	Atomic mass unit	$1.660\ 538\ 782 \times 10^{-27}$ kg
N_A	Avogadro's number	$6.022\ 214\ 170 \times 10^{23}$ mol^{-1}
R	Molar gas constant	$8.314\ 472$ J mol^{-1} K^{-1}
ε_0	Electric permittivity of vacuum	$8.854\ 187\ 817 \times 10^{-12}$ C^2 N^{-1} m^{-2} [F m^{-1}]
0	Vacuum permeability	4×10^{-7} N A^{-2}

Preface

The suspension plasma spray, as one of the favorable methods employed for fabrication of high-performance, highly adhered, mechanically enhanced, chemically stable corrosion resistance oxide and non-oxide ceramic thermal barrier coatings, is a fast-growing route in academic and industrial research and development circles. This method has solidified its applications due to special microstructures formation abilities and numerous controlling parameters to create coatings with micro- and nano-sized phase distribution, vertical cracks or columnar structures, uniformly distributed porosity, and evenly spaced cracks oriented normal to the substrate surface and alike. Along with plasma spraying specification, understanding what affects the rheological aspects of ceramic suspensions, the flow behavior, and how this flow behavior can be optimized for suspension plasma spray applications is essential for successful formulation and performance. This book is focused on the study of colloidal rheology of ceramic suspensions in different aspects to present a profound understanding of flowability effects on coating parameters. The book focuses on a brief introduction to basic concepts of thermodynamics of hot oxidation and oxidation (Chapter 1), the general review of thermal barrier coatings (Chapter 2), the main principles of colloidal science and the concepts of rheology (Chapter 3), the disciplines of suspension stabilization (Chapter 4), and the suspension-property dependency of suspension plasma spray thermal barrier coatings (Chapter 5). The role of main rheological behavior of ceramic suspensions on plasma spray coating physicomechanical and chemical properties with microstructural profiles is reviewed in detail. The aim of the book is to provide both the fundamentals-practical aspects and the performance of the dispersions, and it should serve as a valuable text for those scientists in the industry who deal with the formulation of ceramic suspension of plasma spraying application, and also as an introduction to those researchers and developers in research and industry.

February 2022, Navid Hosseinabadi
Hosseinali Dehghanian

1 The Fundamentals of Hot Corrosion and High-Temperature Oxidation

1.1 INTRODUCTION

In this chapter we will illustrate the fundamentals and basics of high-temperature corrosion and high-temperature oxidation. Different types of high-temperature corrosions as well as the different methods of corrosion prevention will be discussed in detail, and the consecutive relations and educations will be discussed as well. Then thermal barrier coatings (TBCs) will be introduced as a practical physicochemical resistance coating. After a brief introduction to TBCs, mechanisms of barrier will be discussed. To design and fabricate high-performance TBCs for high-temperature facilities, different viewpoints have to be brought, which include thermal barrier performance and bonding characteristics of the coating with sublayer.

1.2 FUNDAMENTALS

After the industrial revolution in the eighteenth century, with the invention of the steam engine, new pioneer industries encountered a new class of degradation that no one had experienced up to that date. Since the consequences of high-temperature corrosion were not obvious, it was a surprise to engineers and scientists. But nowadays, this type of corrosion can be influenced by the performance and lifetime of devices that most people deal with every day. For example, car engines (pistons and cylinders, catalyst carriers), airplane engines, powerhouses, reactors, petrochemical and gas processes industries, etc. The technical importance of some of those applications was discovered around the beginning of the twentieth century, and that was why high-temperature corrosion started to gain attention in literature at that time. The first investigation on high-temperature corrosion was started with a paper by Langmuir working on the temperature and mechanisms of oxidation of Tungsten between 800 K and 1,200 K [1]. The oxidation rate law on high-temperature corrosion was extracted by Tamman in 1920. In his silver/iodine system, the blurring layers on the metal rise according to the following formulation:

$$\frac{dy}{dt} = \frac{p}{y} \qquad \text{and} \qquad y = 2pt \tag{1.1}$$

DOI: 10.1201/9781003285014-1

Tamman used *y* for the layer thickness and *p* as the constant, which is related to diffusion coefficient *D* of iodine in silver/iodine. He described the following relation for the concentration gradient:

$$2p = aD\frac{dx}{dt} \tag{1.2}$$

In 1926, Dunn showed the aforementioned parabolic equation only applies for the oxide films whose properties remain constant by consuming time. He then realized that there must be a more fundamental relation in oxidation and corrosion. So he presented the expression for the variation of diffusion rate with temperature on the basis of thermal agitation in solids. Until 1979, many investigators had worked on oxidation and corrosion at high temperature. In that time, the role of mechanical aspects on high-temperature corrosion was shown in an overview paper [2]. From that time until now, high-temperature applications have surged rapidly due to improved thermodynamic efficiency and alloy improvement, single-crystal and directional castings have increased significantly, but coating processes have started to expand rapidly, especially in gas industries.

The improvement of strength and mechanical properties of base metals was the first reason to coat them. Another reason was to avoid the penetration of heat to base metals with thermal barrier coatings (TBCs), which were made of ceramic with very low thermal conductivity.

1.2.1 High-Temperature Corrosion and Oxidation

Depending on the temperature degree, there are three degradation processes: oxidation and type I and type II corrosion.

1.2.1.1 Type I

It returns to the penetration of sulfur ion to the metal substrate, which leads to the formation of stable oxide layers, and since temperature seems to occur with this type at almost 700–950°C, the reaction between these layers and moving sulfurs causes catastrophic progressions in the interface of layers and molten metals. Although no alloys are resistant to high-temperature corrosion, there are certain alloys which may postpone the degradation process. Superalloys have been developed for high-temperature applications, but there is a contradiction between the corrosion resistance of metals in high temperature and their strength properties [3–5].

1.2.1.2 Type II

This type occurs between 600 and 800°C, which implies sulfates formation from the substrate at some partial pressures. This phenomenon forms low melting point particles that degrade and prevent the corrosion of protective layers.

1.2.1.3 Oxidation

As the temperature rises, the importance of oxidation increases. This type of oxidation is used to determine the reaction between metal and air or oxygen in the absence of water. Material oxidation takes place above 950°C. The rate of process strongly depends on ions' movement activity in the oxide layer and grain boundary. The low dissociation pressure of aluminum and chrome leads to the formation of protective uniform, stable oxide layer. Because of the volatile nature of chromia, aluminum is added in severe conditions to protect the material from degradation. In the case of thermal cycling, the oxide layer is exfoliated due to thermal stress induced. So corrosion resistance can be improved by adding the reactive elements, such as hafnium and yttrium to coatings and alloys. These elements stick to grain boundaries with alumina and diminish the oxidation rate [4–6].

From the thermodynamic viewpoint, the activity of metallic elements, the partial pressure of oxygen in the environment, and the relative affinity to oxygen control oxide formation in the alloy or coating.

In contact with oxygen, oxides of all elements in the alloy will be formed in the condition where oxygen pressure in the gas is higher than the equilibrium pressure for each element. So thermodynamically stable oxide phases can be formed, which results in selective oxidation of specific elements in the alloy or coating. During the formation of different oxide films, only the thermodynamically stable ones remain and some transient phases disappear. The different flakes that form on the surface of the alloy in the oxidation atmosphere contain different stable oxides. The formation of these flakes varies considerably from one alloy to another.

With investigations on kinetics, it can be seen that all stable phases are formed. Then, diffusion processes, such as displacement reactions, begin. The velocity of oxygen penetration in alloy in comparison with element diffusion plays a key role on the configuration of the oxide layer from continuous to dispersed oxide films. Of course, the dispersed oxide films are preferable in working conditions [7]. If oxygen diffuses faster than metallic elements, the volume fraction of oxides may almost be the same with the elements. On the other hand, when oxygen penetrates at a slower rate than do other elements, the volume fraction of the oxide can be more than that of the alloy. In this situation, lateral growth of the film is considered and a uniform, continuous oxide film is formed.

In the case of reaction kinetics, test condition should be the same as the application under consideration, but unfortunately, application conditions are often not precisely known and, even when known, can be extremely difficult to establish as a controlled test. True simulation testing is usually impractical because the desired performance period is generally much longer than the length of time for laboratory testing that is feasible.

Theoretically, the continuous film may eventually disrupt due to cracks from thermal stresses. After, the disruption oxidation process leads to the formation of more stable oxide phases than before, and then the alloy is gradually discharged from the elements which are selectively oxidized.

1.2.1.3.1 Oxidation Mechanisms

Considering the general reaction of oxidation $M + \frac{1}{2}O_2 \rightarrow MO$, it is clear that reaction production in interface of MO and M leads to the detachment of ingredients from the anion reactions:

$$\mathrm{M} + \mathrm{O}^{2-} = \mathrm{MO} + 2\mathrm{e}^{-}$$

$$\frac{1}{2}O_2 + 2e^{2-} = O^{2-}$$

And in cathodic reactions:

$$M = M^{2+} + 2e^-$$

$$M^{2+} + 2e^- + \frac{1}{2}O_2 = MO$$

If anions are dominant in diffusing species, reaction at scale/metal interface continues and scale grows inward. Conversely, if cations are dominant in diffusing species, reaction at scale/gas interface happens and scale grows outward.

As all metallic oxides have ionic nature, monoecious atoms cannot diffuse out from the shell.

There are several mechanisms to justify the transfer of ions through ionic solids, which can be divided into two general categories:

- Mechanisms that afford to stoichiometric crystals
- Mechanisms that afford to nonstoichiometric crystals

There are Schottky and Frankel defects in stoichiometric crystals that can be reviewed in literature. As oxides may exhibit marked deviation from stoichiometry, they can even be unstable. Oxides may contain appreciable concentrations of both two cation and oxygen defects. Oxygen defects relate to anions, and oxides relate to cation defects.

The real-time concentration of different defects is also a function of temperature as well as oxygen pressure. Most of the time, oxidation at high temperature leads to the formation of oxide films. The mechanism of oxidation depends on the nature of metals. The oxide state physically, including solid, liquid, and gas, depends on the temperature and pressure of oxygen.

In solid shell, the solid behavior relates to what the oxide looks like. It means if the shell is compact, it acts as a discriminant layer of metal and the environment. If there is enough oxygen near the surface, the continuation of the process is controlled by diffusion. Compacted layers show good protective properties; therefore, to increase the oxidation resistance, the protective properties of the shells should be increased as well.

What has led to the understanding of the mechanism is the theory proposed by Wagner concerning the parabolic oxidation of metals at high temperatures.

1.2.1.3.2 Wagner Theory

Some basic principles should be considered in the parabolic oxidation of high temperature. These bases are for interpreting or improving the oxidation of metals and are not limited to the oxidation of metals, and they can be used for the metal-gas reaction in general as well.

1. It is used for compressed oxide shells, and hence, it is adhered to the surface (with resistance properties).
2. Volumetric ion diffusion or transfer of electrons from a growing oxide layer is a speed controller.
3. Because the penetration from the shell shown is controlling, the reactions that take place at the interface are fast reactions.
4. There is a thermodynamic equilibrium along the shell and also in the interface between oxygen and oxide, and between flow and oxide, this interaction is established.
5. The formed shell is relatively thick.
6. The solubility of oxygen in metal can be neglected.

If a reaction between a pure metal and a series of oxidants takes place, it is assumed that a single-phase, high-density, uniform oxide layer is formed on the surface.
The following conditions are required for this:

- One kind of defects is present.

Thermodynamic equilibrium is present at the metal/oxide and oxygen/oxide interface.
The oxide shell has a slight deviation from the stoichiometric state.

- The solubility of oxygen in the shell is low.
- Oxide layer does not evaporate.
- Other events should not happen, and if they do, it can be ignored.

Using Fick's rules, it can be shown that speed is parabolic. According to this fact, equilibrium exists at the interfaces and thickness does not alter with time, so we have:

$$\frac{\partial}{\partial x} J = \circ \tag{1.3}$$

And this flux depends only on time and is independent of place.

$$\frac{\partial C_i}{\partial t} = -D_i \frac{\partial^2 C_i}{\partial x^2} \tag{1.4}$$

$$J_i = -D_i \frac{\partial C_i}{\partial x} \tag{1.5}$$

From the aforementioned equations and after simplification, we have:

$$\frac{dx}{dt} = \frac{k'}{x} \tag{1.6}$$

Hence, the parabolic pattern would be given as:

$$x^2 = k't \tag{1.7}$$

The growth rate of the oxide layer is proportional to the flux of metal ions.

$$J_{Me} = J_{Mei} = -D_{Mei}\frac{\partial C_{Mei}}{\partial x} \tag{1.8}$$

Me and *Mei* refer to metal and metallic ion, respectively.

When the diffusion constant does not depend on the defects concentration, we can write:

$$J_{Mc} = -D_{Mci}\frac{\Delta C_{Mei}}{X} \quad \text{I} \tag{1.9}$$

The mass balance is established at the oxide/oxygen interface, where Stefan conditions exist.

$$J_{Mc} = C_{Me}\frac{dx}{dt} \quad \text{II} \tag{1.10}$$

By combining the two equations, I and II, the differential form of the law of parabolic velocity is obtained:

$$\frac{dx}{dt} = \left(D_{Mei}\frac{\Delta C_{Mei}}{C_{Me}}\right)\frac{1}{x} = \frac{k'}{x} \tag{1.11}$$

At t=0 and x=0 (boundary condition):

$$x^2 = k'.t \tag{1.12}$$

The results of Wagner's theory support the following two hypotheses:

1. The migration of ions and electrons along the shell controls the process.
2. Thermodynamic equilibrium is established locally at all points of the growing shell.

When the thickness of the oxide layer reaches about 10 nm, a potential difference of almost 1V can be established between the surfaces. In general, different theories have been proposed based on the difference between electrical and chemical potentials (thin-layer mechanism).

When an ion passes through a layer, two forces are applied to it:

$\frac{\partial \mu_i}{\partial x}$ chemical potential

$\frac{\partial \varphi_i}{\partial x}$ electrical potential

Therefore, the desired flux is equal to:

$$J = CBF \tag{1.13}$$

$$J_i = -C_i B_i \frac{1}{N_a}\left(\frac{\partial \mu_i}{\partial x} + z_i F \frac{\partial \varphi}{\partial x}\right) \tag{1.14}$$

C represents ion concentration, B indicates ion mobility, and N_a is Avogadro's constant. F is the forces applied to each particle per mol, which will be presented by:

$$F = \frac{\partial \mu_i}{\partial x} + z_i \boldsymbol{F} \frac{\partial \varphi}{\partial x} \tag{1.15}$$

In this equation, φ is electrical potential and $\boldsymbol{F}$ is Faraday constant. This equation is known as Nernst-Planck equation. The mobility and the constant of self-diffusion are interrelated through the Nernst-Einstein relationship:

$$D_i = B_i KT \tag{1.16}$$

The flux of all three components (cations, anions, and electrons) is written as follows:

$$J_c = -C_c D_c \frac{1}{RT}\left(\frac{\partial \mu_c}{\partial x} + Z_c F \frac{\partial \phi}{\partial x}\right) \quad \text{cations} \tag{1.17}$$

$$J_a = -C_a D_a \frac{1}{RT}\left(\frac{\partial \mu_a}{\partial x} + Z_a F \frac{\partial \phi}{\partial x}\right) \quad \text{anions} \tag{1.18}$$

$$J_e = -C_e D_e \frac{1}{RT}\left(\frac{\partial \mu_e}{\partial x} + F \frac{\partial \phi}{\partial x}\right) \quad \text{electrons} \tag{1.19}$$

According to the law of electrical neutrality, the following relationship must be established:

$$Z_c J_c + z_a J_a - J_e = \circ \tag{1.20}$$

By replacing the aforementioned equations in the last equilibrium relation, we have:

$$\frac{\partial \varphi}{\partial x} = \frac{1}{F} \frac{-C_c D_c Z_c \frac{\partial \mu_c}{\partial x} - C_a D_a Z_a \frac{\partial \mu_a}{\partial x} - C_e D_e \frac{\partial \mu_e}{\partial x}}{Z_c^2 C_c D_c + Z_a^2 C_a D_a + C_e D_e} \tag{1.21}$$

1.2.1.3.3 Diffusion in Solid State

Solid-state diffusion plays a key role in metal oxidation. This oxidation can lead to different kinds of processes:

- The diffusion of metal ions (from metal to surface) through the oxide layer and the reaction with the anion adsorbed on the surface takes place at the interface (like p-type NiO).

- Anions diffuse into the metal through the oxide layer (like p-type ZnO_2).
- There is movement of both the aforementioned processes.
- Oxygen atoms diffuse into the metal from the oxide layer (such as in-through diffusion of titanium).

1.2.1.3.4 Diffusion Mechanisms

a) Vacancy diffusion
b) Interstitial diffusion
c) Interstitial and substitutional diffusion

One of the most important mechanisms is diffusion through vacancy. The atom oscillates and thus can jump to the nearest empty space. This mechanism is predominant in most metal systems, but in ionic oxides, there is a defective complex structure. These defects can either be Schottky and Frankel, which include vacancy defects.

It is easier for metal atoms to jump to a void because the distance is short, but in the case of ions, it is the opposite. The distance is greater because the cationic sites are surrounded by anions (and vice versa), so the distance is longer (jump distance). Small interstitial atoms can easily move to interstitial sites. Cations are smaller than anions, so cations diffuse internally, but anions have a problem moving freely.

In another mechanism, the atom diffuses through *knocking* and can continue in a chain or path of penetration in a chain, occurring in the ionic lattice. The ion moves into a lattice location and, by tapping, moves the other ion to another position or another lattice site. The driving force for the transfer of reactants may be due to the electric field for thin films or, for thick layers, through the chemical potential gradient along the layer.

1.2.1.3.5 Oxidation Kinetics

The speed of linear oxidation reactions can be calculated from the following equation:

$$\frac{dx}{dt} = K_L \quad \textit{that} \qquad X = K_L.t \tag{1.22}$$

If the oxide surface of the metal is not to be protective and the oxide film does not act like a barrier layer, this condition occurs and the oxidation rate remains constant over time.

In this case, the reactions or surface processes in the interfaces are the controlling factors of velocity.

This may happen in many cases, such as:

- The oxide is volatile.
- Oxide forms molten eutectics with its underlying metal.
- The shell cracks or falls under internal stresses.
- A porous and nonprotective oxide layer is formed on the metal surface.

Diffusion through the shell, when the shell is thin, is probably not controllable, i.e., in the early stages of the process, and the speed of diffusion is linear and follows *Fick's first law.* In this case, the interfaces cannot be assumed to be in thermodynamic equilibrium.

In the oxide/gas interface, the process can be divided into several stages: In the reaction, the reactants must reach the surface of the shell and be adsorbed on them, the molecules are adsorbed on the surface, and oxygen is adsorbed on the surface, which finally absorbs the electrons from the oxide lattice to be initially chemically adsorbed and finally attached to the lattice. Removal of electrons from the shell changes the concentration of electron oxide defects in the shell/gas interface. The first step is to control the speed of process if the oxygen pressure is low. Considering oxygen is absorbed on the surface, it goes through the rest of the steps due to other factors. This reaction (first step) takes place on the surface and is linear velocity.

1.2.1.3.6 Logarithmic Velocity or Inverse Logarithm

At low temperatures, when the film layer is thin and about 100 nm thick, such a situation arises. (Such as Cu at 70°C, Fe at 120°C inverse, Al at 20°C.) In this case, the reaction is firstly very fast, then the rate decreases rapidly and reaches a relatively small rate. Compared to parabolic velocity, the slope of the changes is large. Initially, there is no barrier layer, so growth is rapid. In this case, a barrier layer is formed and the speed is limited.

The transfer process is from the length of the controlling layer, and the driving force is the electric fields on both sides.

$$x = k_e \log(at + 1) \tag{1.23}$$

K_e and a are constant.

1.2.1.3.7 Parabolic Law

In the parabolic cases, velocity has an inverse relationship with the second root of time and is found when penetration through the shell controls the velocity.

$$y^2 = K_p t + C_2 \tag{1.24}$$

The ideal oxidation of pure metals controlled by ionic diffusion follows this rule.

Parabolic oxidation laws are generally very common and are observed in the case of thick and sticky oxides. Metals such as iron, copper, and cobalt are oxidized parabolically.

- Temperature is an important factor, and depending on the temperature, one of the aforementioned conditions occurs. In the presence of a temperature gradient, different types of the aforementioned conditions may occur.
- When a clean metal surface is exposed to oxidation, first, the gas is adsorbed on the surface, and then the oxide is formed on the surface in the following two ways:
  1. In the form of a film or shell (depends on the pressure and the substrate and tends to combine with oxygen)
  2. In the form of separate oxide nucleus

The formation of each of the aforementioned two states is a function of the following factors:

- Surface roughness
- Crystal orientation
- Defects
- Surface preparation
- Surface impurities

Metals such as Ta, Nb, and Hf, and possibly Zr and Ti, form oxides in which the penetration of oxygen ions overcomes the penetration of cations so that the shell grows on the metal/shell joint. Shells formed on ordinary metals, such as Fe, Ni, Cu, Cu, Co, etc., mainly grow at the shell/gas interface, because metal cations move out faster than oxygen anions.

Factors affecting oxidation rate are gas composition, temperature, surface polish, purity of gas, heating method, orientation, gas pressure, geometry and thickness of the piece, crystallinity gas flow, metal purity, and growth stresses.

At low temperatures, growth is logarithmic. For example, if the oxidizing gas enters the furnace in the isotherm state, they are different compared to the case where the temperature increases in the presence of the gas. The geometric shape is a very important example. In the case of the sphere, for example, the reaction of the central nucleus becomes smaller. In other words, the area of the interface is getting smaller and smaller. Therefore, when the reaction speed is expressed in terms of mass per unit area, it should be considered, and therefore the area should not be assumed to be constant. This issue has been investigated by Romanki Brockman and Marouk. The speed obtained will be less (calculated) without considering a correction for it.

Increasing impurities in nickel metal (penetration of cations outward) helps the movement of cations, so the thickness of the oxide layer increases. As oxidation progresses, stresses are created in the shell, which are called growth stresses. This is important in systems where the cations are mobile, and growth stresses are created in the shell. Spatial displacement must occur as the cations penetrate to maintain the metal bond to the shell. As a result, cavities are created in the metal/shell joint. These cavities detach the shell from the metal surface. As the oxidation continues and the bond between the shell and the metal disappears and spreads on the surface of the sample, the cation exchange surface (cation supply) decreases over time, i.e., more and deeper diffusion must be done to reach the cation at the shell/gas interface. Hence, the reaction rate decreases. Detachment creates a porous zone between the outer compact layer and the metal.

When the detachment of the shell occurs, due to the fact that the movement of the cations initially causes the growth of the shell, the cations cannot pass by, creating a cavity between the dense layer and the metal surface. Because oxygen has high activity outside and low activity inside, it penetrates into the cavity. Oxide-free surface comes in contact with oxygen; it grows linearly, and a porous layer is formed on the metal surface. If the impurity in Ni increases, the formation of the cavity is accelerated, resulting in a double-layer oxide layer.

Oxidation of pure metals:

- Systems that create monolayers, like Ni, Zn
- Systems that create multilayers, like Fe, Co, Cu
- Systems with significant oxygen solubility in metals, like Ti, Zr, Hf

Single-layer systems: Ni

- **Nickel oxidation**

1.2.1.3.8 Metal-Deficient Oxide and P-Type Semiconductor

As nickel oxide grows out of electrons and cations, it forms a single-phase shell that reacts chemically with adsorbed oxygen on the surface, and Ni^{2+} ions and cations penetrate the surface through cation voids. This operation is through the gradient of chemical potential and electrical potential.

In the case of thin films, electric field growth is applied, and in the case of thick gradients, the chemical potential is the driving force for the transfer of reactants. There is also a thermodynamic equilibrium at the phase boundaries, and the volumetric diffusion of nickel ions controls the velocity, and according to Wagner's theory, the growth rate is parabolic.

1.2.1.3.9 Electron Microscope Observations

The outer layer is dense and has few cavities, while the inner layer is porous, and to prevent this, the NiO layer must have adequate adhesion to the substrate; hence, due to the weakness of this case, two layers are formed.

Oxide formed on the outer surface has a higher chemical potential than the inner surface, and at low oxygen pressure, Schottky and Frankel defects can be major defects in nickel oxidation and show some deviation from stoichiometric conditions.

1.2.1.3.10 Zinc Oxidation (Monolayer)

N-type semiconductive. Metal systems like this are present in the interstitials (cation surplus), and there are electrons in the conduction band. Oxygen forms ZnO at temperatures around 200–400°C. In these temperatures, crystals are needle-shaped (according to Raether theory). In the temperature range of 225 to 375°C, zinc is oxidized logarithmically. At a temperature of 375°C to the melting point, and also in the liquid state, zinc behaves parabolically.

Logarithmic oxidation probably takes place under electric fields (thin films). But parabolic oxidation is under in Wagner's mechanism, in which cations penetrate outward. The formed film is relatively thin (100–1,200°A), so the electric fields have a more important role in the driving force of oxidation than the concentration gradient.

1.2.1.3.11 Multilayer Systems

Iron oxides contain wustite, hematite, and magnetite. Wustite does not form at less than 570°C, so oxidation will be two layers below it and three layers above it (Fe_2O_3, Fe_3O_4, FeO). FeO is a p-type semiconductor with a metal deficiency that can exist in a large stoichiometric range and can exist in the range from $Fe_{0.95}O$ to $Fe_{0.88}O$

at 1,000°C. Since at a temperature of 1,000°C there is no stoichiometric state, the concentration of ion network defects will be high; defects include cationic vacancies and electron defects. Diffusion is done cationically and through a cationic vacancy.

Fe_3O_4 (magnetite) is a p-type. It has divalent ions in octahedral domains and trivalent ions in tetrahedral domains, so ions can diffuse through both octahedral and tetrahedral domains. Therefore, they have an inverted spinel.

Fe_2O_3 is an n-type semiconductor and can be in two forms: α-Fe_2O_3 with hombohedral lattice, and γ-Fe_2O_3 with a cubic one. The growth of Fe_2O_3 is associated with the diffusion of cations outward and anions inward.

1.2.1.3.12 Oxidation of Iron Alloys

It is oxidized in three layers at a temperature of 700–1,200°C. It is the inner layer of wustite, which makes up 0.95% of the thickness. The middle layer is the magnetite, which is 0.045% thick, and the outer layer is Fe_2O_3, with 0.05% thickness.

1.2.1.3.13 Cobalt Oxidation

Two types of oxides are formed: CoO (p-type) (internal) and CO_3O_4 (external). CoO is a p-type or cation-deficient semiconductor from which cation travels through cation vacancies and electrons through electron cavities. Above 500°C, there are inherent defects of the Frankel type, so according to previous discussions, it can be said that constant changes in velocity with oxygen pressure are more complex. At temperatures of 920–1,300°C, dense CoO oxide is formed by diffusion to the outside of the metal, and due to the use of platinum markers, these markers are placed in the shell/metal interface. Growth is parabolic, and the variation of parabolic velocity constant is as follows:

$$K_p = B\left(P_{O_2}\right)^{1/n} \tag{1.25}$$

This rate constant changes with temperature as an Arrhenius equation:

$$K_p = B'(P_{O_2})^{1/n} \exp\left(-\frac{Q}{RT}\right) \tag{1.26}$$

Different n has been reported at different temperatures. The velocity constant is a function of temperature and pressure. For example, at temperature of 950°C and 1,300°C, n could be 3.4 and 3.96 respectively.

Marwak and Pazuki concluded that in addition to Frenkel defect, other defects could be due to deviation from the stoichiometric state.

Bridge et al. showed that with the formation of a two-layer constant shell, velocity is independent of oxygen pressure, because in the previous state, CoO is porous but Co_3O_4 is dense at a temperature of 750°C, and within ten hours, Co_3O_4 layer is formed. (Co_3O_4 is denser than CoO, and CoO grows faster than Co_3O_4.)

In the case of copper, two layers of CuO and Cu_2O are formed. Until the formation of the constant Cu_2O, the rate velocity is a function of oxygen pressure and will be independent of oxygen pressure with the formation of the CuO layer (similar to cobalt).

1.2.1.3.14 Systems with Significant Oxygen Solubility in Metal

Titanium oxidation:

There are several oxides in titanium in different temperatures, including Ti_3O_5, Ti_2O_3, and (α-β) TiO.

In the temperature range of 600 to 1,000°C, oxidation is parabolic. This growth rate is a function of two processes:

1. Growth of oxide shell
2. Dissolution of oxide in metal

The same manner is true for zirconium and hafnium.

Alloy Oxidation:

Mechanical properties at high temperature: hardness, fatigue resistance, creep resistance strength, and ductility at high temperatures.

1. The most important properties for alloys at high temperatures is mechanical properties.
2. There should be proper corrosion resistance to the environment.
3. They must be easily built and equipped.

Design criteria:

1. Self-metallic or base metal (mostly FCC metals used, such as CO, Ni, Fe)
2. Added alloy elements, such as Cr, Al, Si

The investigation of alloys is very complex because existing alloying elements show different combinations for reacting with oxygen, which is determined by the free energy of their oxide formation, and different ions will have different mobility in the oxide (the diffusion deception of different metal ions varies along the oxide layer). Ternary and multinary oxides may be formed.

Then, solubility in the solid state may exist between conventional oxides that change with temperature and thicknesses.

Dissolution of oxygen into the alloy may lead to the formation of oxide deposits of one or more metals beneath the surface of the metal, also known as internal oxidation.

Internal oxidation is never desirable because external oxidation is controllable and in the internal type there is more damage and there is a tendency to convert the internal type to external.

Internal oxidation is the process by which oxygen penetrates into an alloy and causes one or more alloying elements to precipitate.

Necessary conditions for internal oxidation:

1. ΔG^0 formation of dissolved metal oxide (BOy) for an oxygen model should be more negative than ΔG^0 formation of the base oxide.

$\Delta G^0_{BO_y} < \Delta G^0_{AO_{y'}}$

B will oxidize faster than *A*.

2. ΔG^0 for the reaction $B + \frac{y}{2} O_2 \rightarrow BO_y$ is negative.
3. The base metal has the necessary solubility for oxygen.
4. No surface layer at the beginning of oxidation should prevent the dissolution of oxygen inside the alloy (it should not be too compressed, to allow oxygen to pass through).
5. The concentration of the dissolved element in the alloy should be a certain limit. For example, in Cu-Al, Al should not be too high, because internal oxidation becomes external.

Process mechanism:

1. Oxygen can be formed on the surface of the metal or through any layer that has formed on the surface (dissolution of oxygen).
2. Oxygen diffuses into the field (penetrates to form oxygen oxide).
3. The critical product of solubility is suitable for the nucleation of sediments in the impact of the reaction, and a critical value is created on three buds.
4. Sediment growth should be done.

This process is due to capillary characteristics, so it is called *Ostwald Ripening* in internal oxidation kinetics.

Consider a binary alloy A-B with a small value of *B*, because otherwise a stable oxide layer is formed on the surface. A stable oxide must also be formed for this type of problem to occur (internal oxidation). The position of the medium for oxidation *A* must also be low; otherwise, a stable oxide layer formed on the surface will prevent oxygen from diffusing inside. Therefore, we consider a quasistable state—that is, the amount of dissolved oxygen in the internal oxidation region changes linearly.

1.2.1.3.15 Precipitate Morphology

As the formed precipitate may be very small and not visible through a light microscope, its morphology can be affected by mechanical, electrical, magnetic, etc. properties. Overall, oxide particle size can be determined as a result of competition between nucleation and growth. The higher the nucleation rate and the lower the growth rate, the finer the particles.

1.2.1.3.16 Temperature Effect

The effect of temperature on growth rate is greater than its effect on germination, and therefore, with increasing temperature, the particle size becomes larger.

1.2.1.3.17 Transition from Internal Oxidation to External Oxidation

There must be a certain concentration of B atoms, below which internal oxidation occurs; otherwise, a BO-blocking layer forms on the surface, which prevents further oxidation. Therefore, in the design, it can be considered that the transfer from internal to external oxidation should be the basis of the work, and therefore the soluble elements (such as Cr, Al) should be more than the critical value. This type of oxidation is called selective oxidation.

Critical N_B concentration:

To obtain critical N_B, the Wagner model is used to transfer internal oxidation to external. According to the model:

$$Y = \gamma (D_0 t)^{1/2} \tag{1.27}$$

$$\gamma = \left(\frac{2N_0^{(s)}}{N_B . \vartheta} \right)^{1/2} \tag{1.28}$$

And considering Fick's second law and the law of mass conservation, it can be concluded that by adding a second soluble element (C), it can be oxidized without changing the value of B. Therefore, the second mechanism of N_0 is achieved.

1.2.1.3.18 Classification

Alloys can have two components (base metal, alloying elements), so they can be divided into two categories:

I. Base metal with noble elements and alloying with active elements
II. Base metal with active elements and alloying with active elements

As must be mentioned, noble alloy elements cannot be considered in this division due to the formation of a protective layer on the surface.

In accordance with group I, Au, Ag, and Pt as noble base metal, with Cr, Co, Ti, Al, Ni, Br as active alloying elements. An important example of this class are Pt-Ni alloys, in which $\Delta G_{NiO} < \circ$ and Pt is a noble metal. There is low oxygen solubility in platinum; hence, the oxidation conditions would written as:

$$N_B \geq \frac{\pi g^*}{2\vartheta} N_0^{(s)} \frac{D_0 V_m}{D_B V_{ox}} \tag{1.29}$$

In this equation, D_B is diffusion coefficient of B in alloy, V_{ox} is molar volume of oxide BO_υ, and V_m is molar volume of base metal. By considering g^* as below critical, N_B can be obtained:

$$g^* = f\left(\frac{V_{ox}}{V_m} \right) \tag{1.30}$$

Where f is defined as molar fraction of BO_υ.

1.2.1.3.19 Transfer Conditions from Internal to External Oxidation

- The concentration of soluble element must be increased.
- The amount of oxygen inside the alloy should be reduced.
- Without changing the composition of the alloy, if cold work is done, the base metal can be brought closer to the surface. As grains become smaller, the grain boundaries increase. Hence, a stable oxide layer is formed on the surface.

By increasing the percentage of chromium to 5% Ni, 5% Cr, an inner oxide duplex layer is formed at the bottom of the outer oxide (NiO) layer. This layer is in fact the combination of two porous layers.

The binary inner layer is crude in the form of NiO, which creates additional cationic voids in this region, which increases the mobility of Ni ions in this layer, called the doping effect, which increases the oxidation rate relative to pure nickel. As the inner shell moves toward the metal, the NiO_2-encapsulated Cr_2O_3 oxides form a secondary $NiCr_2O_4$ phase inside the NiO by a solid-state reaction:

$$NiO + Cr_2O_3 \rightarrow NiCr_2O_4$$

As the percentage of chromium increases, the oxidation rate increases due to the increase in cationic voids and the increase in Ni mobility. With the formation of $NiCr_2O_4$, this layer acts as a barrier against diffusion (diffusion block). It prevents the diffusion of nickel outward, so it is expected that with the oxidation rate of the active field, the active alloying elements in these systems, two or more oxides are formed: the base, the alloying elements, etc.

The principles of this class are discussed through examples Fe-Cu, Fe-Si, Fe-Cr, Ni-Cr, Cu-Be in literature.

We discuss three of the aforementioned systems that are more efficient: Ni-Cr, Fe-Cr, Fe-Si.

1.2.1.3.20 Ni-Cr Alloy System

Ni-1% Cr alloy contains an outer layer of NiO oxide (because Ni is active and the amount of Cr is low). The amount of transient is reduced by factors that encourage selective oxidation. Any factor that encourages the formation of Cr_2O_3 shortens the conversion region and reduces the instability region. These factors include increasing the percentage of chromium, reducing the oxidant pressure, and cold work on the alloy (causes the soluble element to be placed on the surface).

- In some areas, the phenomenon of elemental depletion occurs in the case of the element in which selective oxidation is to occur (chromium depletion).

Therefore, it can be concluded that the alloy containing Ni-50% Cr can consist of two phases: Cr_2O_3 and Ni.

- Chromium depletion depends on chromium concentration and shell growth rate, where tangled penetration of this alloy is reduced. This is when the percentage of chromium reaches a certain value. Therefore, by increasing the percentage of chromium to 10%, a complete outer Cr_2O_3 shell is formed outside at a temperature of 1,000°C. In this case, the speed reaches a certain value.

Transition mode is created by diffusion $NiCr_2O_4$ block. The steady state is the result of the formation of Cr_2O_3 (formation of Cr_2O_3 below NiO). Ni initially has a higher ionic mobility than chromium, so in alloys with very large amounts of NiO and $NiCr_2O_4$, a significant amount of Cr_2O_3 can be formed before a continuous layer is formed, because Ni ion mobility is higher. This region is formed for all alloys containing two or more components, and all components oxidize to form stable oxides. Cr_2O_3 is also more stable than NiO. Finally, the oxide is a more stable Cr_2O_3.

The nature of transient oxidation is often based on the selective oxidation process, in which the most stable oxide forms a series of mixed acetates, which is $Fe(FeCr)_2O_4$, and the oxidation rate constant decreases. A transient zone is formed due to the formation of a barrier, and at high percentages of Cr, the selective oxidation of Cr_2O_3 occurs.

There are two types of active ions. Fe^{2+} and Cr^{3+} are usually due to the fact that they do not prevent the formation of oxides for very long periods of time. Fe^{2+} ions are more mobile and always form a layer of FeO, followed by Fe_2O_3 on the surface.

When the percentage of chromium exceeds a critical value, the outer-layer Cr_2O_3 is formed (20% Cr).

- In the case of stainless steels (12% Cr at least), which are designed for corrosion resistance in aqueous solutions and for oxidation conditions at high temperatures, a thin layer of FeO is formed on Cr_2O_3, which indicates even very low diffusion of Fe to the surface, but this issue in Ni-Cr alloys is not observed.

1.2.1.3.21 Fe-Cr Alloy System

Due to the fact that the outer shell is formed faster, it can be considered that the thickness of the oxide layer is less.

$$FeO + Cr_2O_3 \rightarrow FeCr_2O_4$$

Iron oxides combine with chromium oxide. At lower chromium levels, both iron and chromium oxides are formed on the surface. Some chromium dissolves in the FeO phase, but this solubility is limited due to the stability of the $FeCr_2O_4$ spinel.

With increasing amount of chromium, Fe^{2+} ions are blocked by $FeCr_2O_4$ islands, and as a result, the FeO layer is almost much thinner, because the diffusion of Fe^{2+} is prevented by blocks. With more chromium, the amounts of spinel increase and the mobility of Fe^{2+} ions decreases. In other words, the spinel blocks increase. These islands can act as a marker indicating the initial position of the crust/metal in the interface.

1.2.1.3.22 Fe-Cu Alloy System

When Fe-Cu alloys are oxidized, Cu does not enter the oxide phase and remains in the metal, enriching the metal with copper. Solubility of copper in Fe reaches the desired temperature. As a result, copper is rejected and a second phase of enriched Fe is deposited from the copper at the interface, where the shell/metal precipitates. As a result, its melting temperature decreases. When the oxidation temperature is lower than the melting temperature of copper, the second phase is solid and, due to the low solubility of Fe, can be considered as a barrier to the penetration of iron, and iron oxidation will be slower if the oxidation temperature is above the melting point of copper.

There is also a limit to the percentage increase in chromium, because in systems with 25% chromium, a phase (a brittle phase) is formed. As a result, they are not usable in practice, so NiCr alloys perform better. Because a long layer of FeO is formed in Fe-Cr, but not in Ni-Cr, they are also comparable in number of spinels.

1.2.1.3.23 Fe-Si Alloy System

The oxide in this system is SiO_2, which, with the oxides of Fe_2O_3, Fe_3O_4, or FeO, can form iron silicate (Fe_2SiO_4). At lower Si values, SiO_2 forms on the surface of the alloy (apparently as very fine particles distributed on the surface) because instead of being surrounded by a progressive background, iron oxide goes to the surface and accumulates there. In this case, it combines with FeO and produces fayalite (Fe_2SiO_4). As the particles get larger, they are trapped by the shell. These islands are located in the FeO layer. The molten phase can diffuse through the grain boundaries. (*Red brittleness* phenomenon is also observed in the rolling of these alloys.) Copper can be introduced as an alloy element (to prevent corrosion) or as an impurity. The red brittleness phenomenon can occur through Bi and Sn elements, like copper. This phenomenon can be prevented by controlling the amount of copper.

If the part is exposed to a higher temperature, the risk can be reduced by reducing the time.

1.2.1.3.24 Sources of Stress Production

There are two sources of stresses: growth stress and stress of thermal origin (due to differences in shell and metal expansion coefficient). Growth stress is due to reasons that include difference in oxide volume and oxidized metal volume. In combined changes in alloy and shell, stress is related to recrystallization, depending on the samples.

Thermal stress is in proportion to the temperature difference, cooling and heating rate.

1.2.1.3.25 Introduction to TBCs

Background

Thermal barrier coatings are used in some applications in which the working temperature is so high that it damages the surface of the part. In mechanical parts whose working temperature is very high, to prevent the phenomenon of creep and fatigue very soon when it is created, due to high temperature, they use protective coatings, called thermal barrier coating, TBC for short. There are usually several layers, each layer providing some of the properties we want. For the first time, Mexican researchers have used a material in ash to make a composite nanocoat that is resistant to high temperatures. This coating can be a breakthrough in the aerospace industry and in air turbines, as we know the temperature of the inlet gases to the turbines is very high and the turbine blades are subject to severe mechanical attacks; this high temperature provides conditions for fatigue and creep. The temperature in these areas reaches 1,000°C, which causes microstructural deterioration in these turbines. This deterioration eventually leads to changes in the thermal and mechanical properties of the turbine and reduces the energy efficiency of the turbine. Gas turbines, cutting tools, and molten-metal storage plants are used.

In this project, researchers sought to build advanced nanocomposite-based materials that could protect the superalloy structure. To do this, the researchers used ceramic ashes containing different nanoparticles to make nanocomposites. Mullite can also be used. Mullite is a substance found in ash that has high thermal and chemical stability.

In turbines, they reduce the temperature and the heat transfer rate to the blades by various methods; one of these methods is blowing cold, compressed air to the side of the turbine blades, which has its own side problems, which are not discussed here.

Today, due to the industry's need to increase the efficiency and useful life of turbines, engineers are looking for a material as a coating that has the following conditions:

- High melting point
- Low heat transfer rate

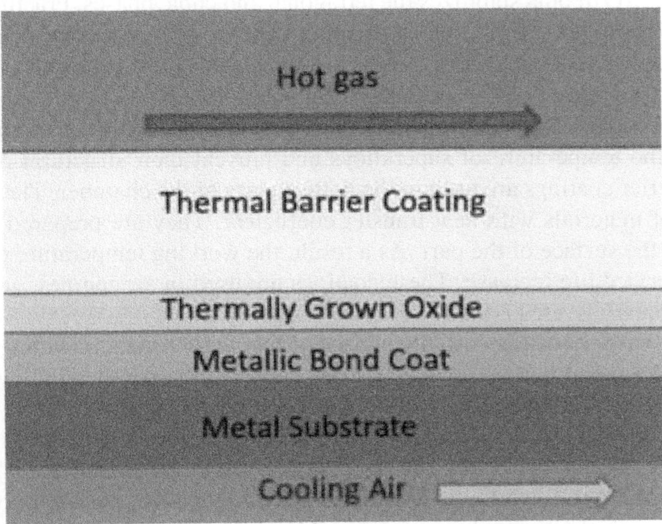

FIGURE 1.1 A common representation of hot oxidation against TBC.

- Very high temperature expansion coefficient, similar to the blade material (for expansion and contraction of equal size)
- Lightness of these covers, to reduce the lame mass of the blades
- High adhesion to the metal surface
- Reluctance to chemical reaction and oxidation
- Low strain tolerance
- Phase stability

Gas turbine blades are affected by destructive factors, such as oxidation, hot corrosion, fatigue, creep, and wear, due to operating conditions. In the production of turbine blades, superalloys are mainly used, which are expensive, and in order to reduce the mentioned destructive factors, suitable coatings on the surface of the turbine blade have been widely used. The use of thermal barrier coatings (TBCs) based on zirconia ceramics on the surface of gas turbine blades increases their power and efficiency, as zirconia has almost the lowest thermal conductivity compared to gasses. Increase the inlet without increasing the temperature under the layer so much.

Thermal barrier coatings (TBCs) are formed on gas turbine blades. A typical TBC coating consists of two layers, an intermediate layer with a combination of (Co, Ni = M) MCrAlY that is resistant to oxidation and hot-dip corrosion. Yttria-stabilized zirconia (YSZ) is formed on MCrAlY, and its main role is to insulate the metal substrate against high temperatures. Using these coatings, the metal surface temperature is reduced, and therefore the temperature can be higher. Applied in turbine operating conditions, it increases efficiency. Pure zirconia is converted from monoclinic to tetragonal at a temperature of about 1,000°C, and during subsequent cooling, tetragonal transformation to monoclinic occurs, which is accompanied by a volume expansion of 3–5% and is destructive. Addition of stabilizing compounds such as Y_2O_3 to zirconia stabilizes the tetragonal and cubic phases. Fracture and separation will result in ceramic coating. In order to improve the hot corrosion resistance of TBC coatings, a laser glazing process (laser glazing) and creating a dense layer without porosity have been proposed as a suitable solution.

Thermal barrier coatings are used to protect parts of high-temperature engines. To reduce the temperature of superalloys and prevent their structural destruction, thermal barrier coatings are used in the hottest parts of the chamber. These coatings are made of materials with heat transfer coefficient. They are prepared to transfer less heat to the surface of the part. As a result, the working temperature of the parts and their service life increase. These coatings are used in air engines, gas turbines, industrial materials, etc [7–9].

Thermal barrier coatings usually consist of two main layers, including a ceramic top coat and a metal bottom coat, which are deposited on the metal alloy under the layer. Sublayers are generally nickel- and cobalt-based superalloys. Of course, in some cases, special steels have been used as a substrate for these coatings.

The underlayment provides the necessary oxidation resistance for the metal substrate. This metal overlay is usually either an intrusive aluminide or an overlay coating with the general composition MCrAlY (Co and Ni = M). In modern coating systems, the substrate is generally one of the MCrAlY group coatings.

This coating, at high temperatures, forms a thermally grown oxide (TGO) layer at the joint of the metal, and the ceramic layer protects the surface of the superalloy from oxidation. The top cover provides thermal insulation conditions. This ceramic layer has low thermal conductivity and is usually made of zirconia, ZrO2. Stabilized zirconia is used to prevent fuzzy transformation of zirconia and to eliminate its volumetric changes. Usually, yttrium-stable zirconia coatings with YSZ by weight percentage of Y_2O_3, the rest of ZrO_2, are mostly used as thermal barrier coatings.

The TGO layer that forms with increasing temperature is due to the oxidation of the MCrAlY coating. This oxide layer continuously covers the surface of this coating and, as a protective layer, increases the oxidation resistance of the coating. The thickness and structure of the TGO layer are effective in stress level, stress distribution, and determination of TBC coating stability. On the other hand, the formation of alumina layer as TGO leads to high strain energy in this layer. The stress caused by the overgrowth of TGO is due to the increase in the volume of alumina caused by the oxidation of aluminum in the substrate. This strain energy can lead to TGO fluctuations, crack formation and propagation, and eventually, metallization of the upper ceramic coating. Also, due to exposure to oxidation conditions, a change in the phase of aluminum oxide is observed.

The most important requirement for a thermal barrier coating is its low thermal conductivity. The best available material to meet these conditions is ceramics, especially zirconia.

Ceramics are compounds between metallic and nonmetallic elements and often contain oxides, nitrides, and carbides. For example, some common ceramics include aluminum oxide (or alumina, Al_2O_3), silicon dioxide (or silica, SiO_2, and silicon nitride Si_3N_4), and commercial ceramics composed of clay minerals, such as glaze, cement, and glass. Considering mechanical behavior, ceramic materials are relatively rigid, and their hardness and strength are comparable to metals. Also, ceramics are usually very hard.

FIGURE 1.2 Exhaust system with thermal barrier coating.

Ceramics have been very suitable for this purpose so far, but the disadvantage of ceramics is their strong ionic conductivity. This conductivity causes the underlying layer, which is the metal that makes up the turbine blades, to be severely oxidized and lose their mechanical properties, including their strength. On the other hand, the adhesion of the ceramic layer to the underlying layer due to oxidation of the material reduces the adhesive and cuts the ceramics into pieces.

Engineers use a layer of aluminum between the ceramic layer and the metal that makes up the turbine blades to solve this problem. This layer of aluminum is converted to aluminum oxide, which has two very desirable and unique properties:

1. Aluminum oxide has a low heat transfer coefficient and in turn acts as a thermal barrier insulation.
2. Aluminum oxide is a nonconductive ion and prevents ions from reaching the substrates.

In this coating, for metallurgical reasons, we need plasma spraying.

Plasma is an excited gas, often known as the fourth state of matter, formed by the ionization of gas, which is usually argon in this process and is a combination of nitrogen and hydrogen gases. This method is one of the most widely used methods in the industry due to the high speed of coating. Coating of various ceramics is used on the rollers, and its function is that the plasma gun consists of a copper anode in the form of a hydrophilic and a tungsten cathode and a coating nozzle.

The procedure is, first, a potential difference is established between the anode and the cathode, and then the gas between the anode and the cathode is ionized and the gas is heated by establishing an electric current, and its volume is increased. The gun is removed; if the coating is in powder form, the coating particles are fed into the flame at the exit of the gun, and if it is in the form of a wire, it enters the plasma part with the help of a few rollers after the particles enter the flame. These particles melt or semimelt, depending on their size, and strike rapidly on the surface of the substrate to form a coating.

In this method, different types of powders of advanced materials, such as ceramic, cermet, intermetallic, and powders, are melted at a temperature between 10,000 and 20,000°C and sprayed uniformly on the surface of the part by the carrier gas; however, the surface of the part is not very hot and does not suffer from heat stress. In this way, a surface will be created with a uniform coating, with excellent adhesion and quality, and the resistance of the part against environmental erosion factors will be increased many times. The advantages of this method are the wide range of applicable coatings and also the very good quality of these coatings after spraying, but the high cost and the possibility of oxidation of the powder due to the very high process temperature are its disadvantages.

According to US studies, the sales market for high-performance ceramic coatings and their after-sales service was estimated at $1.1 billion in 2004 and $1.6 billion in 2009.

1.2.1.3.26 Improvement of Thermal Protective Coating

Although thick thermal barrier coatings can meet our thermal needs, the accumulation of residual stresses in these thick coatings is problematic.

Therefore, optimization of ceramic materials can be achieved in two areas:

- Chemical modification by optimizing the amount of yttrium stabilizing
- Microstructural modification by coating density optimization

1.2.1.3.27 Chemical Modification

For example, the result of chemical modification in the maximum life of a thermal barrier coating by changing different amounts of yttria in zirconia is quite evident when the yttrium concentration reaches 6–8% by weight. The highest amount of nonequilibrium (nontransformation) phase of the zirconia tetragonal polymorph occurs.

Recent advances in this field include the application of a semistable zirconia coating by 7–8% yttrium weight on a metal bond band of Ni or CoCrAlY alloy.

The bonding coating should be applied with plasma vacuum spray (VPS) to prevent the formation of internal alumina deposits that are thought to reduce the adhesion strength of the coating to the superalloy substrate. The adhesion of the bonding coating is also controlled by the concentration of yttrium. In this way, it reduces the stress on the interface by minimizing the difference in the electron density of the capacitance between the substrate and the bonded coating. In addition, yttrium can increase the electron density of the joint surface capacity, thereby increasing the bond strength of the joint surfaces.

Currently, the top layer of PSZ is basically sprayed in APS mode. Since the test coatings are applied by VPS, here PSZ coatings, the PSZ layer has a threefold increase in the amount of Young's modulus compared to coatings applied in APS mode. For this reason, VPS coatings have a higher thermal stress than APS coatings.

The second aspect of chemical modification goes back to the replacement of the yttrium stabilizer with other elements, but few studies have been done in this area. Using ceria instead of yttrium, the microhardness and crack resistance in stable PSZ are significantly increased by 15 mol% CeO_2 and higher.

1.2.1.3.28 Microstructure Modification

Microstructure modification refers to the control of porosity or the progress in the grading of coating systems. The distribution of porosity and fine cracks in the material is the cause of tolerance to thermal stresses and residual stresses. The effect of porosity and joint surface of droplets (Splat) on the effective properties of zirconia coating is monitored by a small scan of differential neutron angle (SANS), SEM and OOFE, and nondestructive detection methods.

Other expected functions of thermal barrier coating include improvement in thermal barrier performance and oxidation resistance, as well as resistance to erosion and corrosion.

Recently, research has been conducted on the application of explosive spray coatings to achieve proper reset of YSZ-based TBCs on NiCrAlY-bonded coatings on nickel-based superalloys. These coatings have a low heat transfer of about 1 to 1.4 (at 1,200–2,000°C) and show excellent shock resistance against thermal shocks, in addition to the ability to work in the heat cycle four hundred times (each cycle is as follows: heating to 1050°C and cooling to room temperature with compressed air), followed by two hundred cycles to 1,100°C, and quantification by water widens the cracks. Possible destruction by shredding is covered by several segregation factors that limit the life span of TBC:

- Crushing due to stress during the thermal cycle
- Zirconia instability due to incremental transformation of cubic monoclinic
- Erosion and microcrushing due to the collision of fine particles in hot gases
- Chemical corrosion by reaction with fuel impurities
- Oxidation of the graft coating due to hot corrosion
- Diffusion of gaseous impurities and accumulation of high-stress phases

To prevent the destruction of the thermal barrier coating during operation by the aforementioned mechanisms, over a long period of time, and high combustion temperature and thermal cycle, the following approaches have been investigated, respectively:

- Penetration of wing temperature abrasion-resistant materials into porous coatings, such as Ni3A or CVD-SiC
- Reduction of surface porosity by applying a top layer of very fine zirconia powder
- Improvement of PSZ applied by plasma spray with slurry with 1 mm zirconia particles and secondary sinter
- Addition of Si to MCrAlY to improve erosion and corrosion resistance
- Laser remelting of the surface (laser glazing)
- Reaction laser melting with alumina
- Plasma spray (PZT contains low amounts of yttrium, which leads to the expansion caused by the tetragonal transformation to monoclinic, neutralizing the thermal contraction)

REFERENCES

[1] Evans AG, Mumm DR, Hutchinson JW, Meier GH, Pettit FS. Mechanisms controlling the durability of thermal barrier coatings. Prog Mater Sci. 2001;46(5):505–53. doi: 10.1016/S0079-6425(00)00020-7.

[2] Richerson DW. Modern ceramic engineering: properties, processing, and use in design. New York: Marcel Dekker; 1982.

[3] Park SY, Kim JH, Kim MC, Song HS, Park CG. Microscopic observation of degradation behavior in yttria and ceria stabilized zirconia thermal barrier coatings under hot corrosion. Surf Coat Technol. 2005;190:357–65.

[4] Qian G, Nakamura T, Berndt CC. Effects of thermal gradient and residual stresses on thermal barrier coating fracture. Mech Mater. 1998;27(2):91–110. doi: 10.1016/S0167-6636(97)00042-2.

[5] Ouyang JH, Sasaki S. Microstructure and tribological characteristics of ZrO2-Y2O3 ceramic coatings deposited by laser assisted plasma hybrid spraying. Tribol Int. 2002;35:225–34.

[6] Han Z, Xu B, Wang H, Zhou S. A comparison of thermal shock behavior between currently plasma spray and supersonic plasma spray CeO2–Y2O3–ZrO2 graded thermal barrier coatings. Surf Coat Technol. 2007;201(9–11):5253–6. doi: 10.1016/j.surfcoat.2006.07.176.

[7] Ghasemi R, Shoja-Razavi R, Mozafarinia R, Jamali H. Comparison of microstructure and mechanical properties of plasma-sprayed nanostructured and conventional yttria stabilized zirconia thermal barrier coatings. Ceram Int. 2013;39(8):8805–13. doi: 10.1016/j.ceramint.2013.04.068.
[8] Xu H, Guo H. Metallic coatings for high-temperature oxidation resistance. In: Thermal barrier coatings. Cambridge: Woodhead Publishing; 2011.
[9] Hauffe K. The mechanism of oxidation of metals—theory. Oxid Met. 1965;79–85.

2 Suspension Plasma Spray

2.1 AN OVERVIEW

Surface engineering involves the application of traditional or modern heat treatment technologies or other surface treatments, such as a variety of coating methods on sensitive engineering materials and components, to achieve a composite material with properties that do not exist in the material itself. It is often seen that different surface technologies are applied to predesigned and prefabricated engineering parts. Surface engineering is the design and construction of a part knowing what kind of surface treatment or surface heat treatment is to be performed on it. In fact, lower production costs, longer storage intervals, increased material recyclability, and reduced adverse environmental impacts are the main reasons for the trend toward surface engineering techniques.

Among surface treatment methods, *thermal spray* is a general term for a group of coating techniques used to deposit metallic or nonmetallic materials on various surfaces. These processes are divided into three main categories: flame spray, arc spray, and plasma spray. High thermal energy density and short spray distances lead to the development and use of plasma spray in industries. This method is one of the most widely used methods in the industry due to the high speed of coating. In fact, plasma spray is a suitable method for creating refractory and corrosion-resistant coatings with suitable quality; for example, it is used to cover all types of ceramics on rollers, and its function is that the plasma gun consists of a water-cooling copper anode and a tungsten cathode and a nozzle of coating material [1]. The procedure is, at first, the potential difference between the anode and the cathode is established, and then the gas between the anode and the cathode is ionized; by establishing the electric current, the gas is heated and its volume is increased. In this case, the gas is forced out of the muzzle of the gun. If the coating material is in powder form, its particles are fed into the flame in the muzzle of the gun, and if it is in the form of a wire, it enters the plasma part with the help of several rollers. After the particles enter the plasma flame, these particles melt or semimelt, depending on their size, and quickly hit the surface of the substrate to form a coating. Ideal conditions are those in which the temperature of the particles on the surface is equal to their melting point.

Plasma spraying has an advantage over some coatings, such as metal spray, in that unlike combustion processes, it can spray very high melting point materials, such as refractory metals, like tungsten, and ceramics, like zirconia.

Plasma spray coatings are generally much denser, stronger, and cleaner than other thermal spray processes, with the exception of HVOF, HVAF, and cold spray processes. Plasma spray coatings are probably the most widely used among thermal spray coatings and applications, making this process the most diverse [2].

DOI: 10.1201/9781003285014-2

The best classification for different plasma spraying methods is as follows:

1. Atmospheric plasma spray
2. Vacuum plasma spray
3. suspension Plasma spray

2.1.1 Atmospheric Plasma Spray (APS)

Economically, this is the most common type of process, atmospheric plasma spray in the air. Powder particles can react with the air atmosphere, which limits the choice of sprays, because it creates oxidants inside the coating. The main applications of this method are in the creation of wear-resistant, corrosion-resistant (fluid, gas), and thermal barrier coatings based on oxide ceramic materials. Other common coating materials include metals and some alloys, especially non-oxidizing alloys. However, due to the quality of the coating and the lower coating cost of this process, it has become widely used in the industry for non-oxide ceramic materials. The porosity of plasma spray coatings in ambient atmosphere is 1 to 5%. The spray distance between the substrate and the plasma burner is 100 to 150 mm, depending on the material and spray parameters. The resulting microstructure usually contains a large number of microcracks, cavities, and inclusions. Disadvantages in the coating allow the penetration of molten streams of metals or molten salts in the environment, leading to the possibility of corrosion of the joint cover and, ultimately, failure and destruction of the part.

2.1.2 Vacuum Plasma Spray

This process takes place in a chamber with reduced pressure. The coating process begins after evacuating the pressure chamber less than 0.1 mbar and filling it with a neutral gas at a pressure of 50 to 400 mbar. In this method, it is possible to clean the surface of the substrate, especially oxide layers, and the substrate can be preheated, both of which cause better adhesion. The temperature of the substrate in this method is very high because the possibility of cooling the substrate through convection is reduced due to the drop in chamber pressure [3]. However, the high temperature of the particles improves the adhesion strength of the coating because in this case the penetration is increased and the stresses due to cooling are reduced; however, care must be taken not to increase the temperature of the substrate by a critical value.

Spraying distance in this process is usually between 250 and 300 mm, and the porosity of the resulting coatings is less than 1%, while the thickness of the coatings is usually between 100 and 150 micrometers. Reducing the chamber pressure to about 0.1 mbar causes most of the oxygen to escape from the chamber, thus allowing the use of oxidation-sensitive materials in both the substrate and the coating material.

2.1.3 Suspension Plasma Spray

Spraying submicron powders or even finer powders can also be a health concern at work. Hence, some efforts have been made to agglomerate nanoparticles and submicron. To overcome the aforementioned challenges associated with the production of microstructured coatings, submicron/nanopowder powders can be suspended in an aqueous or organic solvent, and such a suspension can be used as a raw material for thermal spraying. Therefore, this process only requires a stable suspension of powdered material that must be prepared in a suitable aqueous or organic solvent. Plasma spraying has been the spraying method of choice for the vast majority of suspension coating efforts due to the availability of a high energy flux required for the additional solvent removal step on the suspension. The aforementioned approach forms the basis for the incremental suspension plasma spray (SPS) technique, which has already been shown to be capable of producing coatings with a variety of microstructural properties, from very porous to very dense, columnar, vertical cracking, and so on [4,5]. And new microstructures are not usually obtained by conventional thermal spraying using powdered raw materials. Such microstructures are diverse for a wide range of distinct applications, including thermal barrier (TBC) coatings in gas turbines and air engines, electrolytes for solid oxide fuel cells, coatings for wear protection. Biocompatible coatings are useful. For implants, there are electron-emitting coatings, and so on. Of course, like conventional powder coatings, the properties of SPS coatings are controlled by a set of spray processes and related raw materials and parameters.

Another type of such raw material for plasma spraying involves the use of suitable precursor salts in the form of acetate, nitrate, alkoxides, oxychlorides, etc., which can form the desired coating particles in place.

The role of suspension properties in quality control of SPS coatings is particularly significant. Preparation of suspensions includes precise control of suspension properties, such as solid load, solvent type, viscosity, etc.

The challenges associated with preparing suspensions with all the desired properties increase when dealing with multiple powder mixtures [6].

Suspension plasma spray (SPS) is a relatively state-of-the-art technology that fabricates coatings with micro- and nanoscale characteristics by using ultrafine particles (less than 5 micron). To facilitate the use of these ultrafine powders, the particles are suspended in a solution that provides them a critical velocity to carry them into the plasma plume and make a coating. Suspension feedstocks can cover a range of material compositions, including ceramics and metal alloy blends, because the powder feedstocks used are so fine that the resulting coatings have a wider range of structures that can be reached, including columnar, segmented, and fully dense.

Suspension plasma spray is a process that enables the use of thermal spray feedstocks too small for conventional plasma spray processes. These feedstocks come in the form of a slurry, with micron- and submicron-sized particles suspended in water or any other solvents. In fact, the concentration of the particles in the slurry can be controlled and ranges from 5–80% by weight. This allows for the best combination of coating microstructure and precipitation rate to be gained. During the process, the thermal spray slurry is pumped to the outlet of the thermal spray burner and injected into the thermal spray jet. When the particles enter the plasma medium, the droplets fragment and the liquid phase evaporates, leaving ultrafine particles moving and hitting the substrate at an accelerated velocity. By using these ultrafine particle sizes, uniform coatings are fabricated as thin as 30 microns thick [7].

Suspension plasma spray coatings usually have finer characteristic sizes than conventional air plasma spray coatings, whose raw materials are of a higher order. As a result of the use of very fine-particle raw materials, the coating microstructure can be controlled to create microstructures similar to the physical deposition of electron beam vapor (EB-PVD): completely dense or columnar.

By controlling particle size, velocity, particle temperature, and application conditions, the following different microstructures can be formed depending on the need. There are four main types of plasma spray microstructures.

Columnar. Columnar coatings are formed when small particles with medium momentum flow at a low angle at the substrate surface and form single porous vertical columns. These high-strain coatings often have columns that are described as "feathered" and can perform similarly to EB-PVD columnar structures.

Segmented. Segmented coatings are formed when particles with higher momentum are applied to the substrate surface at an impact angle of ninety degrees or closer to it. Coating segmentation is formed by creating vertical cracks created by the high surface temperature of the coating and the sedimentation rate. This type of erosion-resistant microstructure resembles a dense vertically cracked thermal barrier (DVC) coating.

Dense. Like fragmented coatings, dense coatings are produced by high-momentum particles that strike the substrate at an angle of approximately ninety degrees. However, the bed surface temperature and sedimentation rate are controlled to prevent vertical cracks.

Porous. Porous coatings are like column coatings without column formation. This type of microstructure is very porous and therefore has a very low thermal conductivity; hence, it has the highest thermal isolation.

In addition to the coatings mentioned prior, it is possible to produce functionally graded coatings to be developed with a combination of characteristics. For example, a 7% by weight yttria-stabilized zirconia coating (7YSZ) can be combined with a split gadolinium zirconate coating to provide a high-strength, strain-resistant coating that is also resistant to calcium-magnesium-aluminosilicate (CMAS).

2.2 ADVANTAGES OF SPS

TBCs fabricated by suspension plasma spray (SPS) have been shown to create a porous columnar microstructure that can combine the advantages of APS and EB-PVD coatings together, which are low thermal conductivity and high durability, with relatively low production costs. SPS is a much cheaper technique than EB-PVD for the production of columnar TBCs, both in terms of initial investment in equipment cost and in terms of current cost, due to the ability to cover large parts without the requirement for an exclusive compartment with an inert medium. Lower costs with SPS indicate that SPS can not only be used as an alternative to EB-PVD but can also be used for different parts in gas turbines, including components that have not used EB-PVD due to their high cost. In addition, many of the existing infrastructures for APS can be used to make it a potentially accessible technology. An SPS top coating microstructure is to be a columnar microstructure because this microstructure increases TBC durability and strain tolerance improvement. High porosity and generally lower column density than EB-PVD result in decreasing the thermal conductivity and simultaneously reducing the coating lifetime.

Suspension plasma spray process is a coating process that can occur in many industries, including car industries, aerospace and marine industries, medical applications, electronics, petrochemical, etc.

There are some common applications which are found in the power generation, aerospace, and marine industries. In these cases, thermal barrier coatings (TBCs) generated by the plasma spray are applied to turbine components that encounter hot working temperature to protect them against corrosion, wear, erosion, and creep.

By looking deeply into the aforementioned industries, we will investigate the related properties and the applications of SPS, giving us the horizons to differentiate it with other methods of coatings.

2.3 CAR INDUSTRIES

Limited reserves of fossil fuels are declining several times faster than their formation. When these fossil fuels are burned, greenhouse gases, such as carbon dioxide (CO_2) and nitrogen oxide (N_2O), are formed and released into the atmosphere, leading to greenhouse effects and climate change on Earth. Combustion engines produce 27.7% of greenhouse gases in the United States (2017), and in Europe, according to the European Environment Agency, they are responsible for 27% of greenhouse gas emissions. Therefore, methods of using fossil fuels in a sustainable and environmentally friendly manner should be reviewed and implemented. One of the most accessible ways is to increase the efficiency of engines. Thermal barrier (TBC) coatings applied to thermal injection combustion engine components, especially for diesel engines, have been considered throughout the years to improve engine efficiency. The advantages of ceramic coatings are oxidation resistance, stable microstructure, high resistance to thermal cycle, and reduction of heat loss in the engine during operation, which can potentially increase engine efficiency and reduce fuel consumption. However, some catastrophic phenomena, like low and high cycle fatigue, diffusion of combustion gases in the TBCs, etc., lead to surface cracking, microspallation, and increased rate of crack growth, making coating unsuitable for combustion engine application.

The thermal spraying method used in the past was conventional plasma spraying, which usually resulted in the production of TBC with high porosity and coating thickness. Recent studies, however, have shown that relatively thin coatings with a columnar microstructure, with coatings that follow transient chamber gas temperatures, can provide better performance for diesel engines. The use of these coatings, developed by Toyota and called thermal wall insulation technology, reduces heat loss through the combustion chamber wall without heating the inlet air and leads to better engine performance. The most important parameter for achieving this goal is the adaptation of coating temperature in each stroke cycle.

Suspended plasma spraying (SPS) has already shown excellent results as thermal barrier coatings (TBCs) for the gas turbine industry and its capacity to produce thin and porous coatings (Reference 15). Plasma spray is a technical suspension that uses powder suspension as a raw material, thus allowing the use of nanomaterials in submicron sizes and coatings with unique microstructure and improved thermal and mechanical properties. The coating produced by SPS can be thin with porosity, with homogeneous distribution, and provide high strain tolerance in the cyclic thermal environment, which can help improve the life of the coating.

Uczak de Goes et al. investigated YSZ and GZO TBCs produced by SPS and assessed whether they bring benefits over the atmospheric plasma spray TBCs in light-duty diesel engine. This work was performed to study coating microstructure, thermophysical properties, and engine tests.

According to the study, APS coatings show a layered structure that indicates the coating is made up of porous and laminated layers that affect each other. This feature influences the thermal properties of the coating because in the layered structure, different types of pores are usually formed; hence, the thermal conductivity of the entire coating is reduced. The downside of laminating pores is the creation of horizontal cracks that can easily spread into the coating and lead to workpiece breakdown. However, the YSZ SPS coatings in this study can be categorized as having a columnar feathery microstructure, while the GZO TBC reveals a more porous microstructure with columns that tend to widen toward the top of the coating. It is observed that the fine pores in the SPS coatings represent almost half of the total porosity.

GZO coatings have a higher porosity, and therefore the material itself has a lower thermal diffusivity than YSZ material. S. Mahade et al. revealed that the lower thermal conductivity of GZO can be related to the differences in the crystallographic structures and distinct arrangement of cations and oxygen vacancies of the two materials and is related to the phonon scattering in the ceramic layers. As the variation of temperature in the real working conditions is not in a steady state, thermal diffusivity must be mentioned. It shows the rate in which the material absorbs and returns heat. By considering thermal conductivity and diffusivity, it can show that the YSZ SPS has better heat transfer in comparison with YSZ APS and even GZO SPS.

By investigating engines, Uczak de Goes et al. showed that the best result was the SPS GZO that had around 0.7% increase in efficiency to the combustion chamber walls as compared to the uncoated piston. The SPS GZO was followed by the YSZ APS coating, and then, the lowest reduction was shown by the YSZ SPS. This leads to improved engine power and efficiency.

2.3.1 Medical Applications

High biocompatibility is the most important circumstance required for new materials. Materials with undesirable effects and side effects are not acceptable for medical and surgical goals. Required levels of biocompatibility depend on the contact time between the biomaterial and live tissue, ranging from few seconds, where the requirements do not need to be tough, to several years, where final biocompatibility is an obligation, for example, in hip joint replacement, bone fracture replacement, etc.

Metallic implants like Ti-6Al-4V are commonly used in biomedical applications, but these kinds of materials do not show any bioactivity to form a bond with the surrounding tissues, which causes a reduction in the strength of the bonding. To overcome this problem, some sort of coating is deposited on the metal surface. Hydroxyapatite as a bioactive material is used frequently for its osteoinduction and osteoconduction properties. However, this material is brittle and cannot be used in highly tensile-stressed locations, like bones.

Otsuka et al. investigated the mechanical and antibacterial properties of titania gray/hydroxyapatite SPS coatings on the surface of Ti substrates for biomedical applications recently. Plasma-sprayed HAp composite contains components such as medium crystalline HAp and TCP, the dissolution of which can increase the deposition of amorphous calcium phosphate in vivo. Suspended plasma-coated composite coatings containing TiO_2 with mixed Magneli phases, such as Ti_4O_7 and Ti_3O_5, typically exhibit photocatalytic activity. The increased amount of TiO_2 proliferates the mechanical properties and surface strength by suppressing strain accumulation at the interface between the Ti substrate and the composite coating. It is noteworthy that HAp/gray titania coating SPS without amino acid complex can provide effective antibacterial properties against *Escherichia coli* due to exposure to visible light. The underlying mechanism of enhanced mechanical and antibacterial properties observed by the addition of TiO_2 in their study is identified. Increased antibacterial properties of HAp/SPS gray titania coating can be attributed to the high number of Magneli phase compounds and high wettability. Small grain size was observed due to the use of submicron particles, which was beneficial for antibacterial activity.

The mechanical properties of suspension plasma spray coatings were also improved by increasing the titanium weight percentage. SPS prevented phase decomposition, which helped increase the mechanical properties. Multidimensional measurements showed the mechanisms of accumulated strain damage and microcracks in the interface layers until failure.

REFERENCES

[1] Tingaud O, Bertrand P, Bertrand G. Microstructure and tribological behavior of suspension plasma sprayed Al2O3 and Al2O3-YSZ composite coatings. Surf Coat Technol. 2010;205(4):1004–8. doi: 10.1016/j.surfcoat.2010.06.003.

[2] Tarasi F, Medraj M, Dolatabadi A, Oberste-Berghaus J, Moreau C. Amorphous and crystalline phase formation during suspension plasma spraying of the alumina–zirconia composite. J Eur Ceram Soc. 2011;31(15):2903–13. doi: 10.1016/j.jeurceramsoc.2011.06.008.

[3] Kumar V, Kandasubramanian B. Processing and design methodologies for advanced and novel thermal barrier coatings for engineering applications. Particuology. 2016;27:1–28. doi: 10.1016/j.partic.2016.01.007.

[4] Ganvir A, Curry N, Govindarajan S, Markocsan N. Characterization of thermal barrier coatings produced by various thermal spray techniques using solid powder, suspension, and solution precursor feedstock material. Int J Appl Ceram Technol. 2016;13(2):324–32. doi: 10.1111/ijac.12472.

[5] Otsuka Y, Hiraki Y, Hakozaki Y, Miyashita Y, Mutoh Y. Effects of adhesives on reliability in interfacial strength evaluation method for plasma-sprayed hydroxyapatite coating. IJAT. 2017;11(6):907–14. doi: 10.20965/ijat.2017.p0907.

[6] Mahmud Abir MM, Otsuka Y, Ohnuma K, Miyashita Y. Effects of composition of hydroxyapatite/gray titania coating fabricated by suspension plasma spraying on mechanical and antibacterial properties. J Mech Behav Biomed Mater. 2022;125:104888. doi: 10.1016/j.jmbbm.2021.104888.

[7] de Goes WU, Somhorst J, Markocsan N, Gupta M, Illkova K. Suspension plasma-sprayed thermal barrier coatings for light-duty diesel engines. J Therm Spray Tech. 2019;28:1674–87.

3 The Science and Practice of Ceramic Suspensions

Scope:
This chapter provides a brief introduction to the fundamentals in colloid science, with a special focus on nonorganic (nonpolymer) suspension preparation. It addresses the physicochemical properties of single colloidal particles, the mutual and simultaneous interactions of media and particles (micron [μm], submicron, and nanosized [nm]), as well as the role of processes at the interfaces (forces) and the structure of the interfacial layer properties of particles. It further explains the viscous, nonviscous, complex particle-media interactions that occur among colloidal particle formation and the subsequent responses. Detailed knowledge of colloidal science is explained, and definitions of possible behaviors are provided.

3.1 GENERAL OVERVIEW

Among many processes which involve nanopowders, complete dispersion of the powders in a media (mostly liquid) either as an intermediary by-product or as the end product is necessary. In either case, the flowability under shear stress properties of these suspensions is essential for the feasibility of these processes, especially in ceramic technology and industries. This flowability of a suspension basically determines its stability and its homogeneity, which are determinative of the progress of the process and the quality of the end product. The basic parameters of the suspension can directly determine the properties of the products. For instance, for achieving considerable mechanical strength level, it is essential to prepare high solids content suspensions in which the packing density results in minimum porosity content in final green and solidified bodies; therefore, it becomes an important parameter dependent on the quality of the suspension.

Special consideration of these flowability properties of suspensions is not only essential in selecting and designing production techniques but also indicates production cost via suspension flowability. The semifluid behavior of these suspensions increases in power with decreasing mean particle size of the solid part, which requires manipulation of the interparticle interactions (solid-solid and solid-liquid interactions) to control their behavior and an optimization of the suspension properties. Understanding these parameters and the effects of related manipulations is the core of colloidal science and the processing process. The colloidal processing requires certain organic/nonorganic additives and conditions to be chosen specifically so that the properties of the powders and liquids under consideration change accordingly. The goal of understanding "The Science and Practice of Ceramic Suspensions" is

DOI: 10.1201/9781003285014-3

controlling the powder particles—media interactions, to stabilize the suspension and maintain applicable flowability while obtaining optimum (highest) solids content.

3.2 DEFINITIONS

The science and practice of particle dispersion has evolved around to main terms: *colloid* and *suspension*. The term *colloid* is generally used to describe particles in microscopic range which are formed with at least one dimension in a $1nm-1\mu m$ size range, mostly as a D_{50} for an assumed spherical entity, in a continuous, deformable interface media. As a basic feature of all colloidal systems, the contact surface area between particles and the dispersing medias grows continuously with decreasing particle size. This contact creates interparticle (or surface) forces which strongly influence the suspension behavior. The main solid/liquid dispersions of colloidal particles create *suspensions*.

Colloidal suspensions, in fact, are heterogeneous materials which consist of a liquid continuum and solid, submicron particles. The small size of colloids intensifies the importance of the interface to the particle's physical behavior. In most scientific documents, suspensions contain particles (isolated, aggregated, or agglomerated) with at least one outer dimension below 100 nm which is fully covered by the media. Such particles are basically aggregates which have evolved during the particle synthesis processes by *precipitation* or *crystallization* and may result from the *aggregation* (*coagulation*) of isolated primary particles or can already be present in the powder which is dispersed in the liquid media. Also, the microscopic interfacial properties directly affect the interaction between particles and are pivotal for the macroscopic suspension behavior. During preparation of colloidal suspensions, the dispersion and stabilization should be considered simultaneously. Due to the fineness of colloidal particles, colloidal suspensions may appear transparent (particles are much smaller than the wavelength of light). The small size of colloidal particles not only affects their visibility but also results in exceedingly slow settling and high diffusibility. Furthermore, the significance of particle–particle interactions, in particular those related to electric surface charges, must be considered. Therefore, while a certain solid material is dispersed into a comparatively huge number of particles, it requires the promotion of mutual collisions.

The main approaches for the preparation of solid/liquid dispersions include *bottom-up* and *top-down* processes, which act similar to most nanopreparation routes.

The first group of routes depend on the "bottom-up" or "buildup" of particles from molecular units, which is also related to condensation method, involving two main processes: nucleation and growth. Therefore, it is necessary at start to prepare a highly concentrated molecular (ionic, atomic, or molecular) distribution of the insoluble substances, then, by changing physicochemical conditions, causing formation of small particles as a suspension by leading to the formation of scattered nuclei toward growth in the particles.

The initial stages of nucleation result in the formation of small nuclei with the highest surface-to-volume ratio and, as a result, being governed by the role of specific surface energy. With the ever-growing and ever-increasing size of the nuclei, the ratio becomes smaller, leading to the appearance of large crystals, with a corresponding reduction in the role played by the specific surface energy. The free energy formation of these spherical nuclei, ΔG, is given by the sum of two contributing free energies: a positive-surface free energy termed ΔG_S which increases with an increase in the radius of the nucleus (r), and a negative-volume free energy termed ΔG_V, due to the appearance of a new phase. The ΔG_S is indicated by the production of the surface area of the nucleus and the specific surface energy (solid/liquid interfacial tension) γ; while ΔG_V is related to the relative supersaturation, which also increases with increasing particle radius r. Overall:

$$\Delta G = \Delta G_V + \Delta G_S \ll 0 \tag{3.1}$$

$$\Delta G = 4\pi r^2 \gamma - \left(\frac{4\pi r^3 \rho}{3M}\right) RT\, ln\left(\frac{S}{S_0}\right) \tag{3.2}$$

Where ρ is the density, R is the gas constant, and T is the absolute temperature in current and initial entropies S and S_0, which shape the schematic trends in *Figure 3.1.*

In the initial stages of nucleation, ΔG_S increases faster with increasing r than ΔG_V, and as a result, ΔG term remains positive and increasing toward a maximum at a critical radius r^*. After which, it decreases and eventually becomes negative. This occurs since the second term rises faster with r than the first term (r^3versus rr^2). When ΔG becomes negative, growth becomes spontaneous and the clusters of

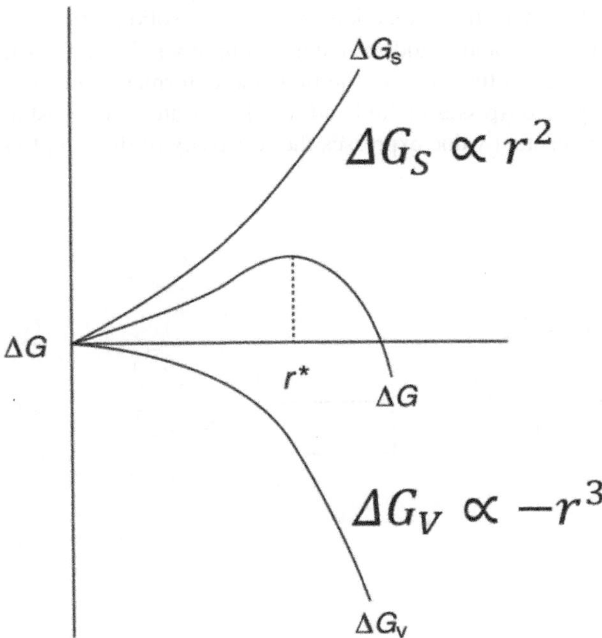

FIGURE 3.1 The schematic Gibbs free energy change during "buildup" suspension formation.

dispersed particles tend to grow rapidly. This spontaneous growth creates the critical size of the suspended particles (r^*) throughout the media. The free energy maximum ΔG^* at this critical radius indicates the amount of energy that the system has to overcome before growth becomes spontaneous. Both r^* and ΔG^* can be obtained by r -differentiating toward zero at maximum, which leads to:

$$r^* = \frac{2\gamma M}{\rho RT ln(S/S_0)} \tag{3.3}$$

$$\Delta G^* = \frac{16}{3}\frac{\pi\gamma^3 M^2}{(\rho RT)^2\left[\ln(S/S_0\right]^2} \tag{3.4}$$

The first group of routes depend on the "top-down" or "up-to-bottom" approaches, which are usually referred to as a dispersion process. In this approach, larger particles/clusters, or "lumps" of the insoluble substances, are subdivided by mechano-physical forces into smaller units. The term *dispersion*, commonly used to refer to the complete process of incorporating the solid particles into a liquid media such that the final product consists of fine particles distributed throughout the dispersion media, can be clearly expressed in this approach as shown schematically in *Figure 3.2*.

Immediately after formation, the physical and flowability (response to mechanical forces) properties of these two colloidal suspension systems will depend on the semiliquid behavior of the liquid, the amount of solids ("solids content"), the size distribution of the solids, the interactions between the solid particles, and the interactions between the solid and liquid, which is commonly referred to as the *rheological* properties. *Rheology* is the study of the flow and deformation behavior of materials in response to applied stresses in fluid and semifluid states. The most important property which as a constant value expresses the resistivity of the fluid to flow at a given

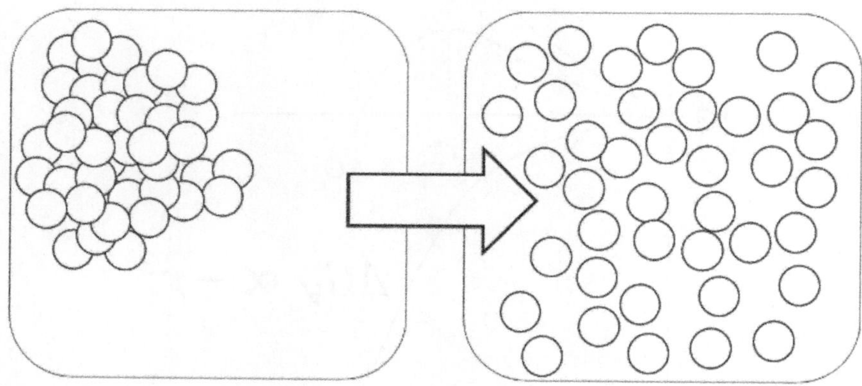

FIGURE 3.2 The schematic response of solid insoluble particles to liquid media during "top-down" suspension formation.

shear rate is *viscosity.* Viscosity (η), or as scientifically known, *dynamic viscosity,* is expressed as the ratio of the external applied stress (τ) to the rate of displacement of the fluid or *shear rate* ($\dot{\gamma}$). This resistivity can also be measured as the resistive flow of a fluid under the weight of gravity expressed as *kinematic viscosity* (ν). The expressions of viscosity are related through liquid density (ρ) as:

$$\eta = \tau/\dot{\gamma}, \nu = \eta/\rho \tag{3.5}$$

In contrast to suspensions with random dispersion of solid particle in liquids, other dual contacts with similar phases can include *emulsion* (colloidal dispersions in which a liquid is dispersed in a continuous liquid phase of a different composition; in most emulsions, one of the liquids is aqueous while the other is hydrocarbon, referred to as an oil); *coalescence* (a process in which two phase domains of essentially identical composition get in contact and form a larger phase domain), *aggregation* (when any of Brownian, *sedimentation*, or stirring causes two dispersed species to clump together, possibly touching at some points, and with virtually no change in total surface area; in aggregation, the species retain their identity but lose their kinetic independence since the aggregate moves as a single unit), and *creaming* (from a density difference between the dispersed and continuous phase, producing two separated layers of dispersion that have different dispersed phase concentrations; basically, it represents the migration of a substance in an emulsion, under the influence of buoyancy, to the top of a sample while the particles of the substance remain separated), which may be encountered in some suspension systems as abnormalities.

The dependency of shear stress (in *Pascal*) and viscosity (in Pascal second [*Pa.S*]) or centipoise (*cP*) versus shear rate (S^{-1}) creates the most common typical behaviors (*Figure 3.3*), as (1) *pseudoplastic with yield stress*, (2) *plastic*, (3) *Newtonian*, (4) *pseudoplastic*, and (5) *dilatant*. The Newtonian suspension behavior depicts a linear variation between the shear stress/viscosity and shear rate while passing through the

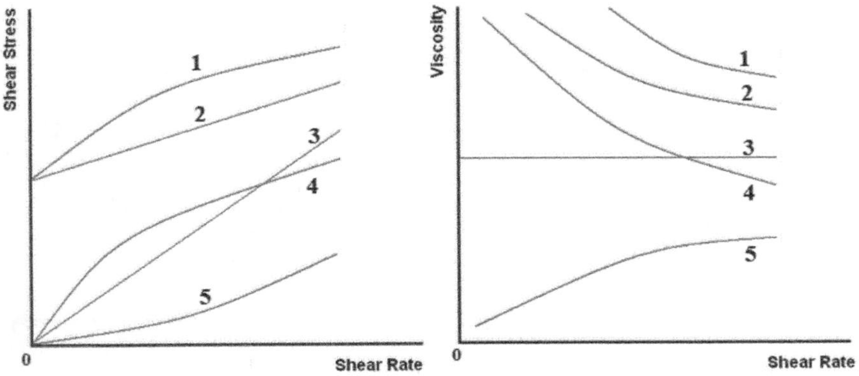

FIGURE 3.3 The most common typical flow curves of shear stress and viscosity versus shear rate.

origin. The coefficient viscosity is a line slope and a constant. For *non-Newtonian* fluids, the coefficient of viscosity is not a constant but is itself a function of the shear rate. In pseudoplastic suspensions, as shear rate increases, viscosity decreases. This behavior can also be called *shear thinning*. The opposite of this behavior is called *dilatancy*. In this case, viscosity increases with the increase of the shear rate. This type of fluid can also be called *shear thickening*. In some colloidal suspensions, the shear rate remains at zero until a threshold shear stress (*yield stress*) is reached, and then *Newtonian* or *pseudoplastic* flows begin. The *plastic* fluid behavior occurs when viscosity is constant. If viscosity is not constant, *pseudoplastic* with yield stress suspension behavior can be detected. The *thixotropy* is a semipseudoplastic flow which is also time dependent, and at a constant applied shear rate, viscosity decreases with time. In this condition, interparticle or intermolecular interactions are influenced by the magnitude of the applied shear and also by the time intervals of shear applications.

Some of the typical stress yielded (τ_y) and nonlinear behaviors compared to linear Newtonian behavior are schematically represented in *Figure 3.4* for (1) *shear thinning*, (2) *shear thickening*, and (3) Bingham plastic in τ and η versus $\dot{\gamma}$. The complex response to increasing shear rate is the main reason for the special attention to these behaviors.

These types of flow behavior can be observed under steady shear depending on suspension composition and the stability of the suspension and are generally characterized as "*viscous flow behavior*." The pseudoplastic or shear-thinning behavior happens while viscosity decreases with increasing shear rate. This behavior can be followed by a yield stress which depends on the strength of the particle network. If the flow behavior is linear above τ_y, the system is referred to as *Bingham* plastic. The *dilatant* or *shear-thickening* behavior occurs when viscosity increases with shear rate. The rheological behavior of highly concentrated colloidal suspensions can often be time dependent. Thixotropic systems exhibit an apparent viscosity that decreases

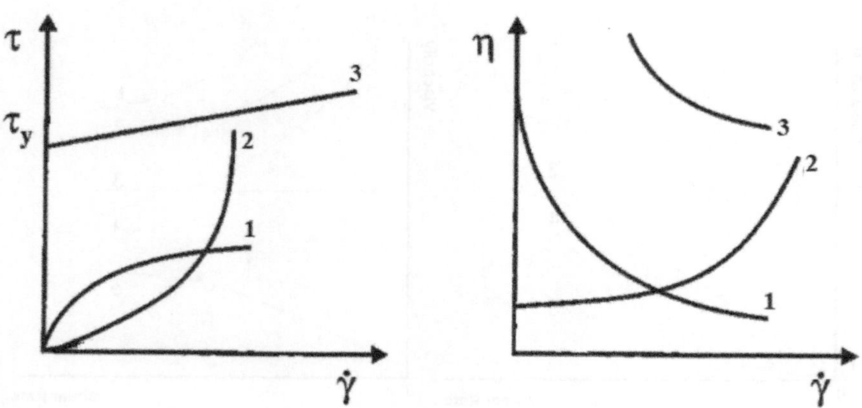

FIGURE 3.4 The rheological behaviors mostly seen in colloidal suspensions.

TABLE 3.1
The rheological models of colloidal suspension based on solids content (Φ).

Model	Parameters	Equation
Batchelor	$\eta - \Phi \precsim 0.1$	$\eta = 1 + 25\Phi + 6.2\Phi^2$
Einstein	$\eta - \Phi \precsim 0.3$	$\eta = 1 + 25\Phi$
Horri	$\eta - \Phi \precsim 0.3 - \text{K(constant}$	$\eta = 1 + 25\Phi + \text{K}\Phi\left(\frac{\Phi}{\Phi_{max} - \Phi}\right)^2$

with time under shear and repeats the original viscosity when flow stops. The opposite behavior is referred to as *rheopexy.*

The perturbations in the structure of liquid media can create perturbations in the rheological behavior. For the colloidal suspension of schematic hard spheres, the equation of the state mostly follows:

$$\eta_s = f\left(\dot{\gamma}, t, \eta_l, \rho_l, R, n, \rho_P, KT\right) \tag{3.6}$$

Where η_s is the suspension viscosity; $\dot{\gamma}$ is the shear rate; t is the time; η_l and ρ_l are the viscosity and the density of the liquid; R, n, and ρ_P are the radius, concentration (number density), and density of the particles; and KT is the thermal energy reflecting *Brownian* motion.

This equation is the basic term indicating the dependence of a suspension's mechanophysical parameter, such as viscosity, to various environmental and intrinsic parameters. Other rheological models for colloidal suspensions consider the charge of particle q, the ionic strength of the suspension I, and the dielectric constant of the media ε. For colloidal systems containing the same type of particles and media at constant temperature, pressure, and shear rate, all variables can be grouped into two groups: solids content (Φ) and interactions between particles (ψ). By considering these two parameters, researchers have developed models for predicting rheological behavior of colloidal suspensions (see examples in *Table 3.1*).

3.3 STRUCTURE OF THE SOLID/LIQUID INTERFACE

The mutual effect of solid/liquid and the quality of the solid/liquid interface are directly related to surface charges. The origin of the charge on particle surfaces relates to a great variety of processes that occur to produce a surface charge. Surface ions are ions that have such a high affinity for the surface of the particles that may be taken as part of the surface, such as A^+ and B^- for hypothetical AB particles. For AB in a solution of a nonorganic salt, such as KNO_3, the surface charge σ_0 is given by the following formula:

$$\sigma_0 = F\left(\Gamma_{A^+} - \Gamma_{B^-}\right) = F(\Gamma_{ANO_3} - \Gamma_{KB}) \tag{3.7}$$

Where F is the Faraday constant (96,500 $C.mol^{-1}$) and Γ is the surface excess of ions ($mol.m^{-2}$). This approach can be extended to an oxide, such as silica or alumina in KNO_3, H^+, and OH^- can be taken as parts of the surface:

$$\sigma_0 = F\left(\Gamma_{H^+} - \Gamma_{OH^-}\right) = F(\Gamma_{HCl} - \Gamma_{KOH}) \tag{3.8}$$

These ions, which determine the charge on the surface of charged particles, are known as "*potential determining ions*." The accepted electrical charge of the particle in the media depends on the pH of the solution. Below a certain pH, the surface will be a positive charge and, above a certain pH, change into a negative charge. At a specific pH ($\Gamma_{H^+} = \Gamma_{OH^-}$), surface is uncharged; this is referred to as the "*point of zero charge*," or *pzc*. The *pzc* depends on the type of the oxide and its reaction toward the media. The surface charge is positive when the pH is below the *pzc*, and negative when the pH exceeds the *pzc*. It is a characteristic property of the particle phase. For instance, an acidic oxide such as silica has an estimated range of pH 2 to 3; for a basic oxide such as alumina, *pzc* is pH ~ 9, and for an amphoteric oxide such as titania, the *pzc* is pH ~ 6. In special cases, some adsorbed ions which are nonelectrostatic toward the surface may "enrich" the surface. These ions should not be considered as part of the surface, i.e., bivalent cations on oxides, and cationic and anionic ions on most oxide and nonoxide surfaces. The schematic charge distribution in positively and negatively charged oxide particles are shown in *Figure 3.5*. The charged surfaces can attract oppositely charged ions, such as NO_3^- or K^+, respectively. The role

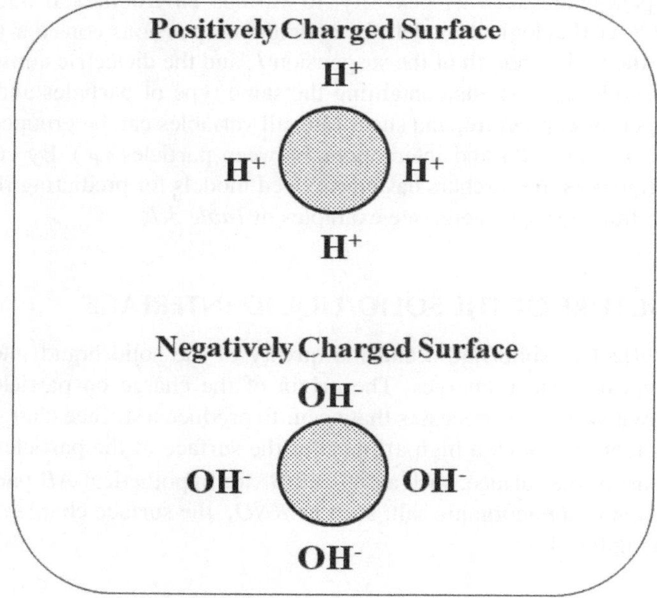

FIGURE 3.5 The charge distribution on oxide particles in contact in media.

of *surfactants*, or *surface-active agents*, in the preparation of suspensions can be summarized by these two methods, basically creating an especially charged surface in suspended particles.

Solid organic particles such as oxides can participate in the replacement of one atom by another of a similar size in a crystal lattice without disrupting or changing the crystal structure during contact with liquid media and create "*isomorphic substitution*" structures. Although there are some difficulties in covalent oxide and non-oxides, it can be easily achieved in *aluminosilicate* (Als: composed of aluminum, silicon, and oxygen, with counter cations with $Al_x^{3+}Si_y^{4+}O_z^{2-}M_{...}^{n+}$ general formula). These structures have the tendency to replace cations inside the crystal structure by cations of lower valency with Si^{4+} and Al^{3+}, like sodium montmorillonite. A typical charge distribution is shown in *Figure 3.6* for an aqueous suspension. The deficit of one positive charge gives one negative charge. The surface of Na-montmorillonite is negatively charged with Na^+ as counterions.

Due to the molecular nature of colloidal interactions in suspensions and the lack of covalent or ionic bond formation, the *Van der Waals* (*vdW*) interactions are the main attractive forces in the system. These bonds originate from the interactions between atomic or molecular oscillations or rotating electrical dipoles in the aqueous

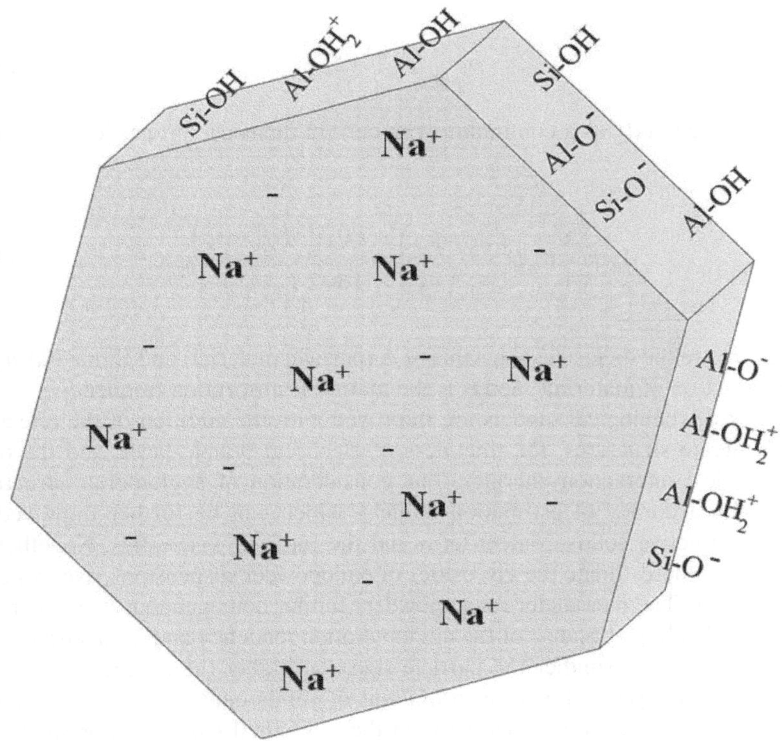

FIGURE 3.6 The charge distribution on oxide Als particle in contact with aqueous media.

and nonaqueous medias and dipole-dipole interactions. If the molecules rotate freely, the dipoles maximize the attractive dipolar attractions and minimize the repulsive forces by aligning the dipoles with respect to each other. Since there is no permanent orientation in isotropic liquids, the molecules correlate themselves according to adjacent neighbor particles, and the net attractive forces occur among the molecules. These forces hold most fluid molecules together. The total attraction potential is calculated on the molecular level with consideration of these forces on a continuum for complex fluid systems. As a result, the *vdW* interaction free energy, V_{vdW}, is equivalent pair-wise to the summation of all atoms in the bodies from the polarizabilities and number densities of the atoms in the two interacting bodies. According to this approximation, V_{vdW} between two approaching spherical particles at *a* distance with media radius can be calculated as:

$$V_{vdW} = -\frac{A_H}{12}\left\{\frac{1}{(x+1)^2-1} + \frac{1}{(x+1)^2} + 2ln\left[1 - \frac{1}{(x+1)^2}\right]\right\} \tag{3.9}$$

Where $x = D/2$ indicates the gap between two adjacent spheres of D diameter, and A_H is the *Hamaker* constant. With the simple approximation ($D \ll a$ or $x \ll 1$), the equation can be simplified to:

$$V_{vdW} = -\frac{A_H a}{12D} \tag{3.10}$$

By treating each body as a continuum with certain dielectric properties, A_H is estimated as:

$$A_H = \frac{3}{4}K_B T\left(\frac{\varepsilon_A - \varepsilon_B}{\varepsilon_A + \varepsilon_B}\right)^2 + \frac{3h\upsilon_e}{16\sqrt{2}}\frac{\left(n_A^2 - n_B^2\right)^2}{\left(n_A^2 + n_B^2\right)^{3/2}} \tag{3.11}$$

Where $\varepsilon_A, \varepsilon_B$ are the dielectric constants of A (particle material) and B (media), n_A, n_B are the refraction of materials, and υ_e is the main UV absorption frequency.

These basic rheological models are improved with the addition of the effects of the agglomerate structures, the thickness of electrical double layer, and the overall interparticle interaction energies. The consideration of agglomerate structures improves the accuracy of estimation models, which can be further improved by adding the effective solids content term and the intrinsic viscosities. Nonetheless, most models underestimate the viscosities of nanopowder suspensions, even when at dilute solutions. The nanometer-range powders inhibit homogeneous dispersion and high particle loadings because of the agglomeration tendency and high surface area of nanoparticles. The smaller the particle size, the higher the number of particles introduced to the system for a given amount of solids content, as a result, creating shorter interparticle distances between the individual particles. For small particles in close proximity, even small disturbances in the system, such as Brownian motion, may result in agglomeration. As the rheological properties of suspensions primarily depend on the interactions in the system, the contribution of the interactions varies significantly for different systems and can be classified into three groups of media-media, particle-particle, and media-particle interactions. The viscosity of the pure liquid media can be determined by media-media interactions. With the addition of small solid particles, these interactions are interrupted, and depending on the interfacial properties between particles and media, different interactions can be introduced. Higher solid particle contents result in particle-particle interactions. The most dominating physical phenomenon in colloidal suspensions is the hydrodynamic drag. This resistance to the motion of a solid particle by the surrounding liquid media, or the resistance to the motion of a fluid caused by solid particles, can be controlled by viscous friction. The light scattering of suspensions with very small particle colloids obeys the *Rayleigh* scattering limit. These physicochemical properties of colloidal particles, their interfaces, and the dispersion media, with the interaction between these particles, the macroscopic properties of the colloidal suspensions, as well as the various characterization techniques, are subjects of colloid science.

The suspended particles experience different physical phenomena, like particle diffusion and particle sedimentation. *Particle diffusion* is explained by its relation to displacement throughout the suspension media. Brownian motion of the scattering centers (i.e., particles) depends on the mean square displacement Δr^2 and is proportional to the translational diffusion coefficient D_t:

$$\overline{\Delta r^2} \sim D_t.t \tag{3.11}$$

The diffusion coefficient is inversely proportional to the translational hydrodynamic diameter $x_{h,t}$:

$$D_t = \frac{K_B T}{3\pi\eta x_{h,t}} \tag{3.12}$$

Also, particle sizing of these suspended particles is related to the quantification of the trajectory lengths (Δr) of the scattering centers for a given time step (Δt). The average displacements Δr of the individual scattering objects for an infinite number of time steps and the mean displacement of a particle in suspension are gathered and measured as a collective average. The measured frequency distribution of the average displacement is a "smeared" projection of the number-weighted size distribution $q_0\left(x_{h,t}\right)$:

$$p\left(\overline{\Delta r}\right) = \int P\left(\overline{\Delta r}, x_{h,t}\right).q_0\left(x_{h,t}\right)dx_{h,t} \tag{3.13}$$

Where the kernel function P depends on the number of time steps. The size distribution $q_0\left(x_{h,t}\right)$can be derived from the distribution of mean displacement $p\left(\overline{\Delta r}\right)$, as the particle diffusion coefficient is inversely proportional to the particle size. For very fine particles, the diffusive transport $-D_t\nabla c$may even outweigh the migration transport, which results in a characteristic, stationary distribution of the local particle concentration.

In *particle sedimentation*, such a state is called sedimentation-diffusion equilibrium, which yields a "barometric" concentration profile for dilute suspensions:

$$c(h_2) = c(h_1).\exp\left(-\frac{\Delta h}{l_g}\right) \tag{3.14}$$

$$l_g = \frac{K_B T}{\Delta\rho V_g} \tag{3.15}$$

Which is described by the characteristic or gravitational length l_g. For high-particle solids content concentrations and long-ranging interparticle forces, the real profile considerably deviates from the aforementioned equation. The sedimentation-diffusion equilibrium in this manner is not just a balance between diffusive transport and settling but can also be considered as an equilibrium between chemical potential and gravity potential and be directly related to residual viscosity.

$$\left(RT.dlnc = -g"\rho V.N_A.dh\right) \tag{3.15}$$

After the start of dispersion during suspension formation, the interfacial energy and variation in its value can be used as a clear indicator of suspension formation. The interfacial energy is a material-specific property and is influenced by temperature and pressure. Its significance results from the fact that its value increases with any increase in the interfacial area. It is an affirmative sign when the droplets of a

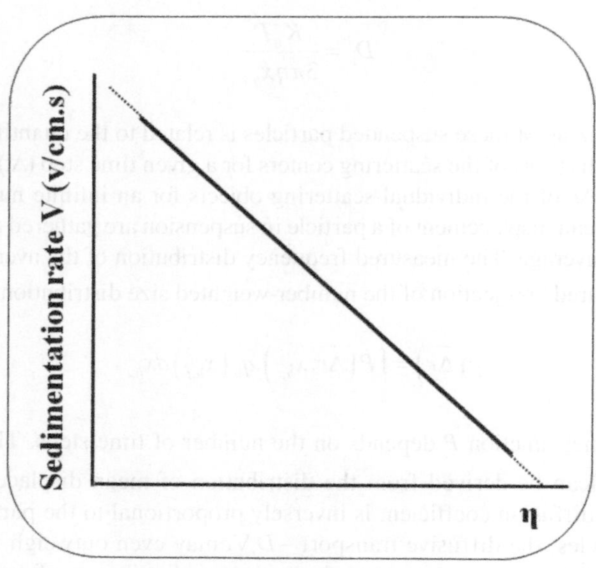

FIGURE 3.7 The relation between sedimentation rate and viscosity of the suspension.

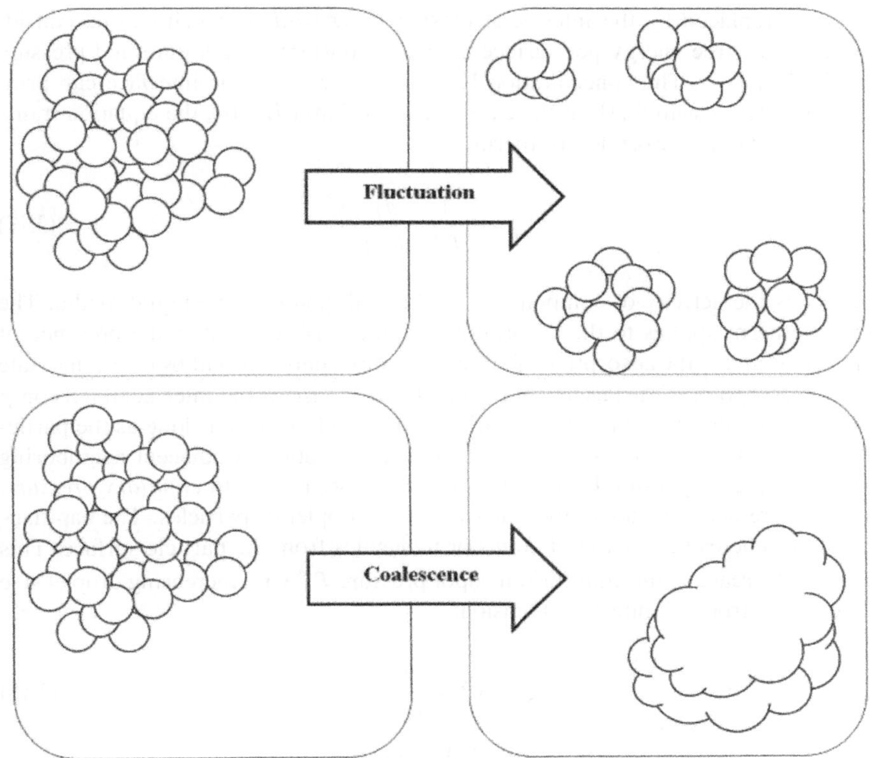

FIGURE 3.8 The changes in free interfacial area during schematic coalescence and fluctuation of particles in the suspension.

colloidal suspension *coalesce* and *fluctuate* (see *Figure 3.8*) and reduce the interfacial area, or when the coarse particles of a colloidal phase grow at the expense of the finer particles. The physical behavior of colloidal particles is largely affected by the properties of the interface between the particle and the continuous phase. The term *interface* does not describe a conceptual two-dimensional boundary but refers to a structured region around the geometric surface of the particle where neither the bulk properties of the particulate phase nor those of the continuous phase prevail. It thus includes the surface atoms of the particle as well as layers of adsorbed ions or molecules, or even the ion cloud that surrounds a charged suspended particle.

Since interfaces are binary or multicomponent systems, their thermodynamics can be described by the Gibbs-Duhem equation:

$$S^{if}\,dT - A_{if}d\pi + A_{if}\,\Sigma``_{i}d\mu_i^{if} = 0 \tag{3.16}$$

Where A_{if}, S^{if}, and μ_i^{if} are the interfacial area, the interfacial entropy, and the interfacial chemical potential of component i, respectively. The interfacial pressure π is frequently replaced by the interfacial tension γ ($d\pi = -d\gamma$), which can be considered as Gibbs free energy per surface area at a constant temperature and pressure ($\gamma = (\partial G / \partial A)_{T,p,n}$). The concentration Γ_i of component i at the interface can principally not be measured [1]. In the case of equilibrium ($dT = 0$), the equation transforms to the Gibbs adsorption isotherm:

$$\Gamma_i = -\frac{1}{RT} \cdot \frac{d\gamma}{dlna_i} \tag{3.17}$$

Where a_i is the activity of component i in the bulk phase of the liquid media. The Gibbs isotherm applies to the adsorption of ionic surface agent in the presence of counterions and to the adsorption of nonionic components. The Gibbs isotherms state that the adsorption of surface agent results in a decrease of the interfacial tension γ. However, the interfacial tension γ will not completely vanish as long as the particulate and fluid phase can be distinguished. The formation of adjacent neighboring particles in the suspension leads to interfacial energy related to *capillary pressure*. This pressure accumulates in the inner part of a droplet or particles. The capillary pressure promotes the release of atoms or molecules from the particle surface. This leads to a decrease of the equilibrium vapor pressure P^{eq} with increasing droplet size and resulting from the interfacial tension:

$$\Delta P = \frac{2\gamma}{R} \tag{3.18}$$

$$ln \frac{P^{eq}(r)}{P^{eq}(\infty)} = \frac{2\gamma . V_m}{\bar{R} . RT} \tag{3.19}$$

The V_m indicates the molar volume, in which $\bar{R}$ is the harmonic mean of the principal radii of curvature.

3.4 ELECTROSTATIC INTERACTIONS

Besides the physical aspects of particles present in the suspension, *electrostatic interactions* spontaneously occur in any system containing charged particles and ions. The natural surface charges of the particles in any media with high dielectric constant result in repulsive interactions between similar particles. *Ionization* or dissociation of surface groups, *adsorption* or binding of ions to the particle surface, or charge exchange mechanisms, such as acid-base type interactions, are the basis of ions present in the suspensions. The interfacial layer formed around the particles in aqueous and nonaqueous medias are referred to as the "*electrical double layer*" *(EDL)* and is shaped by the electrical properties of the particles and the media. The electric double layer consists of subset layers, such as *Stern* and *diffuse* layers [2]. Electric charges on the surface of colloidal particles are the principal behavior of these particles in the aqueous and nonaqueous medias. The electric double layer is formed by the charged surface and the surface-close regions, which exhibits an excess of counterions for charge neutralization. Therefore, it is characterized by the spatial distribution of the co-ions and counterions and by the presence of an electric potential. It depends, to a great extent, on the physicochemical properties of particles and dispersion media. As an example, in an aqueous media, the surface of a ceramic particle interacts with the media and develops a charge on its surface. This interaction builds an interfacial layer between the particle surface and the media. While the interfacial layer thickness (electrostatic double layer) with respect to the particle size for large particles is negligible, it becomes significant in small-sized particles, such as nanopowders. Because the oxide particles are surrounded by their interfacial layer in the suspension, the effective particle size and hence the effective solids content increase and consequently limit the maximum packing density of the particles. The formation of electric layers and their effect on surface electrical potential (ψ) are shown in *Figure 3.8*, with charge distribution at and away from the negatively charged particle surface in contact with an ion- and counterion-filled media for (1) Helmholtz, (2) Gouy-Chapman, and (3) Stern.

This surface charge creates an electrostatic field around suspended particles and consequently affects the scattered ions in the bulk of the liquid. This electrostatic field and the thermal motion of the ions create a countercharged ion and thus orders the electric surface charge. The net electric charge in this screening diffuse layer is equal in magnitude to the net surface charge but has the opposite polarity. As a result,

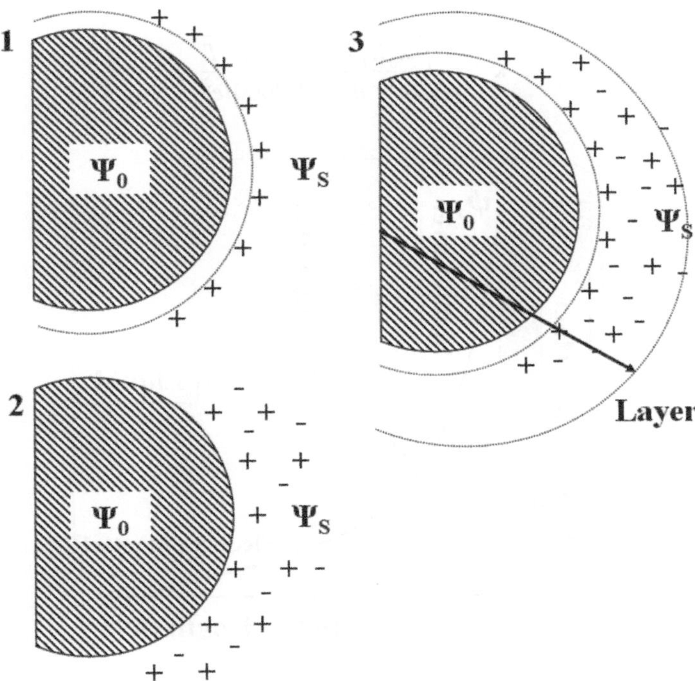

FIGURE 3.9 The changes in free interfacial area during schematic coalescence and fluctuation of particles in the suspension.

the suspension structure is electrically neutral. Distribution of the electric potential in the double layer is shown in *Figure 3.10*, which follows, for a positively charged surface. Electric potential decays almost exponentially, which allows introduction of the "Debye length" as an estimate of the DL thickness. Electric potential drops by approximately 2.7 times at the distance from the surface that equals to the Debye length. Debye length depends mostly on "ionic strength" of the liquid. It is approximately 1 nm at an ionic strength of 0.1 M, and it increases as a reciprocal of the square root of ionic strength (electrolyte concentration), becoming around 10 nm at an ionic strength of 0.001 M. Another important distance within DL is the location of a *slipping plane* associated with the tangential motion of liquid relative to the surface. While the liquid inside of the slipping plane attaches itself to the surface, the liquid beyond the slipping plane moves with the surrounding liquid. The electric potential at the slipping plane is what is referred to as "*zeta potential*" and is measured in millivolts (*mV*), with a maximum absolute value of 100 *mV* in aqueous solutions.

According to this physical phenomenon, the relation between electrical potential and zeta potential can be derived as:

$$\ddot{\ } = \zeta \exp\left(kd_s\right) \tag{3.20}$$

Where the interfacial layer thickness, κ^{-1}, known as the *Debye length*, was derived from the *Poisson-Boltzmann equation (PBE)*:

$$\kappa^{-1} = \sqrt{\frac{\varepsilon\varepsilon_0 K_B T}{2e^2 z^2 n_\infty}} = \sqrt{\frac{\varepsilon\varepsilon_0 K_B T}{2000 I e^2}} \tag{3.21}$$

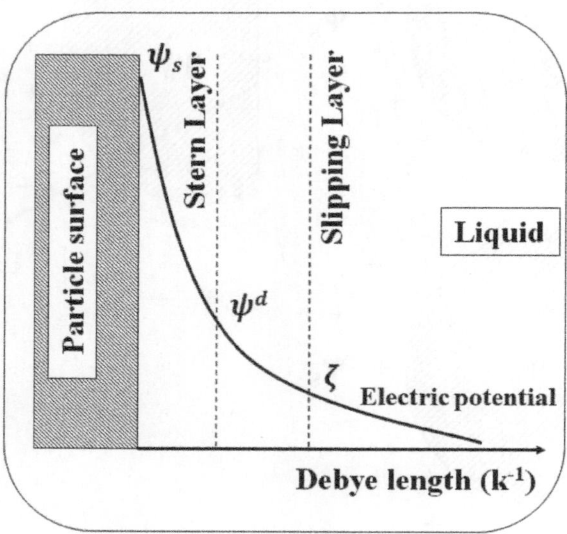

FIGURE 3.10 The position of electric layers along particle surface and the surrounding liquid.

The ionic strength, I, is given by:

$$I = 0.5\sum c_i Z_i^2 \tag{3.22}$$

Where d_s is the double-layer thickness, e is the charge of the electron, z is the valency of the ionic particle, n is the ion concentration, c is the bulk ion concentration, ε is the static dielectric constant of the media, ε_0 is the permittivity of the free space, K_B is the Boltzmann constant, and T is the temperature. As a result:

$$\kappa^{-1} = 0.034\sqrt{\frac{\varepsilon}{n_\infty}} \tag{3.23}$$

The formation of the charged particles in the system introduces attractive and/or repulsive interparticle interactions. The double layer around the particles formed as a result of the interparticle interactions has a profound influence on the rheological behavior of the suspensions. As long as the thickness of the layer, δ, is relatively small, the hard sphere models represent the system accurately. When the ion concentration decreases, the electric double layer becomes thicker and increases the radius of the slide plane. At very low ion concentration, the radius of the slide plane increases and, as a result, increases the resistance of the particles to rearrange and slide against each other in concentrated suspensions and, thus, increases the viscosity according to the following simplified equation:

$$\frac{1}{k} = (\varepsilon KT / 8\pi 2e^2 z^2)^{0.5} \tag{3.24}$$

If the thickness of the layer δ is relatively large, the effective radii of the particles become much larger than the core radius, as does the effective solids content, leading to yield stress and higher viscosity than expected. The effective solids content can be approximated by adding the double-layer thickness to the particle core radius. The effective solids content, Φ_{eff}:

$$\Phi_{eff} = \Phi(1 + \frac{\delta}{R})^3 \tag{3.25}$$

It is due to the fact that some ions from the media adhere on the surface of particles and partially neutralize the surface charge. This layer of motionless ions is another specification for the Stern *layer.* The other ions spread in the liquid media by thermal motion yet are subject to the electric field generated from the charged surface. With growing surface distance, the concentrations of the ionic particles tend to be the equilibrium values of the free liquid media. The region adjacent to the Stern layer with excess of counterions is therefore the *diffuse layer.* In this part of the *EDL*, the ion distribution results from the balance of electrostatic and osmotic forces. The ions may be adsorbed in the Stern layer with or without keeping their hydration shells and may accordingly be located on different planes parallel to the surface (*inner and outer Helmholtz plane—IHP* and *OHP*). The co-ions may be covalently bound to the surface and virtually increase the surface charge, and the structure of the *EDL* may be affected by surface morphology.

The charge separation between the surface and the ions in the Stern and diffuse layer generates an electric field which is described by an electric potential. The potential on the surface is finite and has the same sign as the surface charge. Within the diffuse layer, the electric potential decays exponentially from the diffuse layer potential *wd* to zero in the free media. Following is a schematic representation of the electric double layer with immobilized ions in the Stern layer and mobile ions in the diffuse layer; a further subdivision of the Stern layer is indicated by specifically adsorbed ions located at the inner Helmholtz plane (IHP) and immobilized hydrated counterions at the outer Helmholtz plane (OHP). The OHP marks the beginning of the diffuse layer, where the electric potential exponentially decays with surface distance; the regions between the surface, the IHP, and the OHP are free of charge, and the shear plane or slip plane (SP) marks the transition from stagnant to mobile media:

With the following:

$$\sigma_0 = \sigma_i + \sigma_d \tag{3.26}$$

The concept of the surface of shear, an imaginary surface close to the surface, within which the fluid is stationary, is shown in *Figure 3.12*. The figure illustrates the position of the surface potential ψ_0, the shear plane, and zeta potential (that is close to the Stern potential ψ_d).

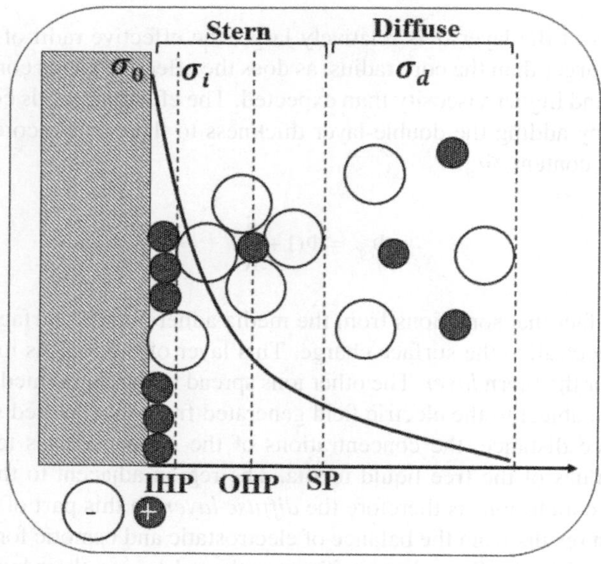

FIGURE 3.11 The schematic presentation of electric layers along particle surface and the accumulated charges.

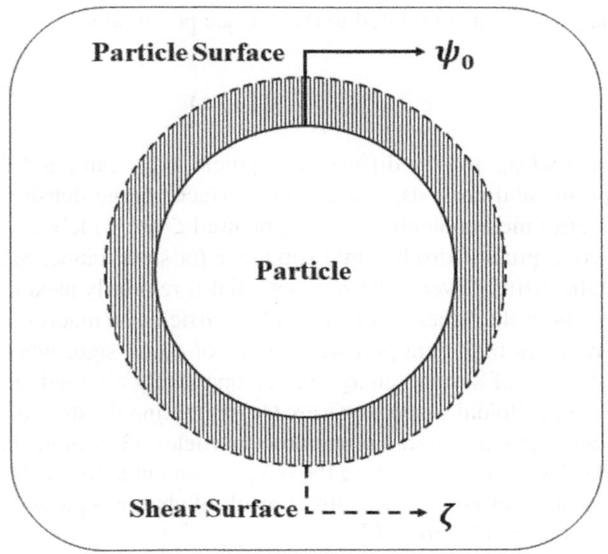

FIGURE 3.12 The schematic presentation of shear planes around a well-dispersed particle in a suspension.

The potential distribution inside the diffuse layer is defined by the *Poisson-Boltzmann equation* (*PBE*). The *PBE* describes a continuous distribution of charge inside the media and on the surface:

$$\nabla\left(\varepsilon_m\varepsilon_0\nabla\psi\right) = -\rho_{e,media} - \sum F_{Z_iC_{n,i}^{\infty}} \exp\left(-\frac{Z_i\psi F}{RT}\right) \tag{3.27}$$

$$\nabla\psi = -\frac{1}{\varepsilon_m\varepsilon_0}\sum F_{Z_iC_{n,i}^{\infty}} \exp\left(-\frac{Z_i\psi F}{RT}\right) \tag{3.28}$$

Where *F*, *R*, and *T* are the Faraday constant, the molar gas constant, and the absolute temperature, respectively. The parameter ε_m is the static permittivity of the liquid, $C_{n,i}^{\infty}$ the bulk concentration of the ionic i, and Z_i its valency. The PBE is frequently reduced to its linearized formulation by the Debye-Hückel approximation:

$$\Delta\psi = -k^2\psi \tag{3.29}$$

Where κ is the Debye-Hückel parameter. For isolated sphere with the surface potential ψ^{iso}:

$$\psi^{iso} = \psi_0^{iso}\frac{a}{r}.\exp\left(-k\left(r-a\right)\right) \tag{3.30}$$

The surface charge σ_0^{iso} can be related to the surface potential ψ^{iso}:

$$\sigma_0^{iso} = \frac{\varepsilon_m \varepsilon_0 \psi^{iso}}{a}(1+ka) \tag{3.31}$$

The surface potential ψ_0 and the diffuse layer potential ψ_d can barely be measured by a direct experimental analysis. Instead, the surface charge density σ_0 (via titration) and the electrokinetic potential or zeta potential ζ are widely used, which correspond to the conceptual hydrodynamic slip plane (or shear plane, SP, between the Stern layer and the diffuse layer). The zeta potential is regularly used to estimate the suspension stability or the adhesion of suspended particles on macroscopic surfaces. The repulsion requires high zeta potential values of equal sign, whereas adhesion occurs in the absence of surface charge or for oppositely charged surfaces. Other applications of ζ in colloidal suspensions are (1) optimizing the dose of dispersion or flocculation agents, (2) maximizing retention of particles, (3) monitoring changes of interfacial properties, and (4) indicating the oxygen content at the surface of ceramic particle. Electrokinetic effects are the direct result of charge separation at the interface between two phases (*Figure 3.13*).

The zeta potential is affected by

- the charge-determining ions, which control the surface charge (e.g., H^+ and OH^- for metal oxides),
- the specifically adsorbed ions which are located in the Stern layer,
- the concentration and valency of all ionic species present in the media, and
- the pH at which the zeta potential is zero (*the isoelectric point* [*IEP*]).

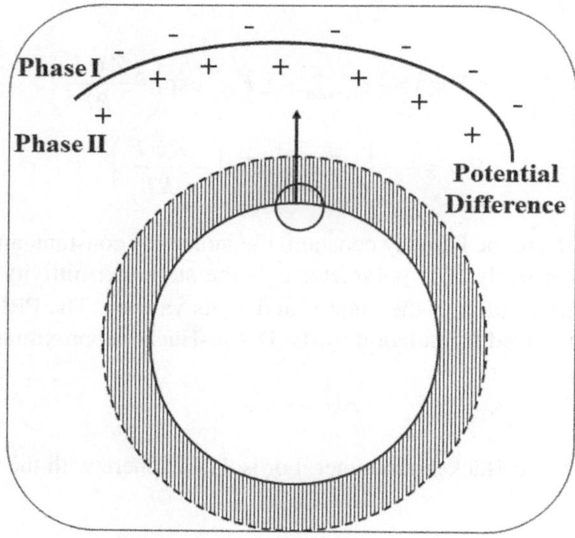

FIGURE 3.13 The schematic presentation of charge separation due to dispersion of a particle in a suspension.

Because of specific ion adsorption, the *IEP* may deviate from the *PZC* (material specific property). As a result, the *IEP* is rather a characteristic parameter of the suspension than a parameter of the particle phase. Similarly, the typical pH dependence of the zeta potential for colloidal suspensions in the presence hydrolysable metal ions is a specification of a suspension. There are three "*points of charge reversal*" (*PCR*), which correspond to the *PZC* of the surface (*PCR*1), to the neutralization of the negative surface charge due to positively charged surface precipitate (*PCR*2), and to the charge reversal of the hydroxide phase (*PCR*3). The second and third *PCR* can only be observed for a sufficiently high concentration of the hydrolysable metal ions and, as a result, for a significantly high coverage of the surface with precipitate.

The second and third PCRs are recognizable in a sufficiently high concentration of the hydrolysable metal ions and for a significantly high coverage of the surface with precipitate. In the case of a complete coverage, the *PCR*3 agrees with the *PZC* of the hydroxide phase. The second PCR depends on the ion concentration and lies above the critical pH, at which the precipitation starts. This pH is approximately one pH unit lower than the critical pH for bulk precipitation. Note that the zeta potential curves of hydrolysable metal ions do not only result from surface precipitation but also reflect the adsorption of the different hydrolyzed ionic species, such as $Al(H_2O)_6^{3+}$ or $Al(H_2O)_5OH^-$, and surfaces with significant solubility, like amorphous SiO_2. In these cases, the precipitate phase may consist of hydroxide and mixed oxides, such as aluminosilicates. For an aqueous system, the effect of such surface charges is limited to the close proximity of the particle's surface. This is because of the high permittivity (i.e., polarity) of water and the high activity of the

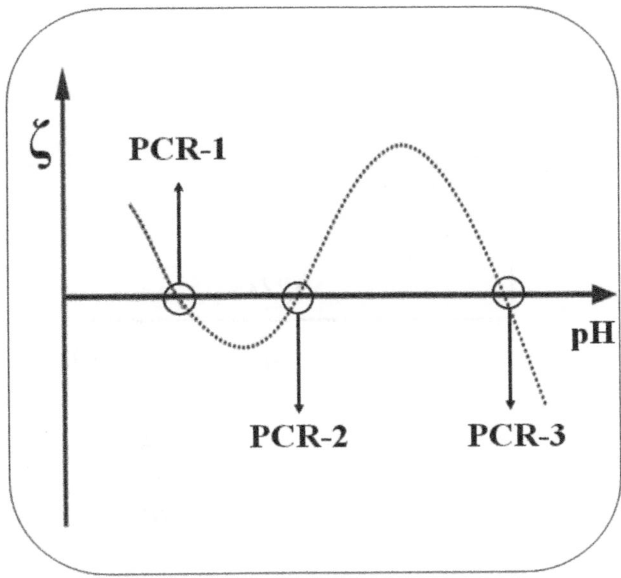

FIGURE 3.14 The schematic presentation of pH dependence of a typical suspension due to surface charges and precipitation of hydrolysable cations.

dissolved ions. As a result, the surface charge is neutralized by an excess of counterions in the vicinity of the surface. This region and the surface can be regarded more simply as two layers of opposite charge and are captured by the term *electric double layer* (EDL) [3]. The surface charge is quantified by the surface charge density σ_0:

$$\sigma_0 = \sum z_i F \Gamma_i \tag{3.32}$$

Where z_i is the valency of the charged surface groups, F is the Faraday constant, and Γ_i is the surface concentration of positive and negatives sites.

The origins for the electric charges at the interface between surface and media can be categorized as:

- Dissociation (e.g., of ionic crystals)
- Acid/base reactions of functional groups (e.g., hydroxyl groups)
- Adsorption/desorption of ionic species (e.g., ionic surfactants/polyelectrolytes)

When (metal) oxides like SiO_2 or TiO_2 are suspended in aqueous media, groups of amphoteric hydroxyl groups are formed, which can be protonated or deprotonated in dependence on the *pH* value via hydrolyzation reactions [4] (see *Figure 3.15*):

$$M-OH \xleftrightarrow{OH^-, K_{a1}} M-O^- + H_2O$$

$$M-OH_2^+ \xleftrightarrow{OH^-, K_{a2}} M-OH + H^+$$

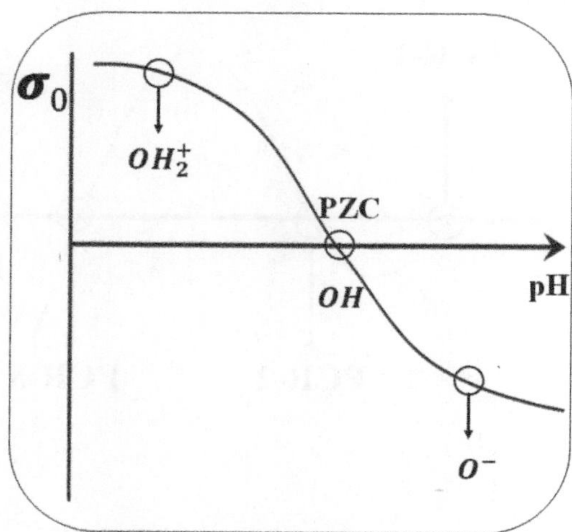

FIGURE 3.15 The schematic surface charge to pH for a typical oxide surface in an aqueous suspension.

The processes of protonation and deprotonation are described by the acid constantsK_{a1} and K_{a2} (acid constants K_{ai}):

$$pH_{PZC} = -\frac{1}{2}\lg\left(K_{a1}.K_{a2}\right) = \frac{1}{2}\left(\rho K_{a1} + pK_{a2}\right) \tag{3.33}$$

The surface distribution of electrical charges results in "*electroviscous*" properties in suspension, which is especially iimportant in ceramic colloidal suspensions. Basically, the presence of the *EDL* affects the hydrodynamic drag of charged particles in a way that the viscosity of the liquid phase might slightly increase. Likewise, the flow through a porous material is apparently retarded when an *EDL* is formed in the pores. These effects and their consequences on viscosity of a suspension is evident in different groups. The *primary* electroviscous effect indicates the retardation of the particle velocity (or the fluid flow in pores) due to the polarization of the *EDL*, which is a consequence of the relative motion between diffuse layers and the particle surfaces. The resulting polarization causes a flux of ions and media that opposes and decelerates the relative motion. This kind of deceleration affects particle sedimentation and viscosity of a suspension. For thin double layers ($\kappa a \gg 1$) and small potentials ($|\zeta| \leq 25 mV$), it can usually be neglected. For concentrated suspensions, the retardation effects observed at interparticle separations larger than 5 *nm* include the retardation effect and are valid for separations smaller than particle size:

$$E_{vdw} = -\frac{A_H}{6}\frac{a_1 a_2}{h\left(a_1 + a_2\right)}\left[1 - \frac{bh}{\lambda}\ln\left(1 + \frac{\lambda}{bh}\right)\right] \tag{3.34}$$

Where A_H is the Hamaker constant, a_1 and a_2 are the radii of the spherical particles at the separation distance of h, b is a constant equal to 5.32, and λ is the characteristic wavelength for the internal molecular motion, which is generally taken as 100 nm. The *secondary* electroviscous effect in concentrated suspensions involves the motion of particles in a crucial zone, affected by the hydrodynamic interaction between neighboring particles, which strongly depends on interparticle distances, i.e., on the suspension structure. This effect is directly influenced by interparticle forces, in particular by the forces that occur when the *EDL* of two particles overlap. As the particle concentration approaches the dense packing values for the charged particles, the double layers around the particles will start to overlap because the small separation distances between them are independent of the repulsive forces, which results in the agglomeration (or flocculation) of the particles. The degree of overlapping can be visualized by calculating the average separation distance S_0 between individual particles:

$$< S_0 >= a\left[(\frac{\Phi_{max}}{\Phi})^{\frac{1}{3}} - 1\right] \tag{3.34}$$

When a suspension contains only a single particulate component, such a double-layer overlap leads to repulsions and thus decreases the particle mobility and increases the suspension viscosity. This behavior depends on the absolute value of the zeta potential and with the ratio of double-layer thickness to mean interparticle distance. The *tertiary* electroviscous effect is related to the dependence of the interfacial geometry on the parameters of the EDL. This can be observed for adsorbed polyelectrolytes, whose conformation on the particle surface (stretched, random, etc.) is determined by surface charge and ionic strength. The *quaternary* electroviscous effect happens as a real change of the liquid viscosity within the EDL and is explained with an orientation of (polar) media molecules in vicinity of the charged surface. This effect is considered to be relevant to a thin region of a few layers of molecules immediately adjacent to the particle surface. The electroviscous effects are presented in *Figure 3.16* with a closely detailed mechanisms in the primary and secondary effects.

The interactions of charged surfaces with ions and molecules are directly related to the coulomb electrical forces. The *coulombic* attraction between the charged surface and the counterions in the media gives rise to ion adsorption on the surface [5]. Additionally, ion adsorption can be promoted by physical forces and chemical bonds and thus may even apply to co-ions. High charge densities and high electrolyte concentration cause strong ion adsorption that coincides with further ion-specific effects. For instance, hydrolyzed metal ions can precipitate on the surface as hydroxide phase. A particular case is the adsorption of surfactants, because they form associated structures at particle surfaces.

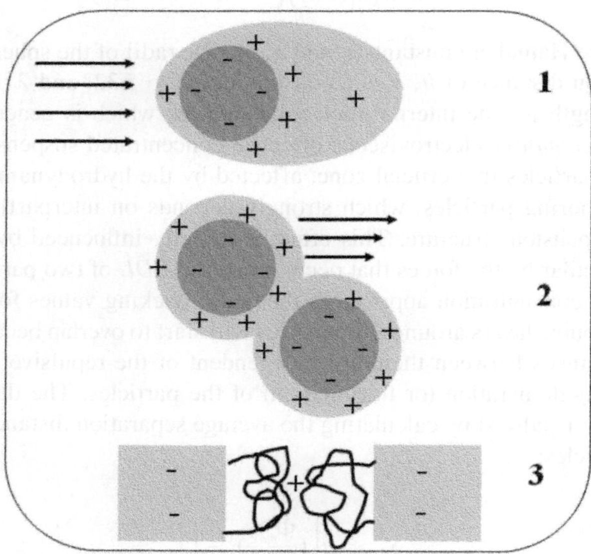

FIGURE 3.16 The schematic representation of the most important electroviscous effects (above) and details on primary and secondary effects (below).

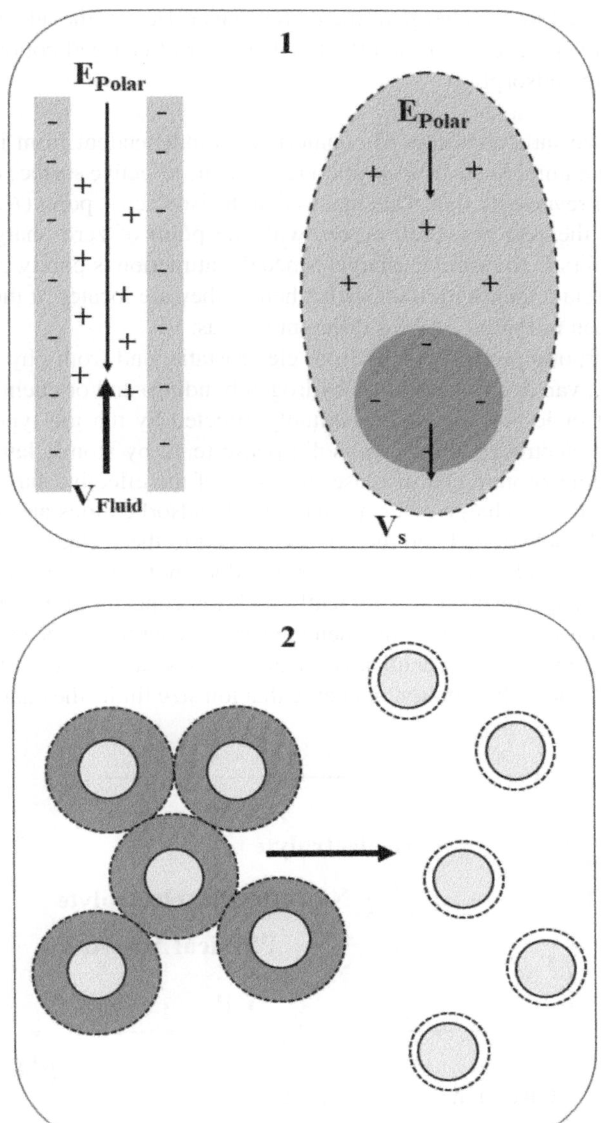

FIGURE 3.16 (Continued)

The other direct effect of charged surfaces happens during the *adsorption* of ions at surfaces.

Although the formation of the EDL is governed by the surface charge and the charge-determining ions, e.g., H for oxides, the interaction with other EDLs is mainly determined by the charge of the diffuse layer. Hence, the adsorption of ions in the Stern layer is crucial for the effective behavior of charged colloidal particles. Therefore, during adsorption:

- The electrostatic attraction of counterions is independent from ion size or chemical nature. Such an adsorption reduces the effective surface charge but can never reverse its sign. That means that the isoelectric point (*IEP*), which refers to the zeta potential, agrees with the point of zero charge (*PZC*), which refers to the surface charge. Since the attraction is purely coulombic, the ions retain their hydration shells; hence, they are located at the *OHP*, as can be seen in Figure 3.17 for different medias.
- The adsorption process results from electrostatic, and from physical adhesion (e.g., van der Waals forces, hydrogen bonding) and/or chemical bonds (covalent or ionic), too; it is inevitably affected by the ion type (e.g., by mass and electronegativity) and solid phase (e.g., by atomic lattice). Such a specific adsorption can increase the value of the effective surface charge or may lead to a charge reversal. Specifically adsorbed ions are partially or completely dehydrated and directly associated to the surface. The physical adsorption is relevant for counterions; it does not affect the IEP but can result in charge reversal at sufficiently high ion concentrations. In contrast, chemisorption can occur even when the surface charge has the same sign as the ion. Consequently, chemically adsorbing ions can affect the IEP. There is no continuous distribution of charge that ion size limits the charge density

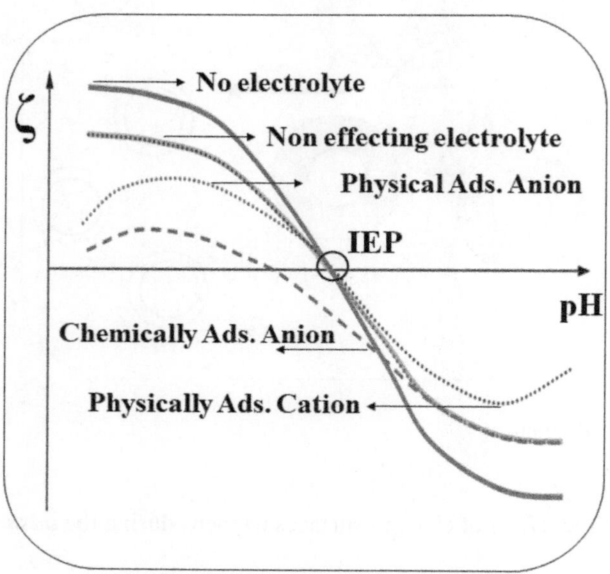

FIGURE 3.17 The effects of electrolytes on cations and anions on trends of zeta potential in different pH ranges.

in the diffuse layer and in the Stern layer, as well as that the electrostatic and steric interactions between ions result in a certain spatial correlation hydration of ions, and surfaces also influence the ion adsorption in the Stern layer. The mobile ions in the diffuse layer interact with the surface not only by coulombic forces but also by van der Waals forces. The PBE will consider the account for the corresponding potential energy additional to the electrostatic energy potential $z_i \psi F / RT$; the nonelectrostatic ion-surface interaction can explain the systematic differences observed for alkali cations, which were previously attributed to hydration. The ion-specific effects are observed for high ionic strengths (≥0.1 M), and the higher the ion's valency, the more likely their occurrence.
- The adsorption process can also include surface complexation, while the individual (hydrated) ions work well for low amounts of adsorbed ions. The adsorption of ionic species, or surface ionization, depends to a large extent on the surface charge, thus on the concentration of the charge-determining ions, and is affected by the nonelectrostatic ion's affinity to the surface and by ion-ion interactions. Additionally, there is a significant influence of ion hydration—in particular for metal cations—which essentially depends on the pH value. Hydration shells hydrolyze with increasing pH, which may result in weakly soluble hydroxides and negatively charged hydroxide complex ions. Hence, the adsorption of ionic species and its effect on the net charge of surface and Stern layer are a function of pH.

The special charge distribution may involve:

- Surface precipitation. If the surface concentration of hydrated multivalent cations or the pH exceeds a critical value, surface precipitation of the respective hydroxide is induced. This qualitatively changes the surface properties and frequently coincides with charge reversal.
- Surface dissolution. The interfacial properties of colloidal suspensions are determined by the chemical reactions (e.g., protonation) and adsorption of solutes. Additionally, the interface can be affected by the dissolution-precipitation equilibrium of the particle phase. The precipitation changes the surface morphology dissolution, which means degradation of the particles, and may result in the loss of the finest particle fractions.
- Dissolution of oxide ceramics. Solubility is a material property; only at the lower nanoscale does particle size matter. The dissolution of oxides strongly depends on the reducing element. Even so, the dissolution of metal oxides can be described by the following general reaction:

$$MO_{x/2(s)} + \frac{x}{2} H_2O \leftrightarrow M^{x+} + xOH^-$$

The released metal ions M are covered by a hydration shell, the size of which depends on the valency of the ion and the size of the nucleus. The hydrate complexes $[M(H_2O)_n]^{x+}$ of multivalent ions can deprotonate (i.e., they act as acids), which affects the chemical equilibrium of the reaction. As a result, the apparent solubility is a function of the pH value. Deprotonation leads to a transformation of the hydrate complexes into hydroxide complexes $[M(H_2O)_n]$. The

neutral hydroxide complexes are hardly soluble. The total amount of dissolved material and its composition can be calculated from the acid-base equilibria of the complex ions and the pH (see *Figure 3.18*). In general, the lower the main group of the metal, the higher the oxide solubility. Dissolution is an equilibrium reaction. That is, the dissolution of the surface is accompanied by the precipitation of the solute oxide species [6].

A similar effect is expected for metalloids and nonmetal oxides, with the dissolution reaction producing acid molecules with oxyanions. Deprotonisation of these acids eventually causes a pH dependence of the solubility, similar to metal oxides. The formation of a gel-like layer on a silica surface results from reprecipitation of dissolved silica and consists of polymeric $-Si(OH)_2-O-Si(OH)_2-OH$ chains. The thickness of this layer is estimated at about 10 Å.

Multivalent ions (such as surface-coordinated ions like Cu[II] and Al[III]) on the dissolution of oxides can effect suspendability of nonmetallic and metalloid (like Si) oxides. The mechanism involves a blocking of the oxide surface groups by such ions. In the case of Al(III), additional effects are the formation of aluminosilicate complexes between Al(III) ions and the silanol (Si-O-H) groups and the precipitation of hydroxide phase at high Al(III) concentrations. In multicomponent suspensions, suspended species from the particulate components participate in adsorption and precipitation on the interfaces and have effects on the suspensions' stability.

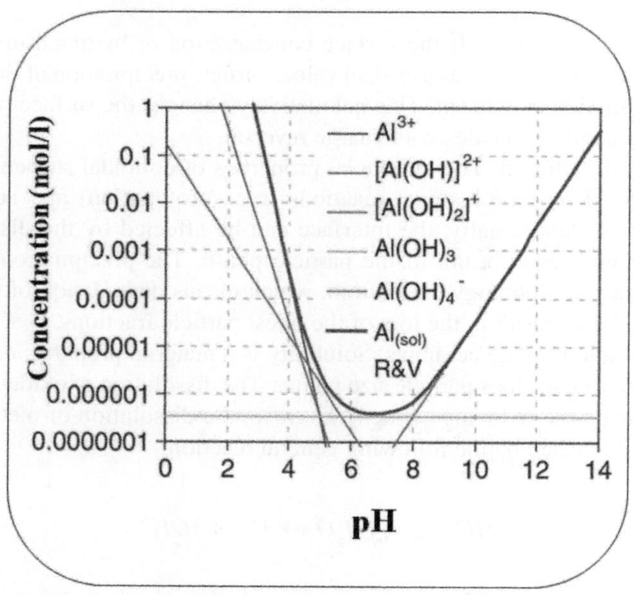

FIGURE 3.18 The Al ion hydrolyzation in aqueous suspensions versus pH and dissolved ion concentrations.

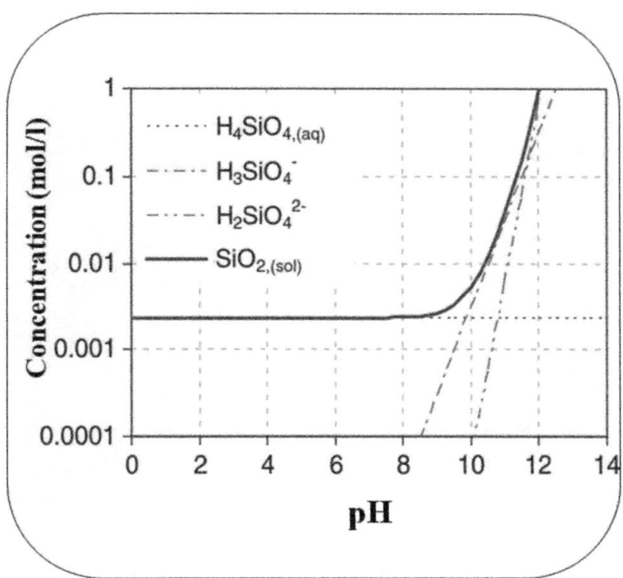

FIGURE 3.19 The Si ion hydrolyzation in aqueous suspensions versus pH and dissolved ion concentrations balanced with orthosilicate acid.

3.5 COLLOIDAL SUSPENSION AND ADDITIVES

The formation of most suspensions is directly related to the adsorption of *surfactants* on charged surfaces. Surfactants are molecules that consist of a nonpolar hydrophobic (e.g., alkyl chains) and a polar hydrophilic part, which can be cationic, anionic, zwitter ionic, or nonionic. The relevance of the polar part for the molecule behavior is quantified in the form of *hydrophilic-lipophilic balance* (*HLB*), for which values above 7 indicate dominance of the hydrophilic groups. The surfactant adsorption is determined by the hydrophobic and electrostatic interaction between single-surfactant molecules and the surface for very low surfactant concentrations. In this situation, there is no interaction between the adsorbed molecules, and the molecular surface area is rather large. By increasing surfactant concentrations, the surfactant molecules start to interact (attraction of hydrophobic parts, repulsion of identically charged groups) and tend to orient into a preferred vertical position. As a result, the effective surface properties change, e.g., the hydrophobicity or the charge density. The interaction of adsorbed molecules (via hydrophobic parts and charge) eventually results in bilayers or molecule aggregates at very high surface concentrations. The key factors for the adsorbate structure include charge and the total electrolyte concentration, which indicates the signs of surface and surfactant. The opposite signs of charge promote the adsorption of surfactants and lead to a neutralization or even a reversal of surface charge. The electrolyte media determines the *EDLs* around the particle and around the surfactant molecules. The high salt concentrations control the coulombic interactions and promote a dense packing of ionic surfactants. A further important factor is

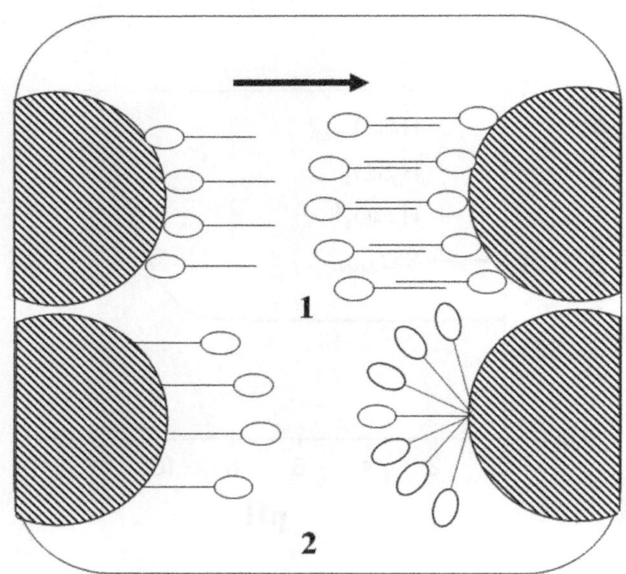

FIGURE 3.20 The schematic structure of the adsorbed nonionic surfactant (with linear hydrophobic head) on a (1) hydrophobic surface and a (2) hydrophilic surface.

temperature, as it affects the configuration of the molecules, the *HLB*, and structure of micelles. The heterogeneous composition leads to amphiphilic properties and interfacial activity, which means surfactants accumulation on interfaces and reduction of the interfacial energy. The surfactant molecules associate to nanosized aggregates and micelles above a critical concentration (*the critical micelle concentration—CMC*). These molecule associations exist in the bulk solution and can be formed by adsorbed surfactant molecules. The adsorption of surfactants on solid particles is affected by surface properties (hydrophilicity/hydrophobicity, surface charge) and by surfactant properties (ionic/nonionic, CMC, HLB). The possible configurations of nonionic surfactants on (1) hydrophobic and on (2) hydrophilic solid surfaces are shown in *Figure 3.20* with the effect on increasing concentration (shown by the arrow).

A similar trend is imaginable for cationic surfactants on negatively charged oxide layers. The formed molecular structure of the adsorbed surfactant consists of monolayers, bilayers, hemi-micelles, and surface micelles.

3.6 COLLOIDAL SUSPENSIONS AND INTERPARTICLE INTERACTIONS

The colloidal suspensions are directly affected by interparticle interactions to the extent that the range and the strength of interactions considerably affect the macroscopic behavior of colloidal suspensions, such as *stability*, *rheology*, and *turbidity*.

The particles in colloidal suspensions interact with each other through hydrodynamic coupling and by a number of attractive and repulsive forces. These interactions are related to the particles' bulk and interfacial properties. The most particular are electrodynamic interactions between the atoms and molecules of the particles (van der Waals interaction) and those interactions that originate from the overlap of electric double layers (double layer interaction). The *Derjaguin-Landau-Verwey-Overbeek* (*DLVO*) theory comprises these two types of interactions for pairs of interacting particles and calculates the corresponding (excess) free energy as a function of particle distance.

The *van der Waals interaction* comprises interactions between atoms and molecules with fixed or induced polarization, which result in a strong attractive force for short distances.

The attractive van der Waals (vdW) forces, the repulsive electrostatic forces, and the steric forces are the operative system forces commonly examined to predict and explain the behavior of particle suspensions, especially when the particles size is in the micrometer scale. The short-ranging and rapid decaying with surface distance are the characteristic of this phenomena. For this reason, it usually suffices to calculate the interaction for minute surface gaps ($h \ll x_i$). For two spheres of diameter x_1 and x_2, with h as the minimum surface distance and A_{123} as a coefficient, which depends on the materials of the particles (1 and 2) as well as on the media 3 that separates them, the vdW is:

$$vdW = -\frac{A_{123}}{12}\left[\frac{x_1x_2}{h.(x_1+x_2+h)} + \frac{x_1x_2}{(x_1+h).(x_2+h)} + 2\ln(1 - \frac{x_1x_2}{(x_1+h).(x_2+h)}\right] \tag{3.35}$$

$$A_{123}(h) = A_{123,s} + (A_{123,0} - A_{123,s}).(1+(h/h_{cr})^{3/2})^{-2/3} \tag{3.36}$$

$$vdW = -\frac{A_{123}x_1x_2}{12h(x_1+x_2)} \tag{3.37}$$

By considering the decreasing contribution of the induced dipoles to the interaction energy with growing distance between the macrobodies, the interaction potential is inversely proportional to the minimum surface distance.

The interparticle interactions can spread into *double-layer interaction*. When two particles approach each other so that their *EDLs* overlap, they experience a repulsive or attractive interaction that results from the electrostatic and osmotic forces between the ions and the surfaces. The overlap of the diffuse layers means a local increase in counterion concentration, which produces a disjoining osmotic pressure. The local increase in counterion concentration produces a disjoining osmotic pressure. In the *overlapping* region, the distributions of potential and ion concentration are changed compared to isolated particles, and the surface charging locally adapts to this new environment. Therefore, the overlap of the diffuse layers of two neighboring particles affects the dissociation and adsorption equilibrium. The surface charge σ_0 and surface potential w_0, as well as diffuse charge σ_d and diffuse layer potential w_d, can experience changes in the overlapping region. For instance:

$$\sigma_0 = S_{reg} - K_{reg}\psi_0 \tag{3.38}$$

To estimate the regulation parameters S_{reg} and K_{reg} and how they can allow for a broad variation of charge-regulation regimes. When the regulation capacity K_{reg} depreciates, the *constant charge* situation is obtained. In contrast, the surface potential remains constant for very large regulation capacities ($K_{reg} \gg \varepsilon_m \varepsilon_0 k$). For an isolated sphere:

$$S_{reg} = \psi_0^{iso} \varepsilon_m \varepsilon_0 \left(k + \frac{1}{a} \right) + K_{reg} . \psi_0^{iso} \tag{3.39}$$

The *interaction potential* is directly related to this overlapping energy. The interaction energy of overlapping double layers can be estimated from the osmotic pressure and the stress tensor or via the free energy of the double layers. The free energy of an electric double layer results from the spontaneous surface charging and from the subsequent formation of the diffuse layer, i.e., from the charge separation. Since the overlap of double layers alters the diffuse layer and surface charge, the related free energy changes as compared to the case of isolated particles. This approach is also theorized by the *DLVO* theory [7]. The interaction between particles usually results from different forces and is quantified as total interaction energy V_t for a pair of approaching particles in dependence on the minimum surface distance h. This theory estimates how fast colloidal particles aggregate or how the particles are spatially arranged in the liquid media. Strong repulsion, thus good suspension stability, is achieved for low electrolyte and high zeta potential values evaluation of the suspension stability via double layer and van der Waals interaction. The stability of colloidal systems is determined by the net interaction potential between particles as the summation of *vdW* and electrostatic potentials. By plotting the net interaction energy as a function of distance, the stability of the colloidal system can be estimated. The *DLVO* is applicable for calculating the net interaction potential between particles as a function of separation distance using the summation of individual forces. The net potential helps to predict the stability and rheological properties of the suspensions. Processing additives are used to stabilize the colloidal systems because they restrict the attractive forces or enhance the repulsive interactions between particles to increase stability. Based on the functional dependency $V_t(h)$, the criteria that allows an evaluation of the suspension stability can be estimated. The shape of the function $V_t(h)$for the total interaction energy depends on the Hamaker function $A_{123}(h)$, the diffuse layer potentials w_d (approximated with the zeta potential), and the ionic strength I (or the Debye-Hückel-Parameter κ). Similar to distance dependency of Gibbs free energy, the interaction energy can be affected by h, as shown in *Figure 3.21*, with $G_t(h)$ and $V_t(h)$ estimated as balances of electrostatic repulsion and attraction.

The curves consist of two local minima and one local maximum, while at vanishing surface distance ($h \ll 1nm$), the total interaction energy tends toward positive infinity due to the *Born* repulsion. The *primary* minimum originates from the van der Waals attraction, which excels the double layer interaction in value for small surface distances. The primary minimum indicates that adhesion is an energetically possible configuration for two interacting particles. The *secondary* minimum at moderate surface distances is neglectable. It can only be observed for strong van der Waals interaction and small Debye lengths. It requires a significant repulsion from

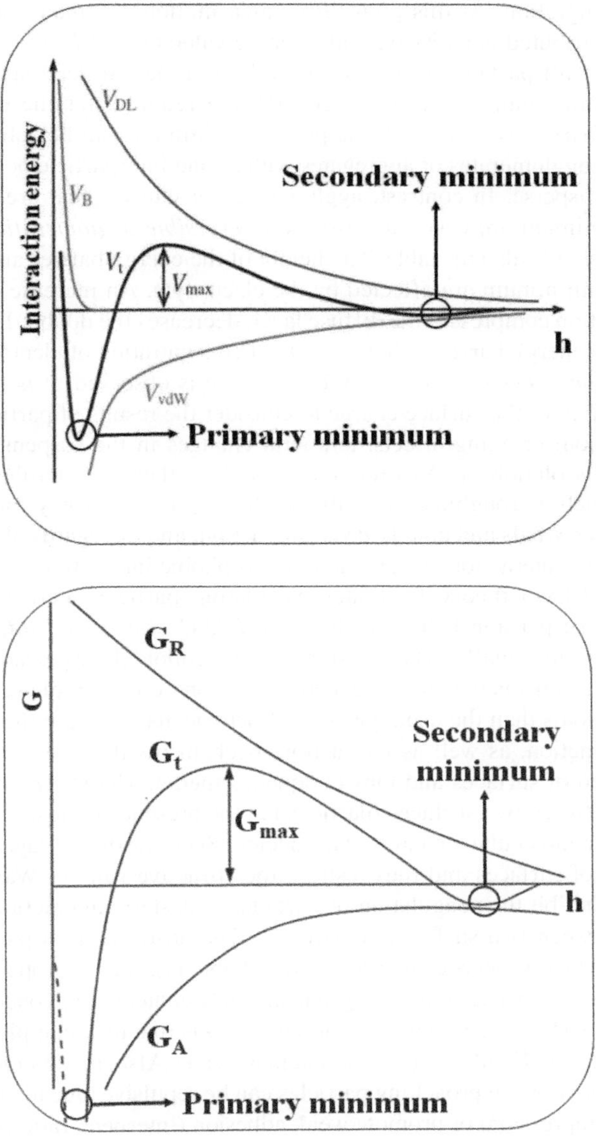

FIGURE 3.21 The schematic variation in interaction energy and Gibbs free energy $V_t(h)$–$G_t(h)$ with surface distance.

the double layer interaction, which is obvious by the local energy maximum V_{max}. This maximum is called the energy barrier because it corresponds to an energetically critical configuration and can prevent two particles from adhesion (i.e., from attaining the primary minimum). The energy barrier is effective at a certain value which exceeds the kinetic energy of the colliding particles, in the absence of shear flow and particle migration. At this point, Brownian motion dominates and the kinetic energy is a distributed quantity with an average value of 1 kBT. In order to prevent coagulation for all particles over a sufficiently long period, leading to suspension stability, the minimum values of 10–20 kBT are required for the energy barrier. If the energy barrier is surpassed, the primary minimum can be achieved, and the particles form agglomerates or aggregates with strong interparticle bonds, which are difficult to redisperse. In contrast, agglomerates or clusters that are related to the secondary minimum are easier to disperse ("*reversible agglomeration*"), although such a suspension is also unstable. The height of the energy barrier and the depth of the secondary minimum are affected by the electrolyte. An increase in the electrolyte concentration compresses the diffuse layer, decreases the double-layer repulsion, and lowers the energy barrier. Above a critical concentration of electrolyte, *the critical coagulation concentration*, quick coagulation is observed. It is not possible to explain the origin of the surface charge to consider the results of particle interaction (e.g., coagulation) or aging effects, temporal changes in the suspension properties (e.g., due to dissolution or Ostwald ripening). This theory simplifies the view on the surface (such as roughness) and the double layer, avoids any interdependence between van der Waals and double-layer interaction, and disregards all other mechanisms of particle interaction (e.g., steric or hydrophobic interaction).

While the DLVO theory facilitates calculating particle interactions and estimation of the suspension behavior, the "*non-DLVO effects and interactions*" are usually relevant for small surface distances. For colloidal suspensions, which are interaction mechanisms which are related to a second colloidal phase, it consists of much finer colloids than the usual particles. Such mechanisms are the steric and the depletion interaction, as well as interaction mechanisms that are somehow related to the hydration of surfaces and ions in aqueous media. The *steric interactions* are interactions between two surfaces (particles) in the presence of adsorbed macromolecules or other molecular or particular species. Such adsorbate layers prevent the direct contact of surfaces and thus restrain the attractive van der Waals forces and can therefore inhibit the coagulation of particle. This steric interaction refers to the interaction between two surfaces or particles. The adsorption on particle surfaces depends on surface properties (charge, hydrophobicity), media properties (polarity, proticity), media properties (charged groups), solids content, and concentration. The condition for a stabilizing effect of adsorbed media is that they completely cover the particles' surface and bridging flocculation may occur. Also, the interaction between adsorbed layers of two approaching particles can be repulsive (due to charged groups or to elastic compression) or promote weak adhesion (interpenetration of the layers).

The depletion interactions are interactions between two surfaces (particles) in the presence of free, i.e., nonadsorbed, micelles, or very fine particles. If the distance between two surfaces h is smaller than the diameter of the media molecules d_M, this region will contain pure media (depletion zone). Therefore, an attractive force corresponding to the osmotic pressure of the bulk solution is acting on the two surfaces. Agglomeration caused by this effect is called *depletion flocculation*, as shown in *Figure 3.22*.

While this attractive depletion interaction prevails for very small surface distances ($h \ll d_M$), there is a repulsive interaction when the distance between the approaching

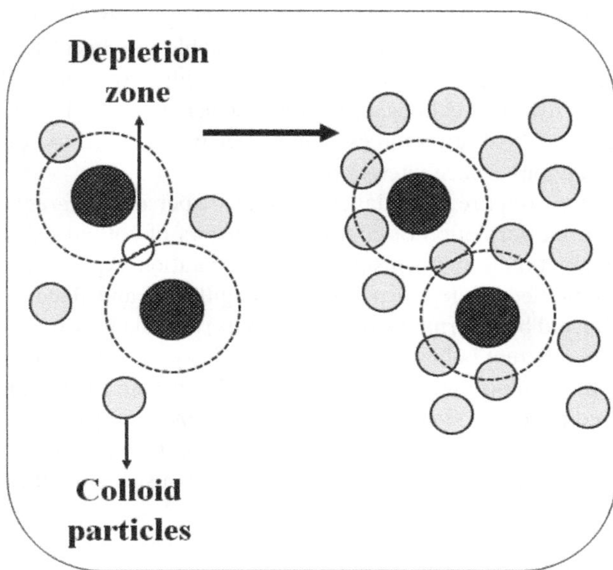

FIGURE 3.22 The depletion interaction between two, with nonabsorbing colloidal particles at low (left) to high (right) concentrations.

surfaces is in the order of magnitude of the macromolecules ($h \approx d_M$) at the start of depletion. The *hydration forces* (especially in oxide ceramics suspensions) indicates the existence of a short-ranging repulsive interaction that cannot be explained by the DLVO theory or the presence of micelles. Since it is observed in particular for hydrophilic surfaces and polar medias, its origin is traditionally attributed to the solvation of the surface and is called solvation or hydration interaction. The hydration interaction decays biexponentially with surface distance with the coefficients L_1 and f_1:

$$V_{hyd} = \pi . x . \left(L_1 f_1 . e^{-\frac{h}{L_1}} + L_2 f_2 . e^{-\frac{h}{L_1}} \right) \tag{3.40}$$

The hydration interactions are common for different oxide materials such as mica, titania, or silica and are absent after the hydrophilization of silica. As an example, in silica-based systems, the influence of ion hydration, the hydration interaction with gel layer, and gel formation on silica surfaces are some of the side effects of hydration in this system. There are several indications that silica surfaces may behave differently as far as the hydration interaction is concerned. While the background electrolyte has a direct effect on this type of interaction for most surfaces, there is no such clear dependency for silica surfaces. This phenomenon is related to the hydration of the surface, which is due to oriented adhesion of water molecules on surface groups via hydrogen bonding and may comprise several layers of water molecules. Such hydration layers have different properties than bulk water, interact repulsively with each other due to dipole interaction, require energy to be destroyed, and have a steric effect. Also, some other effects are responsible for short-ranging repulsion between two particles, such as surface gel layers that reduce the van der Waals attraction, adsorption of hydrated cations, or volume exclusion by the (hydrated) ions. It is believed that that the "hydration forces" can be partly attributed to nonelectrostatic, van der Waals, and ion-surface interactions.

Most suspensions require colloidal *aggregates* dispersion. *Aggregation* is a process that leads to the formation of contiguous clusters of colloidal particles, called aggregates. Aggregation is an intrinsic step in precipitation. Aggregation can also be induced by adding electrolytes or polymers (coagulation and flocculation, respectively) and is affected by external fields (e.g., gravity) and the flow regime. Therefore, aggregation is an inherent behavior of many particle systems with different extents. Aggregation is most important for colloidal particles because their adhesion forces frequently exceed the dispersion forces driving two particles apart. Mostly, the presence or absence of aggregates is interesting and helps in evaluating the stability of colloidal suspensions. During preparation of colloidal suspension, it should be considered that [8]:

(i) Aggregates have a larger surface area than compact particles of the same geometric size, which means higher adsorption capacity, better catalyst performance, and faster dissolution.
(ii) Aggregates are porous, i.e., relatively little solid material is required to occupy a certain space, which is employed for thickening agents. The pores may have been used for the storage of other substances.
(iii) Aggregates are larger than the primary particles. This enhances solid-fluid separation and powder handling. It affects optical properties (turbidity).
(iv) Aggregates have certain mechanical properties which determine the aggregate breakage during dispersion or the erosion of primary particles under mechanical stress. The mechanical properties are also relevant for the elasticity of particle gels.

The aggregate morphology can also be an indicator of suspension stability. The aggregates are neither spherical, homogeneous, nor isotropic. In the course of aggregation, the porosity of the evolving aggregate typically increases. This means that the aggregates' mass does not increase with the third power of their size but follows some noninteger power law (with exponents between 1 and 3). Such aggregates are called *fractal* aggregates.

Aggregation *mechanisms* and *kinetics* are the other affecting parameters on suspension stability. The morphology (i.e., size and structure) of aggregates is closely related to the microprocesses that affect the aggregation process (e.g., particle diffusion and migration, particle adhesion, and aggregate breakage or consolidation). Therefore, it depends on a lot of microprocess and material parameters.

It is commonly agreed that aggregation is the result of *collision* and *adhesion*. Collision means a relative motion between the particles. The major cause for the collision of colloidal particles is their *Brownian* motion. Additionally, relative motion can be caused by laminar shear flow, turbulence, or particle migration (e.g., settling). Hence, collision is primarily governed by the flow regime and by external fields.

Additionally, collision is affected by long-ranging interparticle forces, i.e., by double-layer interaction for colloidal suspensions. In the case of very strong repulsive forces, the collision rate can be reduced by several orders of magnitude. A third factor is particle concentration, and the higher the concentration, the higher the collision frequency, which is directly related to the Brownian motion. In diffusion-limited cluster aggregation, there are weak repulsive particle interactions by which colliding particles and/or aggregates adhere at first contact. Similarly, in the reaction-limited cluster aggregation, there is significant repulsive interaction, and the collision of two approaching particles is very unlikely. As a result, it is more likely to happen after many trials. Apart from Brownian motion, media flow or particle migration can be responsible for particle collision. The adhesion aggregation requires attractive interparticle forces between two particles in contact. Usually, such an attraction is caused by London–van der Waals interaction short-range repulsive forces, e.g., originating from adsorbed molecule layers, it may outweigh the van der Waals forces and impede the adhesion even in the case of small surface distances.

The aggregation at this stage coincides with further processes that affect the growth and morphology of aggregates, such as the breakage of particle bonds, the restructuring within the aggregate (consolidation), the alteration of contact zones due to sintering, or the (slow) formation of covalent bonds between particle surfaces in contact (e.g., siloxane bonds for silica). The repulsive interactions should not be ignored for electrically charged suspended colloids with high surface charges (with a thin double layer). The two basic aggregation concepts are the diffusion-limited cluster aggregation (*DLCA*) and the reaction-limited cluster aggregation (*RLCA*), during which particles and clusters of particles diffuse through the surrounding medium and stick together irreversibly through diffusion or reaction (see *Figure 3.23*). In *DLCA*,

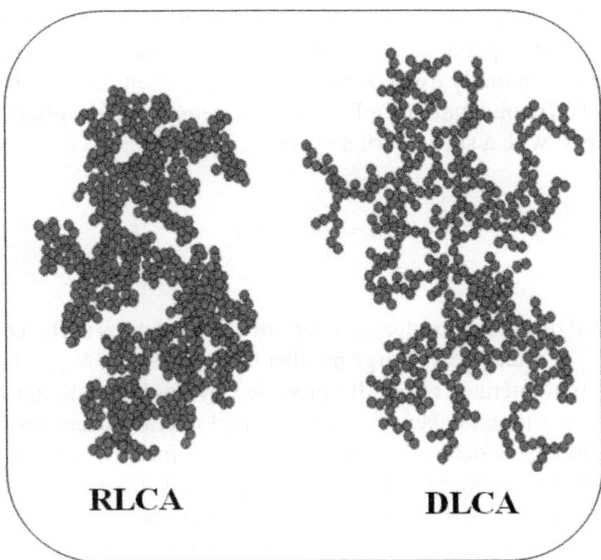

FIGURE 3.23 The depletion interaction between two, with nonabsorbing colloidal particles at low (left) to high (right) concentrations.

a particle is so strongly attracted to a nearby aggregate or another particle that it sticks irreversibly during the aggregation process, and during *RCLA*, the particle doesn't irreversibly stick on its first encounter with another particle or aggregate but instead takes multiple counters in order to stick.

3.7 COLLOIDAL SUSPENSIONS: DISPERSION AND STABILITY

Suspensions of colloidal particles mostly have some agglomeration, which clearly affects the macroscopic suspension properties (e.g., viscosity or turbidity). In order to adjust the particle distribution in suspension, it is necessary to pretreat the powders and destroy the agglomerates and aggregates (i.e., dispersion) beforehand. As a result, the obtained size distribution depends on the employed dispersion techniques, on the microscopic stresses acting on the particles, on the total energy input, as well as on the mechanical strength of agglomerates and aggregates. The state of dispersion in colloidal suspension is not a constant property and changes over time, e.g., due to Ostwald ripening and recrystallisation and more dominantly through coagulation. To ensure the stabilization of a colloidal suspension, special measures should be applied, such as the interfacial properties, so that, e.g., double-layer and steric repulsions are increased. The occurrence of coarse, micrometer-sized agglomerates in a colloidal suspension and the dispersed particles is related to particle synthesis (e.g., precipitation), powder processing (e.g., spray drying), or agglomeration in stored powders and suspensions. The deagglomeration of these agglomerates, i.e., the dispersion, tries the elimination of very coarse agglomerates, which have adverse effects on the suspension quality and the homogeneous distribution of primary particles, or at least at a general reduction of particle size, in order to achieve certain quality requirements (e.g., viscosity or stability).

The aforementioned details clearly indicate the importance of *dispersion of colloidal suspensions*. The dispersion of agglomerates requires a sufficiently high mechanical stress τ_{disp} that eventually exceeds the agglomerate strength σ_{agg}. The ratio of the two quantities, the fragmentation Fa Fa, is thus a measure on the effectiveness of the dispersion process, with a special criterion of:

$$Fa = \frac{\tau_{disp}}{\sigma_{agg}} > Fa_{cr} \tag{3.41}$$

The successful dispersion or deagglomeration also requires sufficiently long periods of stress t_{stress} or a sufficiently large number of stress events N_{stress}. This is because the breakup of agglomerates is usually preceded by agglomerate deformation and/or fatigue processes that weaken the original agglomerate strengths. The strength of an agglomerate σ_{agg} is determined by the adhesive forces F_{ad} among its primary (constituent) particles, its internal structure (e.g., porosity, fractal dimension), and the size of the agglomerate and/or primary particles. In general, σ_{agg} can refer to different types of stress. The dispersion of colloidal suspensions, especially the breakup of colloidal aggregates, increases the total interfacial area and induces changes in the bulk properties of the liquid phase (e.g., salt content, pH) and changes the interfacial properties. In the case of highly intense dispersion processes, like stirred media milling, such changes may be even due to mechanochemical activation of the solid surface. Consequently, it is necessary to monitor and regulate the bulk and interfacial properties of the colloidal suspension during the dispersion process. Otherwise, reagglomeration of the dispersed particles is fairly likely.

The principal aspects of dispersion, such as the impact of agglomerate size and structure, are commonly based on the tensile strength σ_T of aggregates. Tensile strength is related to homogeneous, isotropic, and randomly packed agglomerates. In this case, there is no impact of the agglomerate size but rather of its internal structure, i.e., on the agglomerate porosity ε_{agg} and on the average number of contacts to neighboring particles N_c (coordination number), as well as on the primary particle size x_p:

$$\sigma_T = \frac{1-\varepsilon_{agg}}{\pi} N_c \frac{F_{ad}}{x_p^2} \tag{3.42}$$

This equation is mostly applicable for arbitrarily shaped primary particles, which may be compact individuals or aggregates of even finer particles. The equation can be further refined by considering the interdependence of porosity ε_{agg} and coordination number N_c. Rumpf proposed a simple relationship ($N_c \times \varepsilon_{agg} \approx 3.1 \approx \pi$) which is based on experimentally determined coordination numbers in fixed beds of spheres and thus obtained:

$$\sigma_T = \frac{1-\varepsilon_{agg}}{\varepsilon_{agg}} \frac{F_{ad}}{x_p^2} \tag{3.43}$$

The agglomerate surface is more densely packed, and its interior regions and, as a result, the (tensile) strength of an agglomerate cannot be considered constant throughout the whole agglomerate volume. This fact consequently changes the agglomerate dispersion by external stress, and three fundamental types are distinguishable:

- The erosion of the agglomerate surface, which results in a steady size reduction of the agglomerate size mode and the appearance of a fine size mode and is related to the eroded primary particles or aggregates; erosion is dominant for small stresses in $1 < Fa < 100$ range.
- The fragmentation of the agglomerate, which yields several fragments, the size of which is in the same order of magnitude, and fragmentation occurs at high stress.
- The shattering of the agglomerate, which indicates rupture into a large number of fragments considerably smaller than the original agglomerate; shattering is expected at extremely high stresses ($Fa > 104$).

The type of deagglomeration depends on the type of stress which is applied to the agglomerate. Deagglomeration in colloidal suspensions mainly results from the interaction with the continuous liquid media phase and from viscous forces and pressure fluctuations. The stress on the agglomerates is closely related to the flow field. For example, in laminar shear flow, the pressure and shear forces cause rotating agglomerates. A flow field is laminar when it is governed by the viscous properties which apply very slow fluid velocities or very high viscosities, which is typical for highly concentrated suspensions. There are mainly three ideal types of laminar flow—the uniform flow (e.g., experienced by settling particles), the shear flow (e.g., in rheometers or pipes), and elongational flow (e.g., in nozzles and diffusers)—and only the first two are relevant for deagglomeration. Particles that move in a laminar flow field with velocity gradient $\dot{\gamma}$ experience shear and normal stresses, which vary along/across the surface and induce particle rotation and deformation. The rotation of spheres is stable with an angular velocity, ω:

$$\omega = 0.5 \times \dot{\gamma} \tag{3.44}$$

Whereas aspherical particles or agglomerates rotate in a quasiperiodic or even chaotic manner. This rotation means fluctuating hydrodynamic forces on the surface elements of the particle. The maximum stress for spherical particles can be calculated as:

$$\tau_{disp,max} = 2.5\tau_{lam} = 2.5\eta\dot{\gamma} \tag{3.45}$$

Where τ_{lam} is the shear stress in the undisturbed flow field. The shear stress on the particle is independent from the particle size. The parameter τ_{lam} can be used to calculate the power density P_V, which allows a comparison with other types of dispersion:

$$P_V \propto \gamma\tau_{lam} \propto \eta\dot{\gamma}^2 \tag{3.46}$$

In addition to the P_V, the effect of dispersion, i.e., the degree of deagglomeration, depends on the number of revolutions N_{stress} (i.e., numbers of stress alternations). That is due to the weakening and loosing of particle bonds in the agglomerate and the subsequent agglomerate deformation, which barely happen instantaneously but require some time. The number of rotations is derived from:

$$N_{stress} = f.t_{stress} \propto \dot{\gamma}.t_{stress} \tag{3.47}$$

The *laminar elongational* or converging/strain flow also can exert excessive stretching of the agglomerates due to the fact that the velocity gradient is parallel to the direction of flow. This leads to a stretching of particles by tensile stress and may eventually lead to their rupture. In contrast to laminar shear, there is no permanent rotational motion, and as a result, the particle experiences a quasistatic load. The close study shows that a strong deformation (apparent stretching) of the agglomerates before fragmentation took place. The resulting fragments keep the deformation as long as they stay in the flow field and appear more compact than the original agglomerate. The simulation of agglomerates of monosized particles with uniform adhesive forces indicates that the average number of particles in the fragments anti-correlates with the hydrodynamic stress. That means that increasing stress coincides with a transition from *fragmentation* to *shattering*. The third type of deagglomeration, *erosion*, is barely present in colloidal suspensions. Under basic flow conditions (moderate stirring speed), the elongational regime is more effective to break up aggregates than simple shear flow. The deformation of single droplets in ideal elongational flow in a typical colloidal suspension is much more effective than shear when the viscosity ratio is high ($\eta_{droplet} / \eta > 4$).

The application of *turbulent* flow shows that the dominant stress depends on the agglomerate size, pressure fluctuation, or shear forces. The turbulent flow occurs in any sufficiently rapid flow when the fluid inertia exceeds its molecular friction. It is the typical flow regime when suspensions of relatively low viscosity are dispersed—e.g., by stirring or in nozzles. The turbulent flow is characterized by multiscale eddy structures that unregularly move through the flow field and cause local fluctuations in velocity and pressure. The velocity fluctuations can be quantified by the effective velocity difference $\overline{\Delta u'^2}$ over a distance Δr:

$$\overline{\Delta u'^2} \propto (P_V / \rho)^{\frac{2}{3}}.\Delta r^{\frac{2}{3}} \ (\Delta r > 20 \times l_D) \tag{3.48}$$

$$l_D = (\eta^3 / \rho^2 P_V)^{1/4} \tag{3.49}$$

Where P_V is the volume-specific dissipation rate (power density) and l_D is the Kolmogorov scale of microturbulence. In the viscous subrange, i.e., within the smallest laminar eddies, the velocity difference scales with distance and is affected by the fluid viscosity η:

$$\overline{\Delta u'^2} \propto (P_V / \eta.\Delta r^2 \ (\Delta r \leq 5 \times l_D) \tag{3.50}$$

The deagglomerating effect of turbulence relies on these velocity gradients, which lead to pressure fluctuation, local deformation, and oscillatory motion for particles in the inertial subrange and which also represent laminar shear for particles in the viscous subrange. The stress related to the pressure fluctuation (or turbulent shear) in the inertial subrange can be considered as a kind of *Reynolds* stress:

$$\tau_{ti} = \rho \Delta u'^2 \propto \rho^{1/3} P_V^{2/3} x_{agg}^{2/3} (\Delta r > 20 \times l_D) \tag{3.51}$$

In fact, the deagglomeration in turbulent flows starts with a rapid rupture of large flocs or agglomerates within the inertial subrange and ends up with a comparatively slow size reduction within the viscous subrange. Turbulent flow can be realized in different experimental setups at varying intensity for power densities of less than 100–104 mW / m^3. There is a wide range of achievable size distributions (from a few microns up to a few mm). It is important to understand the scaling laws of dispersions, which require the size dependency of stress and the agglomerate strength. The scaling law for the maximum agglomerate size that stay intact during the dispersion process after a sufficiently long treatment can be estimated as:

$$x_{max} \propto P_V^{-\alpha} \tag{3.52}$$

Where the α power is in the 0.27–0.4 range for most turbulent flows.

The *cavitation* field of intense local agitation of the flow field by collapsing bubbles consists of stresses similar to turbulent flow. This is the phenomenon of bubble formation and collapse in liquid flows of high velocity or in intensive ultrasonic fields and results from the evaporation of the liquid phase or from the release of dissolved gases at zones of low pressure. It occurs when the cavitation number falls below a critical value:

$$N_{cav} = \frac{p - p_{sat}}{0.5\rho u^2} \leq N_{cav,cr} \tag{3.53}$$

Where p_{sat} denotes the saturation pressure of the liquid phase, ρ the liquid density, and u the characteristic flow velocity. In real liquid medias, the ideal critical value ($N_{cav,cr} = 0$) might have positive or negative values. This is because the cavitation is affected by dissolved gases, by the presence of nucleation centers, and by the degree of turbulence. The cavitation bubbles start to grow until they reach flow regions of high pressure ($p > p_{sat}$) and, in the case of ultrasonication, until the bubbles have reached a critical size. In contrast to this growth, the collapse of bubbles is a rapid, instantaneous process which creates shock waves and microjets with diameters of 1 μm to 10 μm and velocities up to several hundred m/s. Particle agglomerates which are exposed to these intensive velocity and pressure fluctuations experience high mechanical forces and possible fragmentations. The cavitation may cause particle-particle or particle-wall impacts and induce deagglomeration. The hydrodynamic stress on a particle generated by bubble collapse can be estimated by:

$$\tau_{cav} \propto \rho c_m u_{cb} \tag{3.53}$$

The quantity c_m is the velocity of the compressional wave (i.e., the sound velocity of the liquid media) and u_{cb} is the velocity of the collapsing interface (i.e., the microjet velocity). The calculated stresses exist only in the close proximity of a collapsing bubble. The dispersing effect of cavitation, therefore, depends on the bubble concentration—or, more exactly, on the number N_{cb} of collapsing bubbles per unit volume and unit time. This quantity, correspondingly, the rate of deagglomeration, increases with rising power input through the flow field.

Ultrasonic dispersion, an alternative to dispersion in flow fields, is the application of highly intensive ultrasonic waves and is probably the most common procedure after severe stirring for suspension dispersion and deagglomeration in so-called ultrasonic baths or with ultrasonic horns (sonotrodes). The former belongs to standard laboratory equipment and is widely used for the preparation and homogenization of suspensions. The stress intensities in such baths are low, which causes some practical limits for the dispersion of strong aggregates of colloidal particles.

Additionally, they facilitate the deagglomeration in the colloidal size range. Sonotrodes are usually optimized for aqueous solutions and suspensions and typically have diameters between a few millimeters and several centimeters [9]. The generated sound fields work with frequencies in the range of 20 to 100 kHz and have a nominal power consumption of 50 W to approximately 1 kW. When operated in a batchwise mode, the induced turbulent field should be intensive enough to ensure the mixing of the complete sample volume, since only a small zone beneath the sonotrode tip actively contributes to the dispersion process. Alternatively, there are flow cells for continuous operation, in which the suspension flows axially onto the sonotrode tip and where the dispersion zone is usually defined by the cell geometry.

The primary dispersing effects of ultrasonication are cavitation and the corresponding hydrodynamic stress induced. Additional effects discussed in literature are particle-particle collisions and the hydrodynamic stress due to acoustophoretic motion of particles. The latter was regarded relevant for high sound frequencies (800 kHz) and coarse particles in the range >10 μm, but simulations for agglomerates of colloidal particles at typical frequency values (24 kHz) showed that this effect is in orders of magnitude less important than cavitation. The maximum stress intensities acting on particles during ultrasonication are considered for energy and for force quantities. Some authors estimated the specific rate of dissipation (power density) based on caloric data, which yielded values in the range of $10^3 - 10^5 \frac{kW}{m^3}$.

The definition of the dispersion zone, the volume where the ultrasonic field significantly contributes to deagglomeration, is intensity-dependent and decreases as the ultrasonic intensity decays exponentially with distance from the sonotrode. A different approach is to consider the hydrodynamic stress τ_{cav} in the cavitation field. For a continuously operated ultrasonic cell with a local dissipation rate P_V of 105 kW/m^3, hydrodynamic stresses up to 100 MPa are yielded. The ultrasonication facilitates the dispersion of colloidal suspensions at relatively high stress intensities. It relies on the presence of cavitation and on a deep penetration of the sound field into the suspension. The latter is guaranteed by low suspension viscosities (e.g., as for aqueous media with low solids content).

The mechanical dispersion of suspension involves stirred media milling. The stirred media mills are commonly used for the comminution of fine solid materials up to submicron or even nanometer regions. This is achieved by pumping the feed suspension through a highly agitated bed of a grinding media (*GM*), which consists of coarse beads (e.g., from glass or ceramic materials) that are much larger (from 100 μm to a few mm) than the feed particles. The stirred media milling is used for fine grinding of crystalline solids, for the disintegration of organic materials, or for the dispersion of colloidal suspension. The continuous agitation of the grinding media leads to frequent and fast collisions between the coarse beads. During these collisions, the feed suspension is squeezed out from the contact region, which causes intense hydrodynamic stresses on the suspended feed particles. However, some of the feed particles remain in the contact zone and are eventually captured by the two beads, thereby experiencing high normal stresses. The second mechanism is mainly relevant for rigid, compact particles, while the first applies chiefly to deagglomeration. The decisive parameters for the final size distribution are the average number of stress events N_{stress} of each feed particle, the stress intensity, and the specific energy input Em. The number of stress events refers either to the rapid flow between the surface of two approaching particles or to the particle capture between two beads in contact. The typical values of the stress intensity lie in a range of $10^{-5} - 10^{-2}\ Nm$, and N_{stress} increases linearly with dispersion time t and rotational speed ω of the stirrer:

$$N_{stress} \propto \omega.t / d_{GM}^{1-2} \tag{3.54}$$

The mechanical dispersion can also be achieved by *disc systems under laminar operation*. The disc systems are agitated vessels where the fluid motion is induced by quickly rotating discs with usually toothed rims. They are frequently employed for the preparation of stable suspensions with high solids contents and are recommended for highly viscous emulsions. Disc systems are used to suspend and homogenize particles in the liquid phase and facilitate their (subsequent) dispersion. The deagglomerating effect of such systems is clearly promoted by high suspensions viscosities, which guarantee the typical, essentially laminar flow field. The tangential velocities of the disc rim amount to 20–30 m/s under typical operating conditions. Decisive for dispersion is the maximum shear rate γ_{max}, which can be approximated by the disc's angular velocity and is usually in the order of 1,000 s^{-1}. The stress on the particles is proportional to the suspension viscosity, which increases disproportionately strongly with the solids content. Best dispersion effects are, therefore, achieved at very high particle concentrations close to the maximum packing, which is qualitatively different to other types of dispersion machines. Conversely, disc systems show a rather-poor performance at low suspension viscosities, which often coincide with low particle concentrations.

The mechanical dispersion through *rotor-stator systems* is the other mechanism for a turbulent flow dispersion. Rotor-stator systems consist of a rotating tool (rotor) and a coaxially fixed wall (stator). The suspension is brought into the thin annular gap between rotor and stator, where the particles experience intensive hydrodynamic stress. Two types of rotor-stator systems can be distinguished. In colloid mills, the suspension flows axially through the conically shaped slit between the two surfaces, which can be either smooth or toothed. The (adjustable) gap width typically lies in the range of a few hundred micrometers, and the rotor operates with a speed of 1,000–5,000 min, which corresponds to tangential velocities up to 40 m/s. Colloid mills are usually operated under turbulent conditions, but laminar flow is also possible. The primary dispersing effect in rotor-stator systems is the hydrodynamic stress in turbulent flow, even though particle-wall collisions may play a role for large particles as well. Hence, the degree of deagglomeration is determined by the specific dissipation rate or power density in the gap between rotor and stator, i.e., the dispersion zone. Maximum values that can be realized lie in the $10^5 - 10^6\ kW/m^3$.

During *high pressure dispersion*, high pressure systems use the kinetic energy of strongly accelerated flow field for the dispersion of colloidal suspensions and emulsions. There are different geometries of the dispersing unit, which can be classified as radial diffusers, counterjet dispersers, and axial nozzle systems. These systems are operated at very high pressures (several hundred up to some thousand bar), which results in extremely high flow velocities in the smallest cross sections. High-pressure dispersion systems are widely used for emulsification below 1 μm, such as high-pressure post feeding (*HPPF*) system.

The parameters that determine the effectiveness of dispersion can be categorized as

- the dispersion time t_{disp} usually normalized by the time period or frequency that the particles are really stressed (t_{stress} or f_{stress}), which yields a dimensionless dispersion time or a number of stress events;
- the power density *PV* (specific energy dissipation) in the zone of dispersion, which is considered to determine the "steady state" of dispersion processes, i.e., for sufficiently large dispersion times ($t_{disp} \rightarrow \infty$);
- the specific energy input (*EV* or *Em*), which either refers to the total suspension volume or to the total particle mass and is employed for relatively short dispersion times, i.e., the "steady state" is not being achieved; and
- the maximum normal and tangential stresses τ_{disp} acting on the agglomerate surface.

The microscopic stresses τ_{disp} on single agglomerates with time, power input, and energy consumption are process parameters. The power density is the characteristic quantity of turbulent flow. It determines the size of the smallest eddies and the intensity of microturbulence. In addition, it is a measure of the "shear intensity" in laminar flows or the "intensity of cavitation" in ultrasonic fields (see prior). The power input *P* in the dispersion zone can be derived from the pressure drop (e.g., in pipes and nozzles). The power density was shown to be the crucial parameter for the "final" size distribution of large micrometer flocs in turbulent flow.

The size reduction of agglomerates is frequently described as an exponential decay:

$$\ln\left(\frac{\overline{x} - \overline{x}_{\infty}}{\overline{x}_0 - \overline{x}_{\infty}}\right) \propto -t \tag{3.55}$$

Where $\overline{x}_0$ and $\overline{x}_{\infty}$ denote the incipient and final mean particle size, with the latter being dependent of power density.

Applying dispersion of pyrogenic powders (ceramic) requires extra effort, as pyrogenic powders are flame-made materials that are composed of fractal aggregates of nanosized particles from amorphous (e.g., silica) or crystalline (e.g., titania) materials. These aggregates are usually further assembled into coarse, micrometer-sized agglomerates. In order to destroy the large agglomerates, which could otherwise depreciate the suspension's homogeneity, heavy mechanical stresses should be applied. Further, for uniformly distributing the aggregates in suspension and creating homogeneity, the fineness of primary particles can be achieved by some type of deagglomeration for instance erosion, fragmentation, or shattering. There is a strong correlation between mean particle size and electric energy density, which implies that upscaling, which is not affected by the ultrasonic power, of ultrasonic dispersion is possible. The strong correlations between the mean particle size and the energy density have been reported for dispersion techniques. They are frequently described by a power law:

$$\overline{x} \propto E_V^{-\alpha} \tag{3.56}$$

Where $\overline{x}$ is some average value of particle size (e.g., a volume-weighted median or an intensity-weighted mean) and E_V denotes the energy density. Alpha values are at the 0.07–0.15 range for 0.1–0.3 μm^3 particle size and 1–10 GJ / m^3 energy density for micrometer-sized oxides. The nature of such fine size fraction depends on the strength of the interparticle bonds which are affected by the conditions of the pyrolysis process.

3.8 COLLOIDAL SUSPENSIONS AND THE PROGRESS OF DISPERSION

The dispersion of agglomerated or aggregated particle systems in a liquid media is often quantified as reduction of particle size. More specifically, this means a shift of the whole size distribution to the lower end of the size axis, a decrease of characteristic size parameters (e.g., medians, means, or maximum sizes), and a decline of the coarse particle fractions. A number of different measurands is utilized in dispersion monitoring:

- The complete particle size distribution profile
- The characteristic average values of the particle size distribution
- The maximum particle sizes
- The weight/concentration of coarse size fraction
- The photometric quantities (*turbidity*) and rheology

The colloidal attractions and flocculated dispersions are the main factors affecting the rheological response of suspensions. A small quantity of flocculated fine particles at a few percent range might be enough to induce pronounced rheological changes. The low shear viscosity in particular can increase enormously and can even evolve into an apparent yield stress below which the suspension does not flow.

As an example, there is a critical solid volume fraction for aqueous alumina suspension. If the solid volume fraction is below the critical value, the suspension shows *shear-thinning* behavior all along. On the other hand, when the solid volume fraction reaches or exceeds the critical value, the rheological behavior changes from *shear thinning* to *shear thickening*. For oxide suspension systems, the solid volume fraction of particles (SVF) formula follows thus:

$$\phi = \frac{m / \rho_{th}}{m / \rho_{th} + V} \tag{3.57}$$

Where m is the mass of alumina powder (g), ρth is the theoretical density of oxide, and V is the volume (ml) of the premix solution in the suspension. The effects of solid load content on rheological behavior of a typical oxide suspension are shown in *Figure 3.24*.

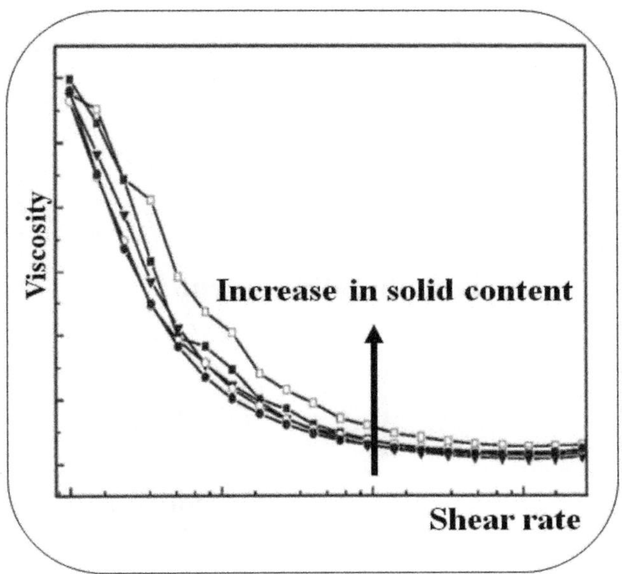

FIGURE 3.24 The viscosity dependence of a typical oxide suspension to the applied shear rate and solids content.

The viscosity at low shear rates is high and decreases as the shear rate increases. Moreover, the viscosity decreases faster at low shear rate. When the shear rate increases to a certain value, the viscosity curve becomes smooth. The entire set of curves shows shear-thinning behavior. A closer study of these suspensions at higher shear rates is represented in *Figure 3.25*. The suspension with the lowest *vol%* solid loading shows shear-thinning behavior in the higher shear rates, and the viscosity increases distinctly. Furthermore, it increases to high viscosity, which indicates that the whole system presents shear-thickening behavior. The viscosity corresponding to the critical shear rates increases with the increase of solid volume fraction, and the increment speeds up over at high contents.

The critical shear rate decreases with the increase in the solid loading of an aqueous alumina suspension and decreases swiftly over certain solids content, which suggests that an aqueous suspension at high solid loading undergoes a transition from *pseudoplastic* to *plastic* behavior at lower shear rate (*Figure 3.26*).

The oxide systems provide a broad operational *pH* range because of low solubility and slightly alkaline isoelectric point. As mentioned, the isoelectric point, IEP, is defined as the *pH* value of an oxide suspension at which the net electric charge of the particles is zero. Because the repulsive forces are created by the surface charge, stable suspensions can be obtained at pH values away from the *IEP* and makes the range of study with oxide particles more flexible.

The particle shape of most oxide systems is presumed spherical for developing rheological relations and equations. For oxide particles of the order of 1 μm or smaller in size (nanoparticles), the Brownian motion introduces an effective force

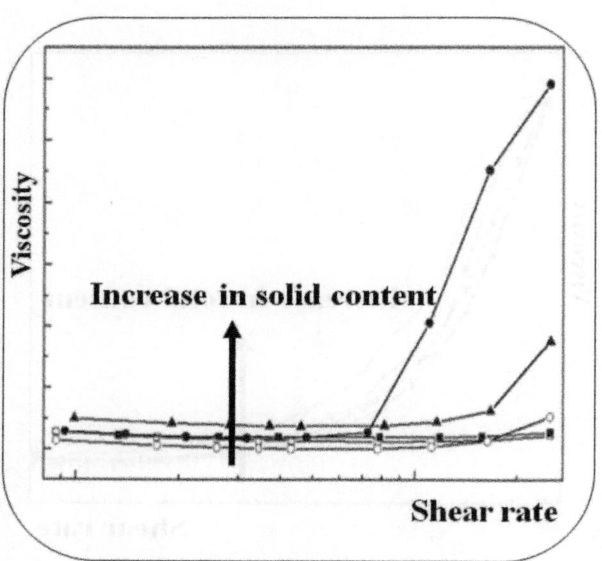

FIGURE 3.25 The viscosity dependence of a typical oxide suspension to the applied shear rate and solids content at high shear rates.

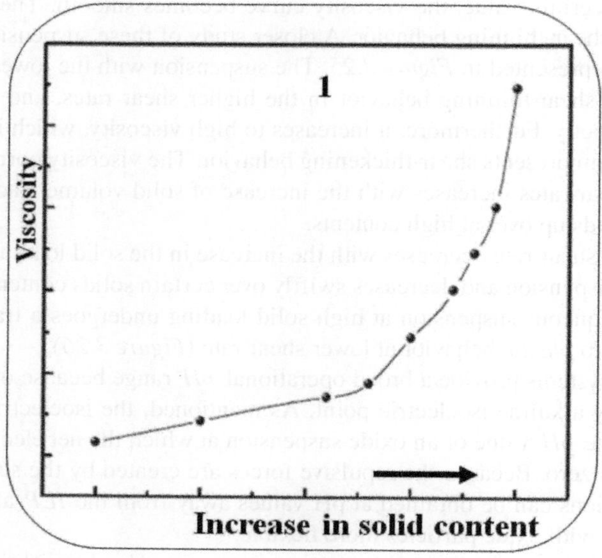

FIGURE 3.26 The viscosity dependence (1) and critical shear rate (2) of a typical oxide suspension to the solids content.

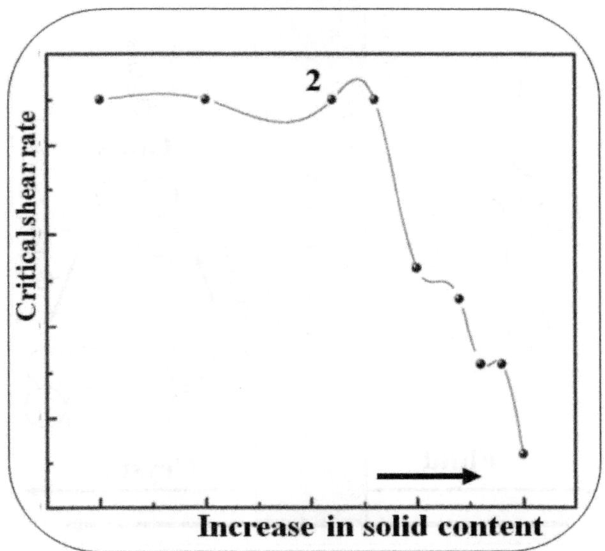

FIGURE 3.26 (Continued)

that acts to keep particles well distributed. This force causes a diffusive particle motion that drives the structure back to its equilibrium state whenever a disturbance occurs. The corresponding time scale, being a relaxation time for particle motion, provides a natural reference by which the time scales of other deformation processes will be indicated. The Brownian motion induces the presence of an underlying equilibrium phase behavior as well as reversible shear and time effects, all of which will be manifested in the rheology of colloidal dispersions. The Brownian hard spheres in colloidal dispersion rheology for dispersions occur according to with the simplest interaction potential and phase diagram of any colloidal dispersion. The dispersion phase comprises of fluid-crystal phase, glass transition, random close packing, and crystalline maximum packing (see *Figure 3.27*), where

$$\phi_{fluid} < \phi_{crystal} < \phi_{glass} < \phi_{closed\ packed}$$

The reduced suspension viscosity should only be a function of the volume of the dispersed solid phase, i.e., the rheological equation of state for the reduced viscosity is $\eta_r(\phi)$. Therefore, the Brownian motion leads to an energy scale k_BT as well as a natural time scale given in terms of the time it takes for a colloidal particle to diffuse a distance a, the particle radius. The energy scale provides a natural scaling factor for the stress as the energy per characteristic volume of a particle, k_BT / a^3, with which a reduced stress σ_r can be defined:

$$\sigma_r = \frac{\sigma}{k_BT / a^3} = \frac{\sigma . a^3}{k_BT} \quad (3.58)$$

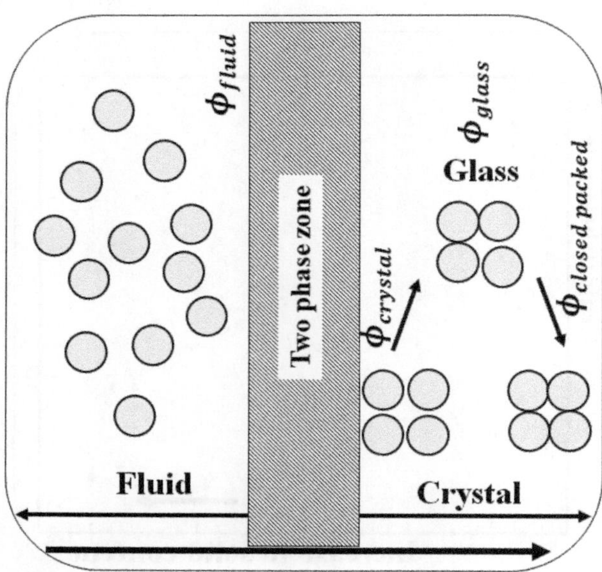

FIGURE 3.27 The schematic hard sphere dispersion phase diagram for a typical oxide suspension.

The reduced stress gauges whether the applied stress is large or small relative to the characteristic stress arising from Brownian motion. The aforementioned indicates that it is a strong function of particle size in order to define a dimensionless shear rate, known as the *Péclet* number. A characteristic time scale for Brownian diffusion is a^2 / D_0, where the diffusivity is defined as

$$D_0 = k_B T / 6\pi\eta_m a \text{ and}$$

$$P_e = \frac{\dot{\gamma}}{D_0 / a^2} = \frac{\dot{\gamma} a^2}{D_0} = \frac{6\pi\eta_m \dot{\gamma} a^3}{k_B T} \tag{3.59}$$

This is the time necessary for a particle to diffuse by Brownian motion a distance a, which is on the scale of its size. The *Péclet* number is the ratio of the rate of advection by the flow to the rate of diffusion by Brownian motion in a dilute dispersion. It defines high and low shear rates as relative to the rate of relaxation by Brownian motion. There is no constant ratio between the suspension viscosity and the media viscosity, and therefore P_e is not simply proportional to the reduced stress. A monodisperse colloidal dispersion can be defined by the particle volume fraction, particle size, media viscosity, and temperature. Hence, we can rationally expect that the dimensionless rheological equation of state should have the form $\eta_r(\phi, \sigma_r)$ and $\eta_r(\phi, P_e)$. There is always some size polydispersity, nonsphericity, and possible surface heterogeneity (i.e., patchiness, roughness). The addition of colloidal particles to a Newtonian suspending media leads to a complex shear viscosity that is of significant practical importance. As for the noncolloidal suspensions, the systematic addition of a solid phase results in a nonlinear increase in the viscosity. At low concentrations, the dispersion viscosity is nearly independent of shear stress. However, increasing the particle concentration leads to marked shear-thinning behavior at intermediate shear rates or shear stresses (*Figure 3.28*). The trend shows how the results for various medias and particle sizes reduce to a master curve when the relative viscosity is plotted as a function of a quantity proportional to P_e. This is of great importance in many industrial applications, as colloidal particles can be used to build low shear viscosity while, because of this shear thinning, they can still be made to flow, pour, spray, or spread as needed with less effort at higher shear rates.

For higher shear stresses, an apparent high shear-limiting viscosity can be achieved over quite a range of shear stresses. This pseudo-Newtonian behavior provides a well-defined viscosity at typical oxide suspensions for coating and processing applications. A rather-remarkable behavior is evident at even higher shear stresses and for higher volume fractions, whereby the viscosity increases significantly with stress. Such shear-thickening behavior is often undesirable for suspension technology and may damage processing equipment or prevent proper materials handling or processing. The forces acting between particles, whether attractive or repulsive, contribute to the shear stresses, and hence to the shear viscosity. The electrolyte addition lowers the viscosity and is most evident at low shear stresses but eventually leads to a viscosity increase. The minimum viscosity level will be the closest to hard sphere behavior. The electrolyte addition has a significantly smaller effect at higher stresses, indicating that electrostatic forces are not the dominant contribution to the viscosity under these conditions.

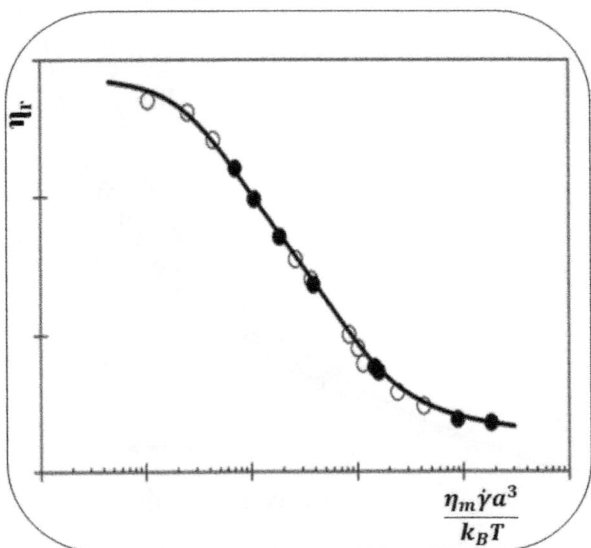

FIGURE 3.28 The schematic viscosity dependency to P_e for a typical oxide suspension.

Similarly, the volume fraction dependence of the limiting low and high shear viscosities $\eta_{r,0}$ and $\eta_{r,\infty}$ is of fundamental importance for colloidal silica dispersions (*Figure 3.29*).

The maximum packing fraction, defined as the volume fraction at which the viscosity diverges, is seen to depend on the shear rate. Most concentrated colloidal suspensions commonly display *viscoelastic* behavior. By applying a frequency-dependent shear stress or strain to a suspension, the shear moduli can be obtained. The complex shear modulus G^* has a real and an imaginary component, as given by:

$$G^* = G' + iG'' \; for \; G' = 5G^*.cos\partial \; and \; G'' = 5G^*.Sin\partial$$

Where ∂ is the phase angle. According to phase difference (∂),

if $\partial < 0$,

the applied strain and resulting stress are in phase, and energy is completely stored, i.e., the suspension is purely elastic (solid-like).

If $0 < \partial < 90$,

the suspension exhibits a viscoelastic response.

If $\partial \approx 90$,

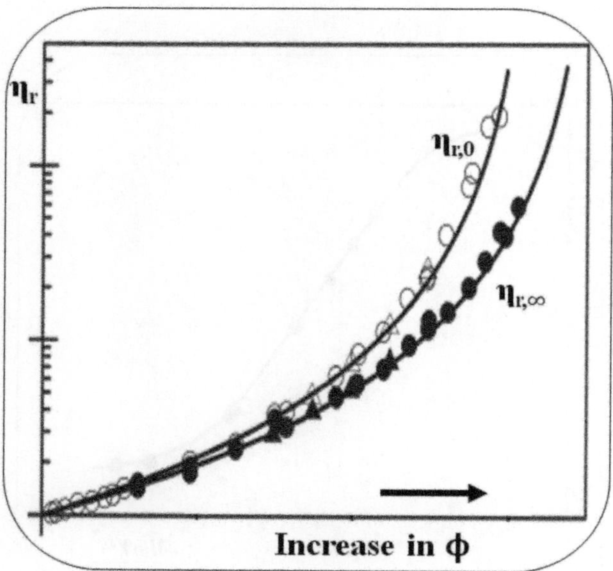

FIGURE 3.29 The schematic low and high viscosity dependency to ϕ for a typical oxide suspension.

the applied strain and resulting stress are fully out of phase, and energy is completely dissipated, i.e., the suspension is purely viscous (liquid-like).

At this situation, a liquid-like response is observed when $G' < G''$ over the entire frequency spectrum (conducting frequency [ω]).

The presence of a relaxation mechanism such as Brownian motion leads to viscoelasticity, where G'' is dominant and the dynamic viscosity ($\eta' = G''/\omega$) follows the same behavior as the steady shear viscosity, namely, a transition from a higher value at low frequencies (corresponding to low shear rates) to a lower one at higher frequencies (and shear rates). The transition regime corresponds to where the storage modulus G' levels off, as shown in *Figure 3.31.*

The Brownian dispersions exhibit very weak, positive first normal stress differences at low to moderate shear rates. However, it can be seen that at sufficiently high shear rates, the first normal stress difference becomes negative. The limited measurements of the second normal stress difference indicate they become negative at higher shear rates.

As mentioned, most nonorganic suspension and the controlling equations are based on spherical particles. Although it explains the basic mechanisms in suspension, *nonspherical* are more likely as real particles are seldom perfectly spherical. This approach assumes particles as anisometric crystalline particles, fibers, and platelets, which constitute two simple shapes that represent typical deviations from sphericity. When such particles are subjected to shear flow, they will, as with spherical particles, be dragged along and rotate. With nonspherical particles, however, the hydrodynamic stresses will depend on the relative orientations of the particles with respect to the direction of flow. The stresses will vary during rotation,

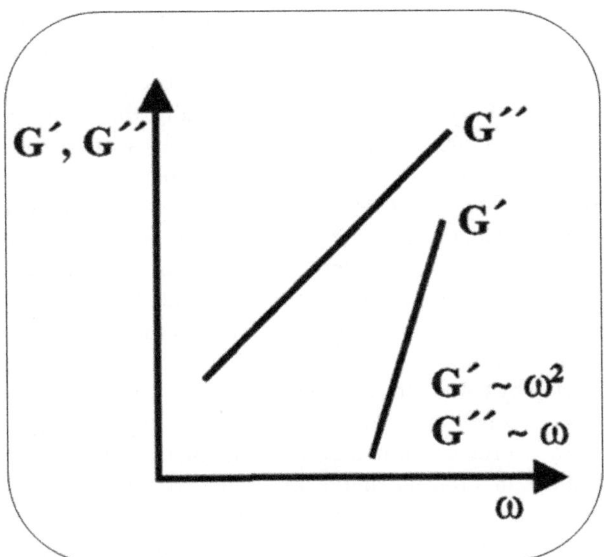

FIGURE 3.30 The schematic representation of oscillatory behavior as a function of frequency for a liquid media of a suspension.

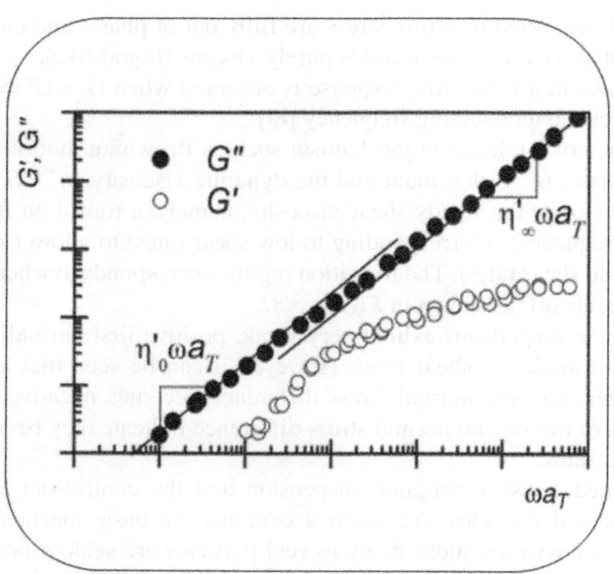

FIGURE 3.31 The schematic representation of viscoelasticity in a Brownian motion suspension.

causing a time-dependent motion of the particle in steady shear flow. The rheology of a suspension of nonspherical particles will depend on particle orientation. As rotation and orientation depend on particle shape, particle motion and rheology will be strongly coupled. The behavior in flow of individual non-Brownian particles occurs with arbitrary shape, consisting of axisymmetric particles, i.e., those with rotational symmetry and asymmetric. More specifically, this includes rods (including fibers), circular disks, and spheroids. All these shapes can be characterized by an aspect ratio, defined as the ratio of the dimension along the symmetry axis to that in the cross direction. The aspect ratio can be larger or smaller than unity; spheroids are then prolate or oblate, respectively. Because of the strong influence of sharp edges on the drag on a particle, cylinders and spheroids with identical aspect ratios (i.e., $L / d = a / b$) will move differently in the flow field. To compare other axisymmetric shapes with spheroids, an effective aspect ratio that results in identical rotational behavior can be used.

The forces exerted by a fluid on nonspherical particles depend on their orientation with respect to the flow field. As a result, isolated, non-Brownian particles in shear flow will change their orientation in a periodic manner but not at a constant rate of rotation. For instance, in a rodlike and disklike rods, the rotation slows down when the symmetry axis is close to the flow direction because the torque on the particle is then at its minimum; for disks, this occurs when the symmetry axis is perpendicular to the flow direction.

The viscosity of a suspension will depend on the orientation distribution of the constituent particles because stress due to the particles depends on their orientation.

In shear flow of Newtonian fluids, non-Brownian particles will describe periodic motions along closed orbits. Hence, their orientation at any time will depend on the initial orientation distribution.

The Brownian motion of colloidal particles will randomize the orientation at rest or at low shear rates, just as relative positions were randomized for Brownian hard spheres. As with spheres, the Brownian motion will eliminate the dependence of the viscosity on initial conditions. The effect of the aspect ratio on the zero-shear viscosity of a hypothetical suspension is shown in *Figure 3.32*. The features of the flow between zero-shear and high-frequency shear are adequately modeled using some relatively straightforward equations. The reasons for such an approach are to describe the shape and curvature of a flow curve through a relatively small number of fitting parameters and to predict behavior at unmeasured shear rates. Three of the most common models for fitting flow curves are the *Cross, power* law, and *Sisko* models. The most applicable model largely depends on the range of the measured data or the region of the curve in *Figure 3.32*, where *right* for η_0 is the zero-shear viscosity; $\eta_\infty$is the infinite shear viscosity; K is the cross constant, which is indicative of the onset of shear thinning; m is the shear-thinning index, which ranges from 0 (Newtonian) to 1 (infinite shear thinning); n is the power law index, which is equal to $(1-m)$, and similarly related to the extent of shear thinning, but with $n \rightarrow 1$ indicating a more Newtonian response; k is the consistency index, which is numerically equal to the viscosity at $1\,s^{-1}$.

It can be seen that viscosity increases with increasing aspect ratio and can become substantially higher than that for spheres. Both the translational and rotational Brownian motions provide a source of dispersion viscoelasticity. As noted,

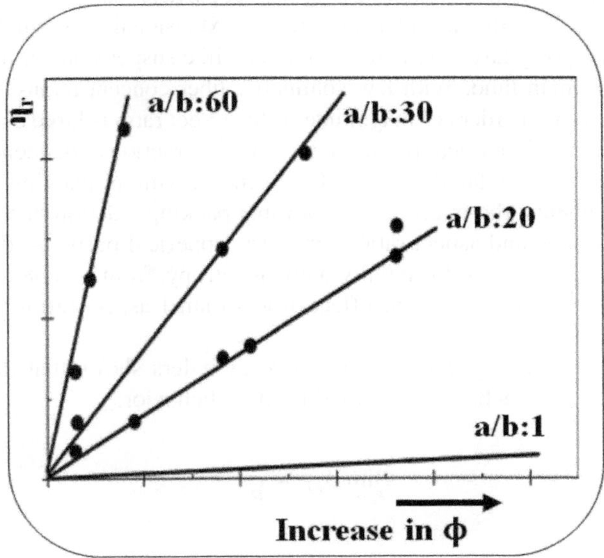

FIGURE 3.32 The schematic representation of viscosity dependency to solids content in nonspherical particles and increasing nonsphericity ration.

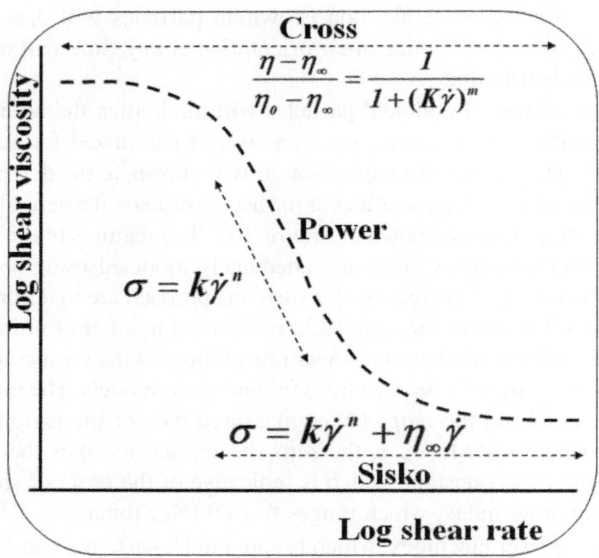

FIGURE 3.32 (Continued)

flow can orient nonspherical particles (note that although spherical particles also undergo Brownian rotation, the flow cannot orient them). The elastic effects can already appear in dilute suspensions due simply to the orientation of single particles.

The effect on the viscosity of adding long slender particles can vary strongly according to the type of flow. In particular, for uniaxial extensional flow, small amounts of fibers can cause a very large increase in viscosity, like suspensions of non-Brownian fibers in a Newtonian fluid. With few volumetric fiber concentrations, viscosity can increase by one or two orders of magnitude if the aspect ratio is large enough.

For suspensions of spherical particles, the volume fraction dependence of the viscosity can often be correlated by considering the maximum packing fraction. For random arrangements, the maximum achievable packing fraction depends strongly on the particle shape and aspect ratio. For nearly spherical particles, the maximum random packing increases substantially with anisotropy, from ~0.638 to nearly 0.74. The trends in *Figure 3.33* show the effect of less than 1 aspect ratios (P_a) (squares) and larger (circles).

The maximum packing fraction for sphero-cylinders shows that the maximum random packing fraction has an empirical limiting behavior:

$$\lim_{P_a \to \infty} \phi_{max} = \frac{c}{P_a} \tag{3.60}$$

Where $c = 5.4 \pm 0.2$ is the average number of contacts. The particle anisotropy can dramatically affect the equilibrium microstructure and phase behavior as well. Shape anisotropy shifts transitions, such as glass and gel lines in the state diagram.

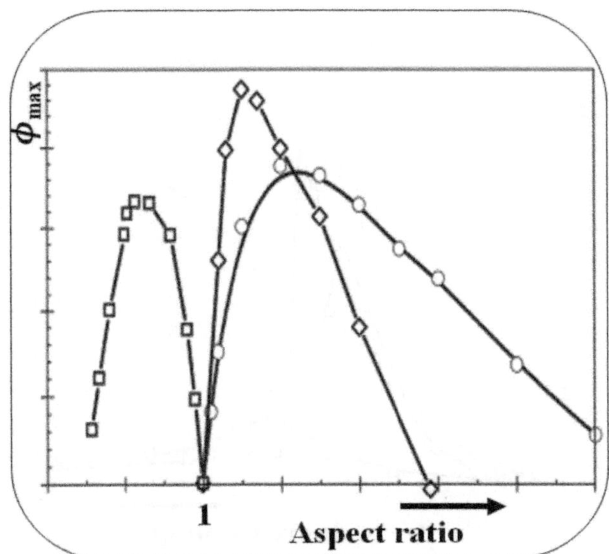

FIGURE 3.33 The schematic representation of maximum solids content dependency to particle aspect ratio.

Deviations from a spherical shape can also lead to liquid crystalline states and additional phase behaviors not possible for spherical particles. At sufficiently high concentrations of these suspensions, an orientationally ordered arrangement of the rods is favored over a randomly orientated state. In this orientationally ordered state, there is a favored direction, termed the *director*, and the rods are aligned about this direction. Notwithstanding the substantial narrowing of the orientation distribution, the relative positions remain unordered. While the positions reflect liquid-like behavior, the orientational ordering suggests a more solid-like nature.

Considering the effect of aspect ratio on the maximum packing and its effect on the viscosity would result in higher viscosities for suspensions of long, slender objects than for spheres at high volume fraction. For Brownian rods, there is, however, a maximum in the relation between zero-shear viscosity and concentration. The rodlike molecules simulate the behavior of Brownian rods with maximum viscosity point (see *Figure 3.34*).

The high aspect ratio particles can interact at very low volume fractions. A "*dilute*" dispersion is typically defined in such a way that the volume fraction regarding the particles is small—for $nL^3 < 1$, where L is the longest length scale of the particles in the dispersion. This length scale is used as it defines the volume that is affected by particles during flow. In this limit, hydrodynamic interactions between particles can be neglected. For high aspect ratio particles, it may require exceedingly low particle volume fractions. The viscous friction experienced by the different particles in the flow field will depend on their orientation. Individual particle contributions can simply be summed to calculate the viscosity of a dilute suspension. This proceeds with multiplication of the contribution to the property of interest for each particle orientation by

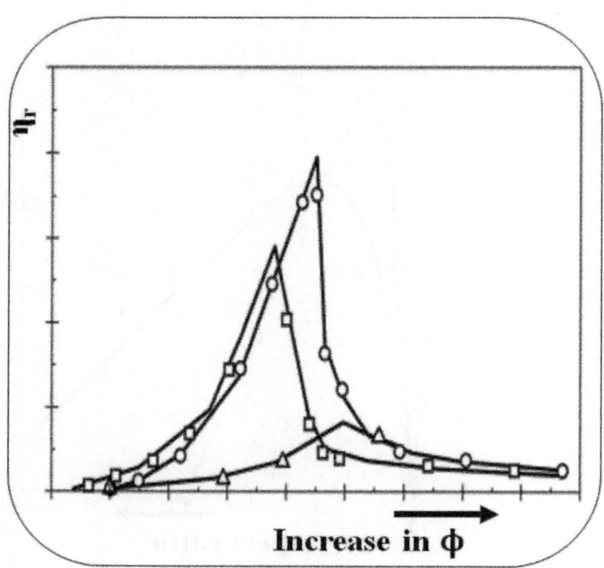

FIGURE 3.34 The schematic representation of relation between shear viscosity and concentration for a suspension of rodlike particles.

the probability of that orientation. Rotational Brownian motion will tend to restore the random orientation and, consequently, induces non-Newtonian behavior.

For non-Brownian particles, the probabilities are determined by the orientation distribution functions resulting from motion along the Jeffery orbits. Brownian motion causes deviations that will depend on the rotational *P´eclet* number. Without Brownian motion, the orbits depend on the initial positions; hence, the rate of rotation as well as the stresses will vary in a periodic manner. With Brownian motion, the orientation will be random at low Pe_r and become oriented with increasing Pe_r. The rheological consequence is a high degree of shear thinning for dilute solutions of long, slender particles. Limiting values for the viscosity of prolate and oblate spheroids with extreme aspect ratios at high and low Pe_r indicate the flowability. The high Pe_r limit implies that $Pe_r \gg (P_a^3 + P_a^{-3})$. The low shear viscosity increases more with deviations from sphericity than do the high shear values; increasing shape anisotropy also increases the amount of shear thinning.

For moderate aspect ratios ($p_a < 15$):

$$\eta_{r,Pe_r \to 0} = 1 + \phi[2.5 + 0.48\left(p_a - 1\right)^{1.508}] \tag{3.61}$$

For $p_a > 15$:

$$\eta_{r,Pe_r \to 0} = 1 + \phi[1.6 + \frac{p_a^2}{5}\left(\frac{1}{3(ln2p_a - 1.5)} + \frac{1}{ln2p_a + 0.5}\right] \tag{3.62}$$

Once more, the details of the shape of the particles are important. For rods, as opposed to ellipsoids, the zero-shear and high-frequency limiting shear viscosities valid for large aspect ratio are:

$$\eta_{r,0} = 1 + \frac{8 p_a^2}{45 \ln(p_a)} \phi \tag{3.63}$$

$$\eta'_{r,\infty} = 1 + \frac{2 p_a^2}{45 \ln(p_a)} \phi \tag{3.64}$$

In this case, "semidilute" is defined as the concentration range where nonspherical particles cannot rotate freely anymore but are still sufficiently far apart for the hydrodynamic interactions between particles to be small. The lower limit corresponds to the upper limit for the dilute region. In this system,

$$nL^3 < 1 \rightarrow \phi > \frac{1}{p_a^2} \tag{3.65}$$

A semidilute suspension requires nonspherical particles act like sphere with diameter L, h the mean spacing between particles, for a free rotating nonspherical particle is $d \ll h \ll L$, for randomly oriented particles, and

$$\frac{1}{p_a^2} \ll \phi \ll 1 \tag{3.66}$$

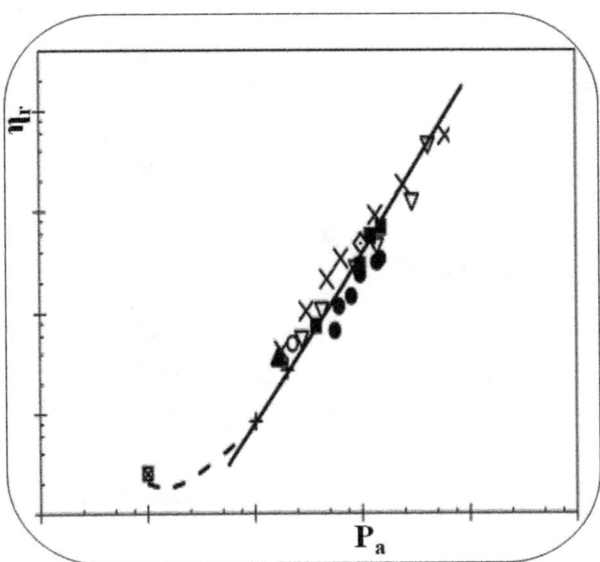

FIGURE 3.35 The schematic representation of viscosity dependency to rodlike particles aspect ratio.

In the semidilute regime of concentrated suspensions of nonspherical particles, the finite volume of the particles is ignored. With increasing volume fraction, the fact that different particles cannot occupy the same space starts to impose nonnegligible constraints on their rotational motion and their orientation. The transition to this concentrated regime can be situated at $\varphi \approx 1/P_a$. The dispersions of platelike particles (such as clays) also show increased ordering and alignment under flow and with increasing volume fraction. However, experiments on clay dispersions suggest that flow alignment becomes independent of volume fraction for concentrated systems. A detailed analysis of the rheology suggests that the platelike particles exist in aligned domains whose size depends on the shear rate, similar to what is observed for dispersions of rodlike polymers. As indicated, the electric charge on the surface of spherical particles can significantly change the rheological behavior of colloidal suspensions. The same applies to nonspherical particles, as can be seen in *Figure 3.36.*

Besides viscosity and shear behavior, *Yield stress*, a constant value of the steady state stress at low shear rates, which can also be called the dynamic yield stress σ_y^d, is an applicable parameter. At stress levels below σ_y^dvalue, some slow creeping motion might still be possible, but the corresponding viscosities can become extremely large. The other useful point is the *nonzero stress level*, below which there is absolutely no flow. A number of materials suspensions does not flow under gravity for very long periods and therefore seems to have a real yield stress. In reality, there seems materials that tended to a limiting shear stress at low shear rates but that nevertheless did still flow at lower shear rates or shear stresses. Nevertheless, slow aging in colloidal gels and glasses is explained in terms of hopping over a barrier between local thermodynamic

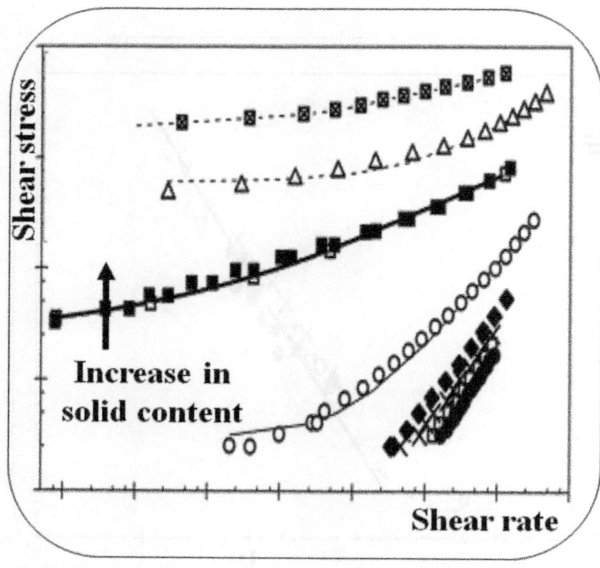

FIGURE 3.36 The schematic representation of shear stress dependency to platelike particles with different solids contents.

minima in frozen-in structures. Therefore, the term *"ideal" yield stress* refers to yielding on a time scale shorter than that of thermally induced barrier hopping.

The yield stress point is in direct relation with compressive yield stress, P_y. The compressive yield stress indicates the collapse of the network under compressional forces. The power law indices that characterize the dependence on volume fraction are often in the range of 3 to 5. The yield stress in shear is substantially smaller than the compressive one. The former is most often slightly less dependent on volume than P_y and G. In some cases, the power law indices for shear yield stress and moduli are quite similar.

While it is considered that the Brownian motion and colloidal interparticle forces give rise to viscoelastic effects, the *thixotropy* must also be considered. When a constant shear rate is applied to some colloidal suspensions, the viscosity can exhibit long transients while viscoelastic features such as normal stress differences are hardly detectable. In this situation, movements such as shaking turns the colloidal suspensions from a gel-like substance into a free-flowing liquid, but when left alone, it will gradually stiffen and return to a gel. The continuous decrease of viscosity with time when flow is applied to a sample that has been previously at rest, and the subsequent recovery of viscosity when the flow is discontinued, is an exact specification of this behavior. *Thixotropy* is reversible, time-dependent, and flow-induced change in viscosity and should not be confused with shear thinning or shear thickening, where the viscosity depends on the applied shear rate (or shear stress). It is also accompanied with a type of nonlinear viscoelasticity, reversible time effects, and including an overshoot stress in start-up flows with substantial recovery of viscosity after a sudden drop in shear rate. The effects of sudden changes between elastic and inelastic shear rates for (a) normal, (b) viscoelastic, (c) inelastic thixotropic, (d) most common response are shown in *Figure 3.37.*

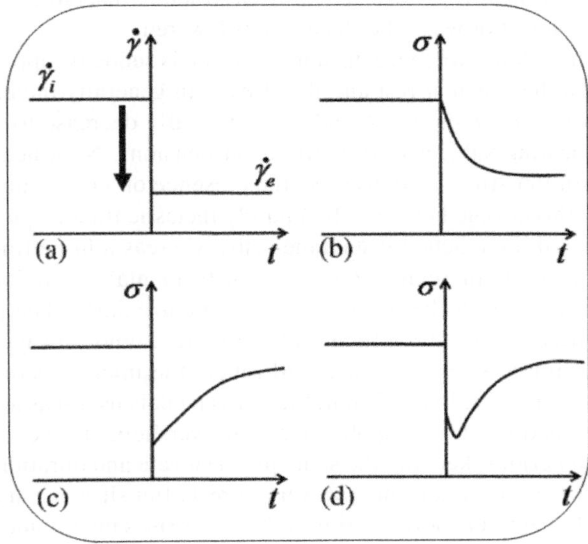

FIGURE 3.37 The schematic representation of effects of shear rate drop on different systems [10].

The behaviors involve stress transients resulting from the sudden decrease in shear rate. (b) A normal viscoelastic fluid responds by means of a stress relaxation. (c) A thixotropic material has a stress evolution in the opposite direction to describe this without recourse to viscoelastic phenomena; a separate class of inelastic or dissipative thixotropic materials is defined. Such materials are typically liquids and gels, where elastic effects such as normal stress differences are marginal. Any elastic effects will decrease with increasing shear rate, contrary to the behavior of viscoelastic polymers. And (d) some degree of viscoelasticity, producing a stress response to a sudden decrease in shear rate, includes both a sudden drop and a gradual stress relaxation, followed by thixotropic recovery.

The sudden changes in shear rate (reversely time dependent) can indicate the effect of passing time periods on colloidal suspensions. This *aging* phenomenon expresses the slow particle dynamics that still occurs in systems. With the application of flow, these aging effects can be reversed, a process which is called *shear rejuvenation* and is reversible in nature. The irreversible effects include *work hardening* or *work* softening, depending on whether the viscosity increases or decreases during flow. The existence of several metastable microstructures shows that a colloidal suspension lacks reversibility within a reasonable time over a certain range of shear rates. These behaviors are applicable for designing suspensions. Some suspensions are especially formulated to exhibit a well-defined time evolution for viscosity recovery after shearing. Special additives known as "*thixotropic agents*" are introduced to induce and control such behavior, such as hybrid amide wax/oxidized polyethylene wax. As an example, ferric oxide suspensions could be transformed into liquids by shaking. This response is the opposite of thixotropy and start-up flow, or a sudden increase in shear rate causes an increase in viscosity over time. The term now generally accepted for such behavior is *antithixotropy*, which requires a structure that builds up under shear and breaks down at rest or when the shear rate is lowered.

The *intermittent* flows, when a constant shear rate is suddenly applied to a thixotropic sample that has been at rest and the stress can generally increase to a maximum, termed the *overshoot stress*, and then gradually decrease to a steady state value, is an indicating subject in thixotropic suspensions. Nonlinear viscoelastic fluids produce similar stress responses, so the presence of a maximum is not sufficient to identify thixotropic behavior. In a purely inelastic thixotropic material, the overshoot stress will be reached instantaneously, whereas a finite time is required in a viscoelastic fluid. In reality, owing to instrumental limitations, it always takes a finite time to reach the overshoot stress (*Figure 3.38*). The peak stress is often reached after a very short time and therefore cannot always be resolved. Nonetheless, the initial stress response will reflect the microstructure existing at the start of the experiment. In particular, the stress peak is useful as a characteristic metric of this structure. In one application, the overshoot stresses are measured after various rest periods, keeping the same preshear rate and duration (an example of intermittent flow) and thixotropic recovery at rest after shearing at a given shear rate may occur. For zero or short rest times, the stress rises monotonically when the flow is restarted. This is a consequence of a recovery of the structure, which is possible as the shear rate is lower than the preshear rate. For longer rest times, a stress maximum develops in the start-up curves and grows with increasing rest times.

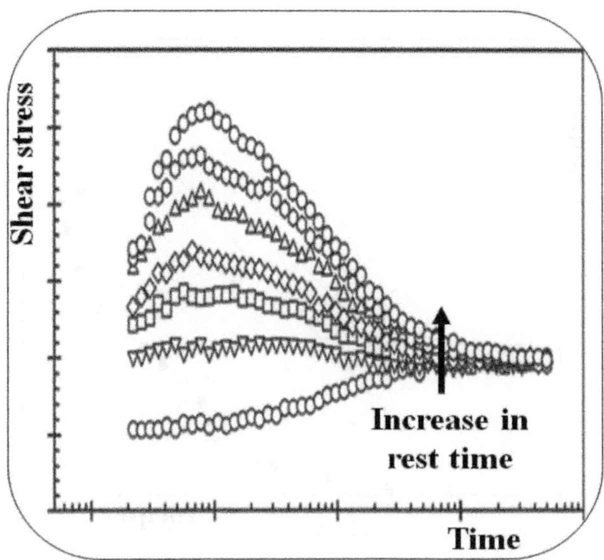

FIGURE 3.38 The schematic representation of intermittent flow, the time dependency, and effect of rest time.

The reversible nature of this behavior manifests in thixotropy *hysteresis* loop. It consists of applying an ascending shear rate ramp, followed by a descending shear rate ramp, starting and ending at rest. The ramp rate and maximum shear rate can be varied. While plotting shear stress versus shear rate, the ascending and descending curves coincide for a Newtonian or shear-thinning fluid. For a thixotropic sample, the curves describe a loop, as schematically shown in the diagram.

Unlike start-up and intermittent flows, where time and shear rate can be varied independently, the shape of the hysteresis loop is the result of the combined effects of shear rate and time. As the shear rate is increased, the microstructure breaks down gradually, thus reducing the viscosity.

The stress transients result from a jump from the steady state stresses at various initial shear rates to a final shear rate. The jump to a lower shear rate level causes a sudden drop in shear stress, followed by a gradual increase to the new steady state. On the other hand, the jump to a higher shear rate causes a sudden rise in shear stress, followed by a gradual decrease. The thixotropic behavior is often associated with a structural buildup or breakdown on a time scale in a way that few cycle conformity might occur over time (see *Figure 3.40*).

In loops, the stress reaches a maximum before the maximum shear rate (not always); such a local maximum indicates that the rate of structure breakdown reduces the stress below the stress increase associated with increasing shear rate. Irreversibility may occur when the total strain during shearing is still smaller than the strain required for yielding. The peak stress then reflects yielding, similar to that observed in start-up experiments. The result of the combined effects of shear rate

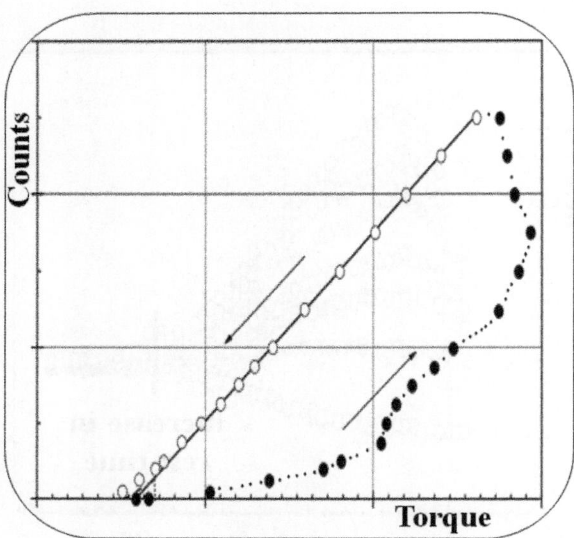

FIGURE 3.39 The schematic representation of loop concept in thixotropic behavior in a suspension.

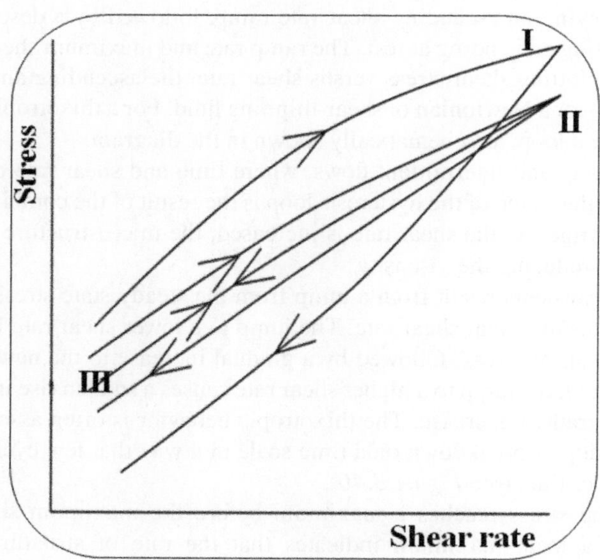

FIGURE 3.40 The schematic representation of loop nonconformity in thixotropic behavior of a suspension for three cycles.

and time increases the loop effect. As the shear rate is increased, the microstructure breaks down gradually, thus reducing the viscosity. The relatively slow rate of structure breakdown in many thixotropic systems results in the structure lagging behind the increasing shear rate and not reaching its steady state value at any shear rate during the ascending-ramp test. The viscosities measured during ramp-up at each transient shear rate are larger than expected, as they are generated by a greater degree of structure than expected at the corresponding steady state shearing condition. During the ramp-up in shear rate, the structure breakdown lags behind the shear rate, leading to transient viscosities greater than those expected at steady state. On the descending branch, the structure continually rebuilds as the shear rate is decreased. The main reason for formation of a hysteresis loop for the measured shear stress is that the structure again lags the stresses, and since the stresses are decreasing, the viscosities are lower than those obtained at steady state. The surface area of the loop has been proposed as a quantitative measure of thixotropy. The result depends on several parameters, like the shear history prior to the test, the maximum value of the shear rate, and the rate of acceleration or deceleration. As shear rate and time change simultaneously in hysteresis experiments, it is difficult to distinguish the role of each parameter in a straightforward manner. Varying the test parameters can provide some insight, but this remains difficult to quantify without reliance on a suitable model. Nonetheless, hysteresis loops can be used as a fast screening test or for comparison purposes. It should be pointed out that the presence of a hysteresis loop is, by itself, not absolute proof of thixotropy.

The *creep* experiments and tests can be used to detect thixotropy. This test is carried out by the application of a constant small stress on a sample that has been at rest for a sufficiently long time, with the strain gradually rising to a limiting value. The final strain level increases with the applied stress. This response is characteristic of a viscoelastic solid, as shown in *Figure 3.40*, for platelike ceramic suspensions.

When the stress is released, the strain will, in principle, recoil to zero as long as the applied stress is lower than the yield stress. After the applied stress exceeds a critical value, the material will ultimately flow. The strain will then increase at a constant rate that depends on the applied stress. A transitional stress region exists between "flow" and "no-flow," where flow is delayed. The delay time decreases with increasing stress. This behavior can be understood on the basis of slow, stress-driven rearrangements in microstructure that weaken the structure and ultimately lead to yielding and flow. The term *rheopexy* typically refers to the increase in viscosity with time of some thixotropic materials held at constant low shear rate (or stress). This results in the reversible shear thickening in stable colloidal dispersions, although in practice shear thickening may also be associated with irreversible particle aggregation and dilatancy. The importance of this phenomenon is, shear thickening in colloidal dispersions often seriously limits formulations for coating and spraying operations, as well as flow rates for pumping concentrated dispersions.

However, shear-thickening dispersions have mechanical properties that can also make them uniquely qualified for specific applications. Indeed, because of their unique rheological response, shear-thickening fluids have been proposed as dampers and shock absorbers. The increase in the rate of shear leads to an increase in

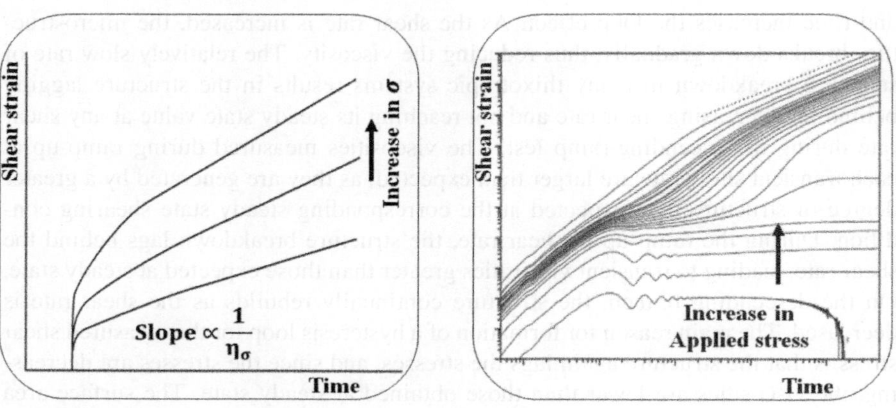

FIGURE 3.41 The schematic representation of creep experiments with increasing applied shear stress (left), and an experiment for a platelike suspension (right).

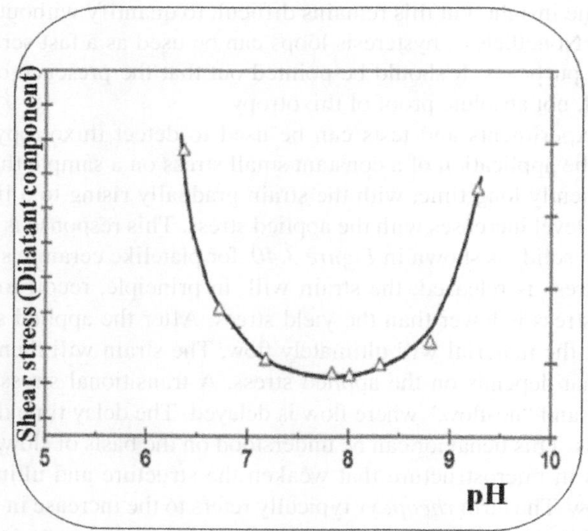

FIGURE 3.42 The schematic representation of pH dependence of a dilatant behavior in a typical oxide suspension.

the viscosity of the dispersion, a behavior known as *inverse plasticity*. The *dilatancy* associated with volume expansion was attributed to packing effects, whereby shearing a dispersion above a critical packing fraction was associated with particles expanding their occupied volume (see *Figure 3.42*).

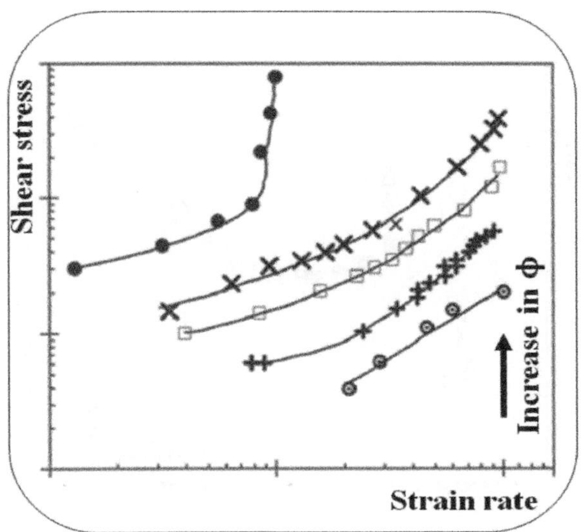

FIGURE 3.43 The schematic representation of shear stress vs. strain rate behavior for a typical oxide suspension with gradual increase in solids content.

The typical behavior of many oxide particle suspensions shows a typical strong shear, especially for the more concentrated suspensions, changing from a linear stress-strain rate trend to showing a critical shear rate at higher solids contents, as shown in *Figure 3.43*. The critical shear rate for shear thickening was observed to decrease with increasing media viscosity, pointing to the importance of the media viscosity. The hypothesis for shear thickening in colloidal dispersions is related to shear-induced aggregation. Vice versa, shear thinning in these suspensions involve the shear-induced breakup of weakly flocculated dispersions, which further increases in shear rate and could result in shear aggregation. Shear forces were presumed to be strong enough at the point of shear thickening to drive particles over the repulsive barrier so that strong, short-range attractions maintain a flocculated state. The point of maximum stability (highest surface charge) corresponds to the minimum shear-thickening behavior. The rheological dilatancy results from a progressive increase in flocculation due to the applied shear.

Shear thickening in oxide suspensions was assumed to be due to an increase in effective particle volume fraction, by accounting for the suspending media trapped within the shear-flocculated clusters. In particular, it was determined and verified experimentally for their systems that the critical shear rate decreases with increasing particle size, media viscosity, and concentration and increases with increasing colloidal stability. Both ascending and descending steady shear flows result in a reversible shear thickening, driven by hydrodynamic interactions. The effective volume fraction of the hydrodynamic clusters reaches a value sufficiently high for the sample to be dilatant. The authors observed the sample to become opaque and to fracture under extreme shear-thickening

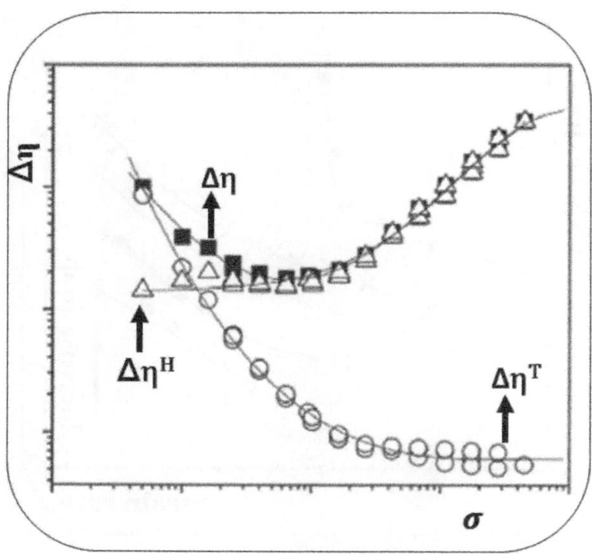

FIGURE 3.44 The schematic representation of viscosity vs. shear stress and indicating the thermodynamic and hydrodynamic trends.

conditions. The accumulation of trends in *Figure 3.44* for *thermodynamic* (T) and *hydrodynamic* (H) viscosity indicates the final viscosity behavior in an oxide suspension. The concentrated colloidal dispersions in these systems can exhibit the same viscosity at two (and sometimes even three) very different shear rates (or shear stresses).

The shear thickening in nonorganic suspensions is used to describe the reversible increase in dispersion viscosity with increasing shear rate and is distinguished from shear-induced aggregation and dilatancy (volumetric increase upon shearing). Although severe rheological responses are associated with shear thickening, it may be surprising at first to recognize that even very dilute dispersions, such that only two particles can interact, will exhibit shear thickening. Many systems show shear-thinning behavior, and at higher rates of shear, the viscosity rises to a value greater than its low shear value. Also, the shear-thickening viscosity is dominated by the hydrodynamic component of the viscosity. For $P_e \sim 0$, as a consequence of diverging lubrication force, the microstructure and rheology change fundamentally at high shear rates. In $P_e > 1$, the shear force is stronger than the characteristic Brownian force acting between particles, and the convective flow dominates the particle motion. The colloidal particles are driven by the shear flow into close proximity such that the lubrication hydrodynamic forces become significant. Because shear flow drives particles together, the probability of finding a neighboring particle increases dramatically along the compression axis of the flow. This transition from a Brownian-dominated regime ($P_e < 1$) to a hydrodynamically dominated one ($P_e > 1$) not only initiates shear thickening but also causes changes in sign for the first normal stress difference, as well as nonmonotonic variations in the normal stress differences and dispersion osmotic pressure. The onset of shear thickening will be accompanied by a change in the first normal stress difference from positive to negative. By comparing the microstructure measured for suspensions as $P_e \to \infty$ to that calculated for high P_e , it is apparent that the Brownian dispersion's microstructure under flow tends toward that observed for the non-Brownian suspension. Such suspensions have negative normal stress differences as they are solely governed by hydrodynamic interactions, so this transition in normal stresses and microstructure is as expected. At the limit of large P_e, the colloidal dispersion will approach the behavior of an ideal non-Brownian suspension. However, the mathematically singular nature of the limit $P_e \to \infty$, due to the singular nature of the lubrication forces, means that $P_e \to \infty$ is not the same as $P_e = \infty$.

The role of nanoscale surface forces becomes critical in determining the limiting high-shear rheological behavior of colloidal dispersions. The effects, such as surface roughness, the molecular nature of the media, surface charges, and adsorbed species, will cut off the shrinkage of this boundary layer and lead to a limiting high-shear viscosity. In these conditions, shear-thickening rheology will be proportional to the suspending media viscosity and will scale, as for suspensions, with the thermal properties of the media. The onset of shear thickening for hard sphere suspensions will scale inversely with the particle volume, where larger particles will shear-thicken. These rates occur because hydrodynamic interactions govern the shear-thickening rheology. The surface forces that prevent particles from approaching sufficiently close to couple through lubrication hydrodynamic forces can mitigate and even suppress shear thickening. These forces need to act at a distance of the order of 10% of the particle radius. The shear thickening will be very sensitive to nanoscale surface forces. The development and structure of the boundary layer at high P_e will depend significantly on all deviations from ideal hard sphere behavior. This behavior can be a dominant interaction in most nonorganic suspensions in aqueous and nonaqueous medias (*Figure 3.45*) with formation of hydroclusters (flow-induced density fluctuations; as the particle concentration is higher in the clusters, the fluid is under greater stress, which leads to an increase in energy dissipation and thus a higher viscosity).

As seen, the particles adopt an organization that permits flow with fewer interparticle encounters. When the shear forces become sufficient, particle motion becomes highly correlated, as particles are pushed into close proximity, and shear thickening results. These correlated groupings of particles are known as hydrodynamic clusters/hydroclusters and are self-organized or flow-organized microstructures formed as a result of the lubrication hydrodynamic interactions in dilute dispersions (*Figure 3.44*). With increasing concentration, local concentrations of particles are created and destroyed in the flow field. It is critically important to note that hydroclusters are not aggregates or coalesced particles, but rather local transient fluctuations in particle density. The hydroclusters are more densely packed than the average dispersion, but they are anisotropic and have very high stresses. The evidences for hydrocluster formation can be collected via the increase in suspension turbidity and scattering dichroism observed in rheo-optic experiments and the extensive measurements by small-angle neutron scattering. The hydroclusters are very broadly distributed and strongly correlated in spatial distribution; the hydroclusters are transient local density fluctuations, which explains why shear thickening does not show

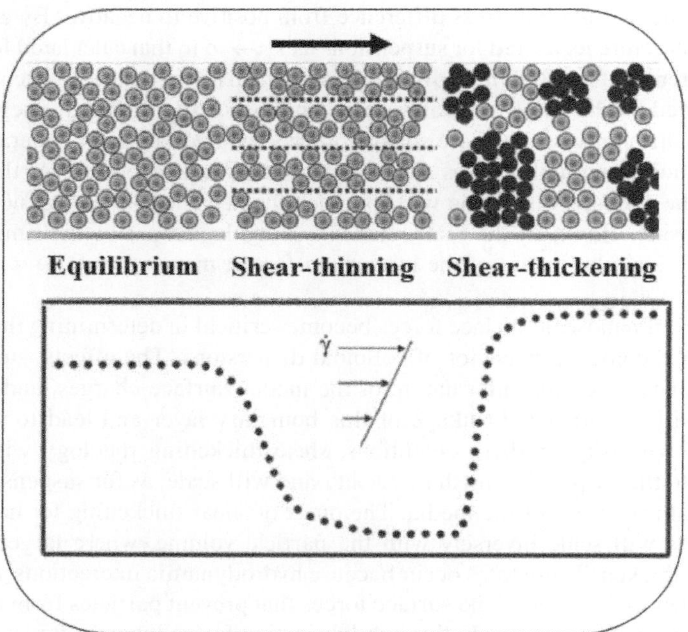

FIGURE 3.45 The schematic representation of viscosity dependence to structure formation in colloidal suspensions.

thixotropy or any significant hysteresis. The particle configurations go under shear flow, in which only particles in locally denser regions of the flow. The density variations are dynamic with particles leaving and joining; they are not rigid, rotating units, and this behavior can be seen from a comparison of the four configurations, each separated by one unit of strain.

REFERENCES

[1] Somasundaran P. Adsorption of surfactants and polymeric surfactants at the solid/liquid interface. Appl Surfactants. 2005:85–113.

[2] Yong H, Jinlong Y. Novel colloidal forming of ceramics. Singapore, Springer Nature Singapore Pte Ltd.; 2020.

[3] Tadros TF. Applied surfactants: principles and applications. Weinheim, Germany, WILEY-VCH Verlag GmbH & Co. KGaA; 2005;1–17.

[4] Tadros TF. Surfactants as dispersants and stabilisation of suspensions. Appl Surfactants. 2005:187–257.

[5] Tadros TF. Adsorption of surfactants at the air/liquid and liquid/liquid interfaces. Appl Surfactants. 2005:73–84.

[6] Jelić RM, Joksović LG, Djurdjević PT. Potentiometric study of the effect of sodium dodecylsulfate and dioxane on the hydrolysis of the aluminum(III) Ion. J Solut Chem. 2005;34(11):1235–61. doi: 10.1007/s10953-005-8016-y.

[7] Zhang C, Jiang Z, Zhao L, Guo W, Gao X. Stability, rheological behaviors, and curing properties of 3Y—ZrO2 and 3Y—ZrO2/GO ceramic suspensions in stereolithography applied for dental implants. Ceram Int. 2021;47(10A):13344–50.
[8] Babick F. Suspensions of colloidal particles and aggregates. Springer Cham, Springer International Publishing, Switzerland; 2016.
[9] Yaghtin M, Yaghtin A, Tang Z, Troczynski T. Improving the rheological and stability characteristics of highly concentrated aqueous yttria stabilized zirconia slurries. Ceram Int. 2020;46(17):26991–9. doi: 10.1016/j.ceramint.2020.07.176.
[10] Mewis J, Wagner NJ. Thixotropy. Adv Colloid Interface Sci. 2009;147–148:214–27. doi: 10.1016/j.cis.2008.09.005, PMID 19012872.

4 The Processes for Stabilizing Suspensions for Ceramic Thermal Barrier Coatings (TBC)

Scope:
The application of suspensions (from semiconcentrated for TBC applications to highly concentrated suspensions for slurry-slip cast applications) requires a basic understanding on suspension stabilization. The stabilization is essential for attributing rheological properties to a suspension (from basic parameters, i.e., η, ζ, and alike, to specially introduced parameters based on application. Although in many cases suspension formation is equivalent to stable suspension, indeed, certain measures must be applied to ensure the stable behavior of the suspension. The stabilization of ceramic oxide/nonoxide suspension in aqueous/nonaqueous media, with micrometer to nanosized particles, with or without the presence of surface adjusting agents, should be considered in different systems of mono- or multiple-type ceramic suspension.

4.1 GENERAL OVERVIEW

As mentioned previously, the suspension preparation and handling of colloidal suspension, or even their characterization, requires a basic understanding of the physical effects that happen on the microscopic level. Some of these effects are directly related to the fineness of the particles because it coincides with a large specific surface area (which corresponds, e.g., to a high adsorption capacity), a considerable curvature of the particle surface (which promotes dissolution), a high number concentration (i.e., significant osmotic pressure and large collision frequencies), small interparticle distances (i.e., strong interactions between scattered waves), low particle mass (i.e., low weight and inertia), or with the fact that the wavelength of light is larger than particle size (i.e., weak optical scattering). A further effect of the small size of colloidal particles is that their diffusive motion is much faster than for micrometer particles. Moreover, the interactions between colloidal particles are decisive for the macroscopic behavior of the whole suspension (e.g., with regard to rheology or coagulation). These interactions may be attractive and/or repulsive and depend on the chemical nature and the material properties of the particles. Most of all, they are affected by the interfacial properties which reflect the physicochemical interaction

DOI: 10.1201/9781003285014-4

between the particulate phase, the solvent, and the solutes (e.g., hydrophobicity, adsorption, surface charging).

The most common and widely used ceramic particle colloidal suspensions happen in *aluminosilicate* systems or *clays*. The clay-water system's extensive applications require a detailed study in ceramic suspension/colloid systems [1]. Clay particles have a platelike morphology commonly consisting of negatively charged faces and positively charged edges when suspended in a polar solvent, such as water. These particles readily undergo cation exchange reactions, swelling, adsorption, and even intercalation of organic species to alter their surface charge, chemistry, and crystal structure. The charged outer surface of these particles clarified and helped the development of colloidal study of ceramic suspensions. The results indicated that binder-free, mono- or multidisperse colloidal suspensions were optimal for achieving the microstructural homogeneity required for functional ceramics. The control of the rheological behavior of ceramic powder suspensions is vital in colloidal processing, not only for easy handling, but also to achieve optimum microstructure in the final product (*Figure 4.1*). There are three main forces effective in ceramic colloidal suspensions of charged particles: Brownian, interparticle, and hydrodynamic forces. At low shear rates, Brownian and interparticle forces dominate. At high shear rates, on the other hand, hydrodynamic forces become more dominant. As an example, monodisperse colloidal systems (e.g., those based on silica spheres) have served as excellent model systems for the study of aggregation, rheological, and sedimentation behaviors. The ceramic suspension viscosity in these systems is proportional to the medium viscosity, multiplied by the effect of the addition of particles. Therefore, the medium viscosity should be as low as possible so that suspensions with high solid

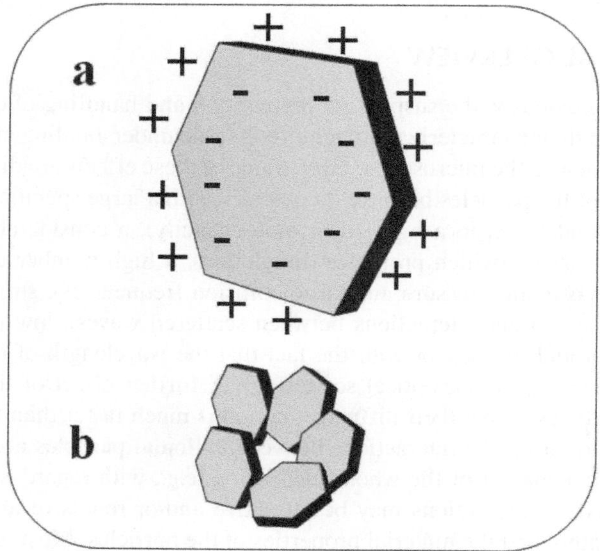

FIGURE 4.1 The schematic of charge distribution around a typical clay particle in a colloidal suspension (a) and aggregated particle network formed to opposite-charge attraction (b).

fraction and moderate viscosity can be achieved. Alumina (Al_2O_3), zirconia (ZrO_2), yttrium-stabilized zirconia (YSZ) [2–4], and titania (TiO_2) are some of the most widely studied ceramic nanopowder systems.

Through control of interparticle forces, ceramic colloidal and suspensions can be prepared in dispersed and weakly fluctuated ceramic unary and binary systems. As mentioned in the previous chapter, the shear-thinning behavior is the most possible behavior, caused by the lack of resistance to interparticle interactions or external forces. The hydrodynamic interactions overcame the interparticle interactions with increasing shear rates and decreased. At high shear rates, the hydrodynamic forces become so strong that hydroclusters are formed and thickening may occur. According to bimodal model, shear thickening is not dominant at high solids contents or when the small particle fraction increase.

The most obvious approaches to increasing the concentration of dispersion with a solids content (concentration) exceeding $\phi_0$is the transition to the polydisperse mixtures. The idea of this approach is demonstrated in *Figure 4.2*, where the smaller particles fill the free volume between the larger particles, thereby increasing the total volume concentration of the disperse phase.

The theoretical analysis of binary low-concentrated ceramic suspensions has led to relations between the viscosity and the ratio of particle dimensions. The main purpose is to determine the largest possible degree of filling depending on the specific and optimal values for parameters such as particle dimensions ratio, the volume content ratio, with lowest values for filling and viscosity.

In the dispersed state, discrete particles that exist in the ceramic suspension repel one another on close approach, provided the repulsive barrier is $\propto K_B T$, which shows

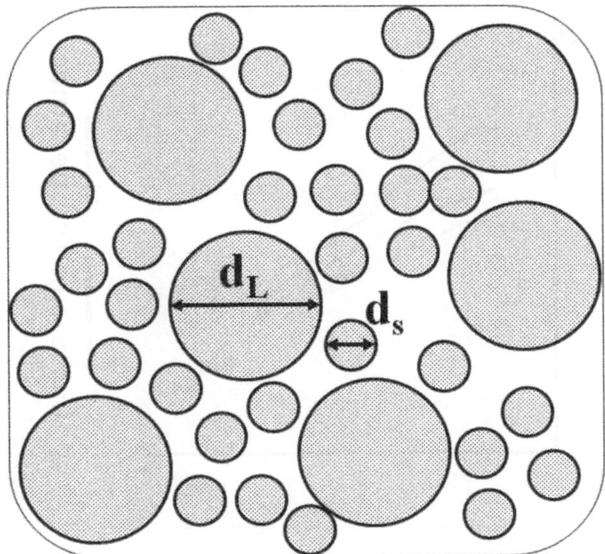

FIGURE 4.2 The schematic ceramic of bimodal particle distribution in a suspension.

the temperature dependency of viscosity at growing shear rate. In the weakly flocculated state, particles aggregate in a shallow secondary minimum, forming isolated clusters (or flocs) in suspension at volume fractions (*Figure 4.3*).

The bases of ceramic suspension dispersion evolved from the fact that equilibrium separation distance exists between aggregated particles. This fact is schematically shown in *Figure 4.4* for completely dispersed, weakly fluctuated, and strongly fluctuated systems. The energy trend curves (potential energy versus distance) comprise of strong repulsion in completely dispersed ceramic suspension, attractive interaction mediums in weakly fluctuated suspension, and complete attractions in strongly fluctuated suspension, with a maximum secondary minimum and primary minimum for the systems, respectively. In contrast, particles aggregate into a deep primary minimum in the strongly flocculated (or coagulated) state, forming either a touching particle network or individual clusters in suspension, depending on their concentration. The trends indicate that any change in surface composition will lead to significant changes in the properties of the ceramic nanopowder suspensions, which is expressed by the DLVO theory.

Three conclusions can be drawn from the DLVO theory: (i) the larger the Hamaker constant, the larger the attraction between particles; (ii) higher surface potentials result in larger repulsive forces; (iii) the higher the electrolyte concentration, the smaller the distance from the surface at which repulsive forces are effective. A large Debye length, κ^{-1}, with a correspondingly high repulsive force predicts that the system is well dispersed and does not necessarily correspond to optimum condition for flow and compaction of highly loaded particle assemblies. The larger the distance, the higher the expected viscosity of the concentrated ceramic particle suspension.

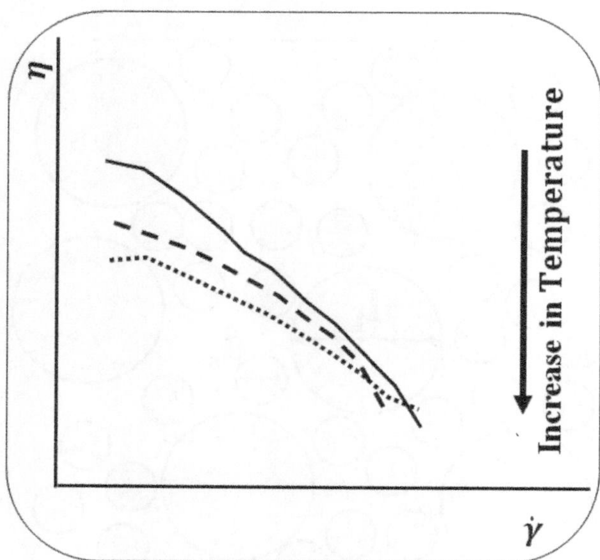

FIGURE 4.3 The schematic non-Newtonian behavior of the ceramic suspension system dependency to temperature.

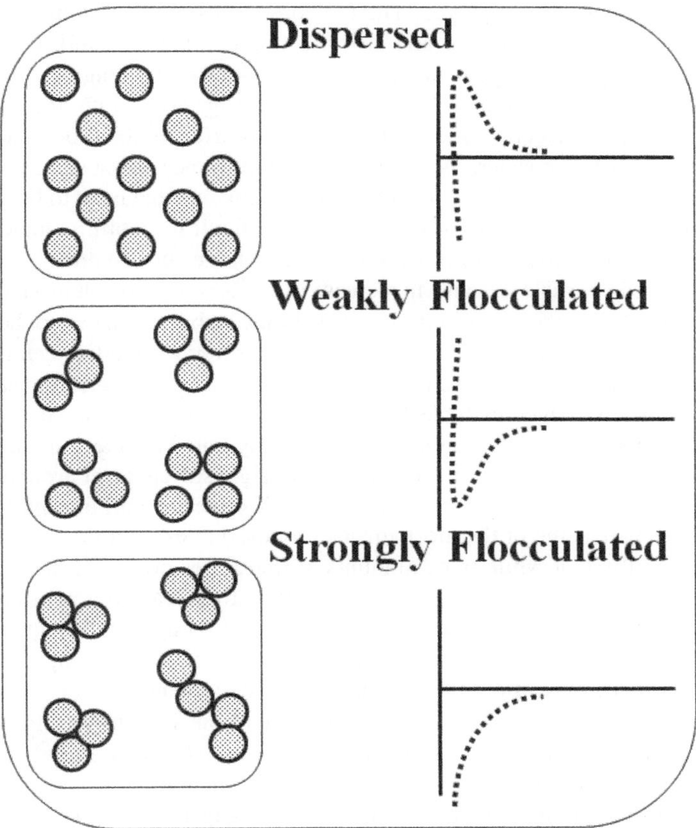

FIGURE 4.4 The schematic non-Newtonian behavior of the ceramic suspension system dependency to temperature.

The viscosity of ceramic suspensions depends on a number of variables, such as particle size, shape, size distribution, surface charge, solution chemistry, solids content, deformability, flow conditions, and others. The viscosity of a suspension is strongly dependent on the solid volume fraction, with the viscosity approaching infinity at a maximum volume fraction, ϕ_m, where ϕ_m relates to the particle concentration at which the average separation distance between the particles approaches zero and the particles pack together, making flow impossible. The relative viscosity is the ratio between the viscosity of the suspension and the viscosity of the monomer solution. The experimental results of ceramic suspension can be fitted to a modified *Krieger-Dougherty* equation:

$$\eta_r = (1 - \phi / \phi_m)^{-n} \tag{4.1}$$

Where ϕ_m and n are used as fitting parameters. The optimum microstructures can be obtained when highly loaded, homogeneously dispersed and stable suspensions with reasonable viscosities are employed. The flow properties of the ceramic suspensions can be controlled via the electrostatic or steric stabilization mechanisms. However, control of viscosity in ceramic nanoparticle suspensions is challenging.

The major stabilization mechanisms can be categorized into three different groups: (i) steric, (ii) electrostatic, and (iii) electrosteric stabilization. The electrostatic behavior of the powders is of particular importance for several reasons: (i) powders exhibit charged surfaces in aqueous environments that need to be controlled in order to stabilize the particles; (ii) it is very difficult to obtain completely pure oxide powders, especially nanopowders, because of large surface area.

The decrease in particle size and an increase in solids content result in a reduction of the zeta potential (ζ), and as a result, the electrostatic stabilization cannot be successful, and it is necessary to impose steric repulsion in order to stabilize ceramic nanopowder suspensions. The zeta potential is the key parameter for indicating suspension stability and, as represented in Table 4.1, can predict the state of a typical suspension.

The secondary hydration forces associated with hydrated counterions adsorbed on the ceramic particle surface and the formation of a rigid solvent layer around the powder particles and the associated repulsive forces effective at small interparticle separations ($<5\,nm$) are some of the mechanisms proposed to explain this phenomenon.

It is noteworthy that while trace amounts of impurities do not result in a shift in the isoelectric point of dilute ceramic suspensions, they lead to a substantial shift in concentrated ceramic suspensions. Therefore, even though the isoelectric points obtained for the ceramic powders in dilute suspensions are similar, impurities on the surface of particles may affect the initial pH and zeta potential of the ceramic suspensions, leading to changes in the dispersion state of the suspensions. For instance, the alumina nanopowder ($55nm$) aqueous suspension ($\Phi : 0.2$) shows fluctuating behavior. The suspension viscosity decreases first and then increases with increasing electrolyte concentration. This trend indicates the competition of two effects of the double-layer compression: (i) decrease in effective solids concentration and (ii) the reduction of the range and magnitude of the electrostatic repulsion (see *Figure 4.5*).

The solids content, similar to any suspension, plays a profound role in concentrated ceramic suspensions because of the strong interparticle interactions and the overlap of double layers. Because ceramic nanopowders have a higher specific surface area

TABLE 4.1
The rheological parameters dependency (ζ) of typical ceramic suspensions.

State of the ceramic suspension	Zeta potential (ζ) range
Fluctuation/coagulation	$0\,to\,\pm 5$
Instable	$\pm 10\,to\,\pm 30$
Moderately stable	$\pm 30\,to\,\pm 40$
Acceptably stable	$\pm 40\,to\,\pm 60$
Highly stable	$> \pm 61$

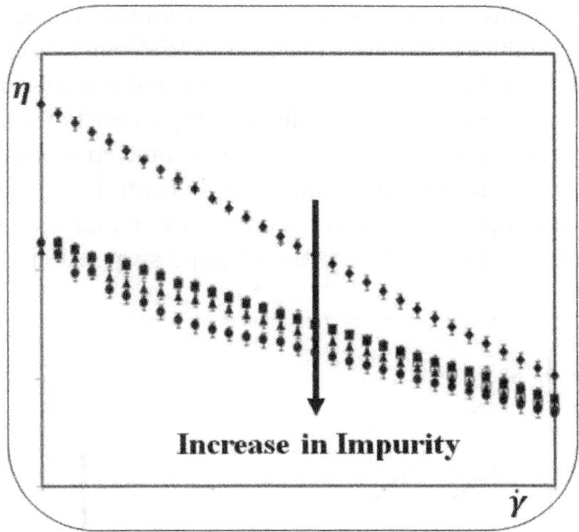

FIGURE 4.5 The schematic behavior of typical oxide ceramic suspension system and the effect of increasing concentration of nonorganic impurity on viscosity.

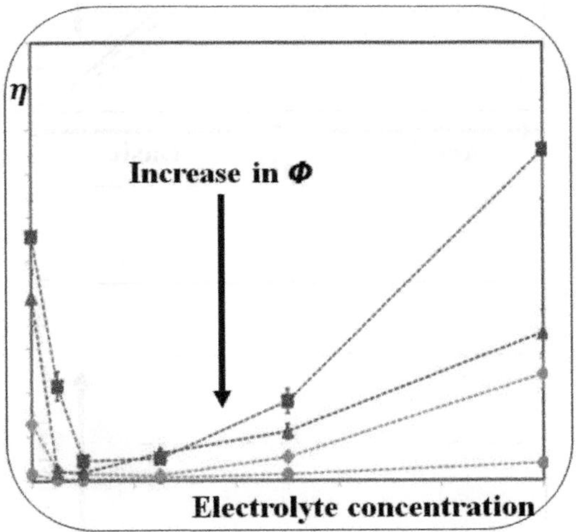

FIGURE 4.6 The schematic behavior of typical oxide ceramic suspension system and the effect of increasing concentration of nonorganic impurity on viscosity of different solids content systems.

than micron-sized particles, the change in double-layer thickness has an amplifying effect on the effective volume fraction of particles (see *Figure 4.6*). The presence of an electrolytic impurity can have fluctuating effects on ceramic suspension behavior. At higher electrolyte concentrations, the powders tend to flocculate due to the screening of the surface charges. The combination of flocs and the increase in effective volume fraction may be caused by the occluded volume of water within the flocs and results in higher viscosities. At higher ion concentrations than the critical coagulation concentrations, anomalous stability of ceramics particles is evident. The non-DLVO behavior, stable suspensions even at the *IEP*, can be explained by the solvation layer formed around the particles by adsorption of polymeric metallic cations. The main focus of most researches is on the stability of particles at much lower ion concentrations. The results show the importance of ionic strength on the stability of ceramic nanopowder suspensions, especially for concentrated systems.

The zeta potential of ceramic powders decreases with increasing indifferent electrolyte concentration (see *Figure 4.7*). If the ceramic suspension is dilute enough to

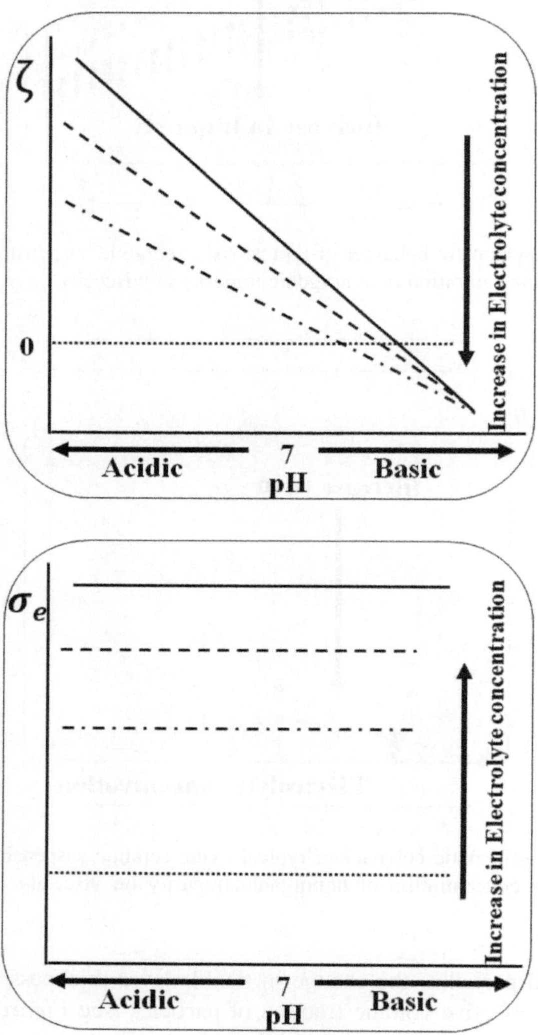

FIGURE 4.7 The schematic behavior of typical oxide ceramic suspension system and the simultaneous effect of pH and electrolyte concertation on zeta potential (left) and electrical conductivity (right).

neglect any interactions between particles, the origin of the zeta potential changes should stem from charging mechanisms of the particles, such as dissolution kinetics or adsorption of ions on the surface. The gradual change in the pH of the media can indicate the surface layer ionic integrity, which, as a result, affects ζ and σ_e even at constant Φ of nanosized oxide ceramic particles.

Increasing ion concentration can decrease the double-layer thickness and reduce the effectiveness of electrical repulsive forces. Two significant features in *Figure 4.8* are the height of the potential barrier, which predicts whether the particles flocculate or not, and the distance at which the net interaction energy is higher than thermal fluctuation caused by Brownian motion.

At very low concentrations of electrolyte, repulsive potentials overlap at large distances, leading to large effective volume fractions. As the electrolyte concentration increases, the closest particle approach, hence the effective volume fraction, decreases. In addition, the repulsive interactions between particles increase because of the increase in surface potential (as measured zeta potential indicates), leading to a higher potential barrier against flocculation. A further increase in electrolyte concentration results in screening of the double layer, and flocculation ensues.

The large surface area of the oxide ceramic nanopowders amplifies the influence of the bound layer on effective volume fraction. According to *Figure 4.9*, the H^+ compresses the bound layer around the particles by releasing a fraction of the water, hence decreasing the effective solids volume fraction. On the other hand, addition of OH^- increases the bound layer thickness, increasing the effective volume fraction and viscosity of the suspensions.

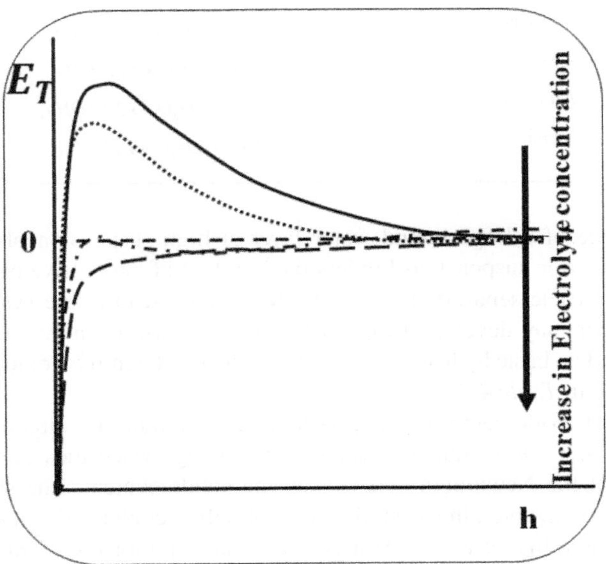

FIGURE 4.8 The schematic behavior of DLVO potential (E) of a typical oxide ceramic suspension system with gradual increase in electrolyte concertation.

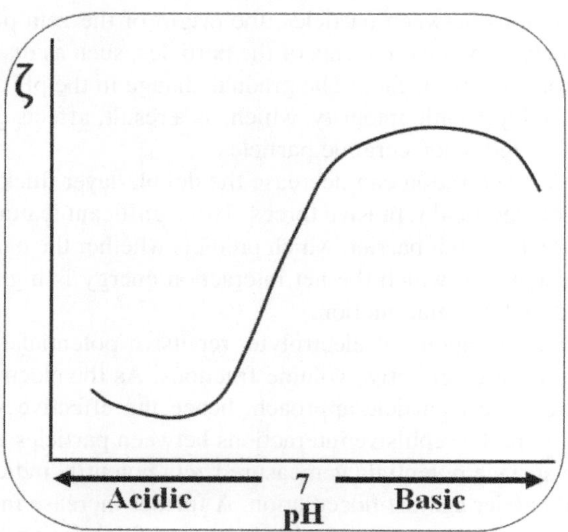

FIGURE 4.9 The schematic effect of a typical oxide ceramic suspension pH on ζ.

TABLE 4.2
The basic hydrolysis reactions of the most common oxide suspensions.

Oxide ceramic suspension	Hydrolysis reactions
Zirconia	$ZrO_2 + 2H_2O \leftrightarrow Zr(OH)_4$
YSZ	$Y_2O_3 + 3H_2O \leftrightarrow 2Y(OH)_3$
Alumina	$Al_2O_3 + 3H_2O \leftrightarrow 2Al(OH)_3$
Titania	$TiO_2 + 2H_2O \leftrightarrow Ti(OH)_4$

The presence of a large bound water layer at high suspension pH for concentrated oxide ceramic suspensions hinders the building of a repulsive barrier because of short interparticle separation distances. Nevertheless, in dilute suspensions, the repulsive barrier may develop at high pH values because of large particle-particle separation [2]. The basic hydrolysis reactions of the most common oxide suspensions are represented in *Table 4.2*.

The study of oxide ceramic nanopowders such as alumina, zirconia, YSZ, and titania (*Figure 4.10*) shows that all suspensions have high viscosities, and almost none of them exhibit the Newtonian behavior at any solids content. The shear-thinning behavior is recognizable almost at all levels of solids content (Φ), and it becomes more prominent at higher levels. At high shear rates and for low solids content systems, shear thickening is dominant.

A comparison of the viscosities of different oxide systems at a constant shear helps in understanding the plateau region of low solids content oxide ceramic suspensions,

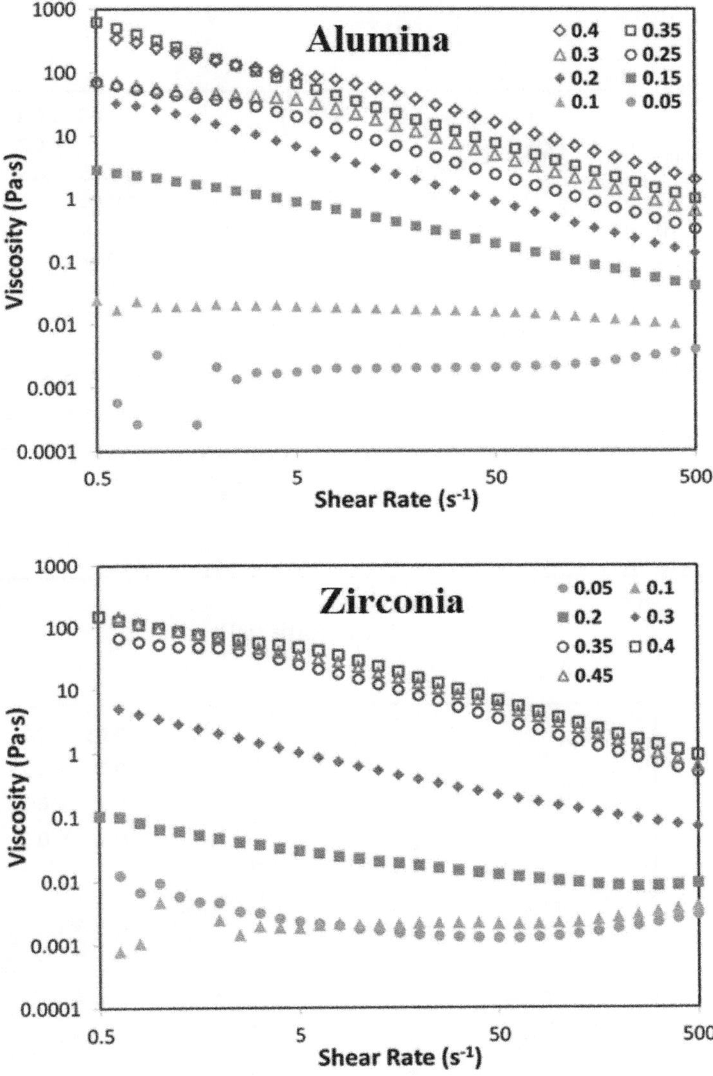

FIGURE 4.10 The effect of solids content in most common oxide ceramic suspension on ζ dependency to shear rate.

where hydrodynamic interactions are most effective and where the interparticle interactions are relatively less effective. As can be seen in *Figure 4.11*, the viscosity of the suspensions increases exponentially with solids content, as can be predicted by the $K-D$ relation. While the increase in viscosities is comparable for zirconia, YSZ, and titania, the alumina nanopowders exhibit the highest viscosities at a different rate of increase than the other systems [3].

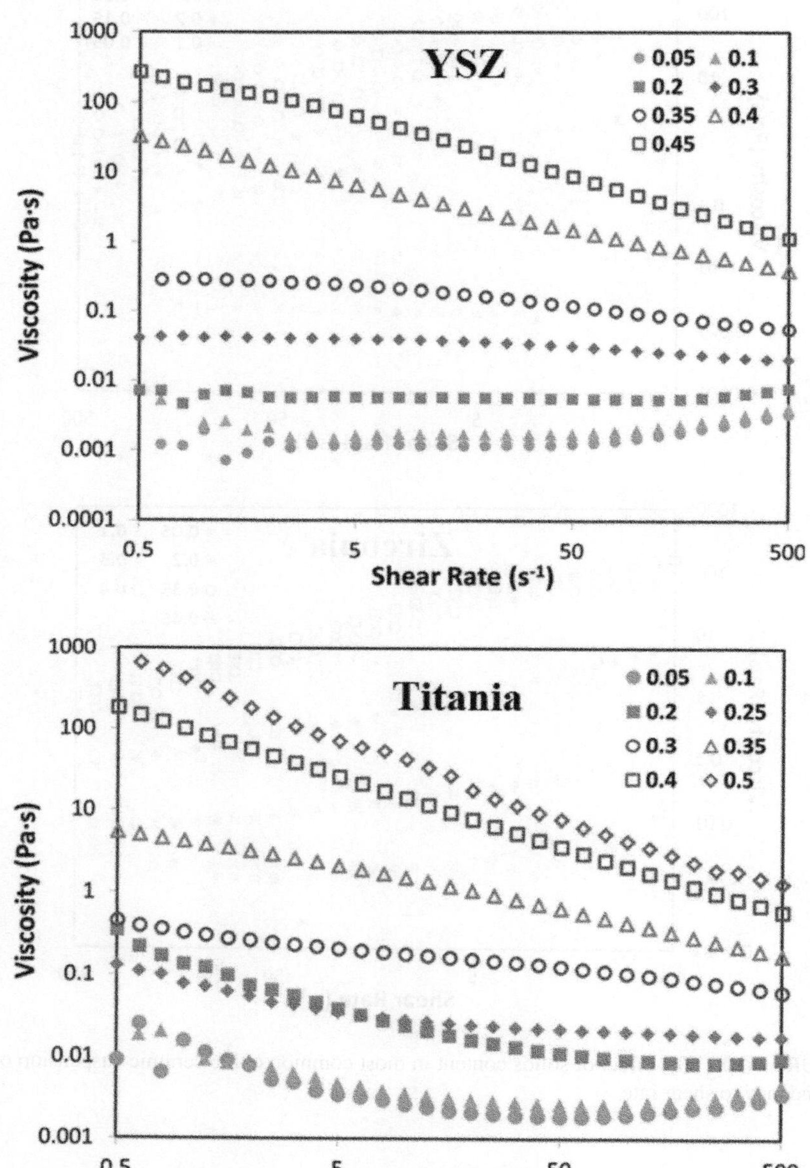

FIGURE 4.10 (Continued)

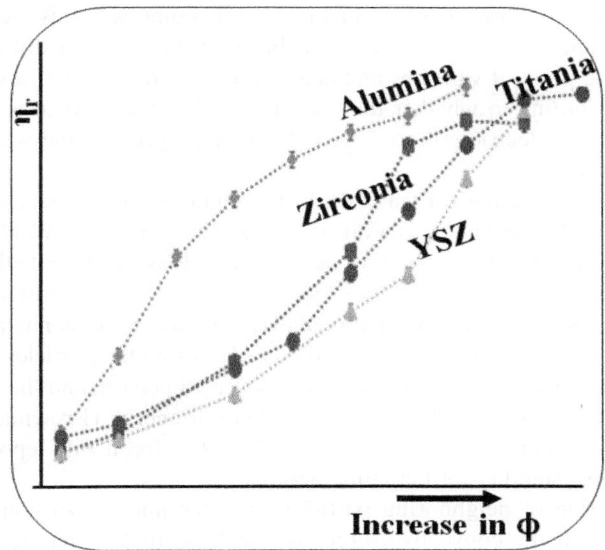

FIGURE 4.11 The comparison of ζ between oxide ceramic suspension with increasing solids content.

4.2 COLLOIDAL STABILITY OF CERAMIC SUSPENSIONS

In colloidal science, the term *stability* especially points to the quality of a substance or a system to remain at the same state. In practice, one needs to specify if "state" refers to the isotopes of elements, the chemical composition, the state of matter, or something else. When talking about the stability of colloidal suspensions or suspension stability, one usually addresses the size distribution of the particles, the homogeneity of the disperse system, and/or the structure of the suspension. A suspension can be permanently stable due to enthalpic or entropic reasons, or it may appear stable within relevant time periods (kinetic stability) because its physicochemical state is changing very slowly. The stability can be manifested and studied in micro- and macroscales.

Microscopic stability is related to size and concentration of the particles and is essentially affected by particle-particle interactions. Very frequently, *stability* is referred to as the absence of particle aggregation (microscopic or colloidal stability), which means that the particles keep their individuality. Therefore, microscopic stability is intrinsically a kinetic stability because aggregated systems are thermodynamically preferred. There are certain criteria for microscopic stability. As defined previously, *microscopic stability* means the absence of particle aggregation (coagulation). A sufficiently high degree of such stability is, therefore, achieved when repulsive forces between particles prevent them from collision. As a result, it is closely related to the surface charge of colloidal particles to their coagulation behavior. *Macroscopic* stability refers to a constant (and homogeneous) distribution of the dispersed phase in the liquid.

The ceramic suspension stability is an isoelectric point, which is found to be one of great importance. As it is neared, the stability of the hydrosol diminishes until, at the isoelectric point, it vanishes and coagulation or precipitation occurs, the one or the other according to whether the concentration of the particle is high or low, and whether the isoelectric point is reached slowly or quickly, and with or without mechanical agitation.

The principal mathematical framework that quantifies the repulsive interaction between electric double layers as well as the attractive van der Waals interaction in the DLVO theory can be applied for stable states. This theory calculates the free energy that is needed to approach two (particle) surfaces from infinity. The decisive parameter with regard to coagulation and in contrast dispersion (stability) is the energy barrier, i.e., the maximum energy required to bring the particles in contact. It depends on the *Hamaker* function, the double layer properties, and the particle size. When the particles move solely due to their thermal energy (Brownian motion), a minimum value of Vmax in the range of 10 to 20 $k_B T$ is frequently reported to guarantee microscopic stability for long time periods.

The distribution of neighboring particles in a ceramic suspension is given by a balance between the convective forces of the flow, hydrodynamic interactions, Brownian interactions, and interparticle forces due to electrostatic interactions. Driven by Brownian motion, colloidal particles will come into contact at a rate that is governed by diffusion. Von Smoluchowski first calculated the resulting rate of flocculation, known as rapid Brownian flocculation, assuming that each binary collision would cause the two particles to stick together:

$$J_0 = \frac{8K_B T}{3\eta_m} n^2 = \frac{3K_B T}{3\eta_m \pi^2 a^6} \phi^2 \tag{4.2}$$

The rate of doublet formation is dependent on the rate of diffusion and is proportional to the square of the particle volume fraction. For real colloidal particles, the flocculation rate will be slightly accelerated by the presence of an attractive interaction and greatly retarded by stabilizing forces, such as electrostatic repulsion [4]. They modified the theory to include an interparticle potential, including hydrodynamic interactions, to yield the exact formula for the rate of Brownian flocculation J. This is often cast in terms of a stability ratio W, defined as:

$$W = \frac{J_0}{J} = 2a \int_{2a}^{\infty} \frac{e^{\phi / K_B T}}{r^2 G(r)} dr \tag{4.3}$$

Where $G(r)$ is a hydrodynamic function that describes the resistance to motion as two particles move toward each other. Note that the potential of interaction can exhibit a large barrier that provides significant stability by retarding the rate of Brownian flocculation. The other convenient approximate form of the stability ratio is cast directly in terms of the energy barrier Φ_{max} as:

$$W = W_\infty + 0.25 \left(e^{\frac{\Phi_{max}}{K_B T}} - 1 \right) \tag{4.4}$$

The W_∞ is the rate of rapid Brownian flocculation or aggregation in the absence of any stabilizing forces. It shows how the stability ratio can increase substantially above that for rapid Brownian flocculation. Higher values could, for example, be achieved by an increase of the surface charge or a reduction in electrolyte concentration [5]. The characteristic time scale for aggregation can be calculated as:

$$\tau_{agg} = \frac{\pi\eta_m a^3 W}{\Phi K_B T} \tag{4.5}$$

The changes in electrostatic potential between colloidal particles for nanosized oxide ceramic suspension are shown in *Figure 4.12*. As observed, the potential decays rapidly with increasing surface separation. Upon the addition of an electrolyte (1:1) the range of the potential is rapidly reduced by screening.

The importance of the shear rate, when the effects of electrostatic repulsion on the viscosity are considered, is also clearly evident in *Figure 4.13*. The difference between yield stress behavior (ever-increasing viscosity as the shear rate approaches zero) and a show of plateau as the shear rate approaches zero is shown on left. The curves for zero-shear viscosity (the viscosity of a suspension when it is effectively at rest) diverge with increasing volume fraction, similar to those for Brownian hard spheres. However, when the electrolyte concentration of the medium is lowered, they diverge at systematically lower volume fractions as a result of the increasing range of the electrostatic repulsion with decreasing electrolyte levels. As with hard spheres, the limiting high-frequency viscosity η'_∞

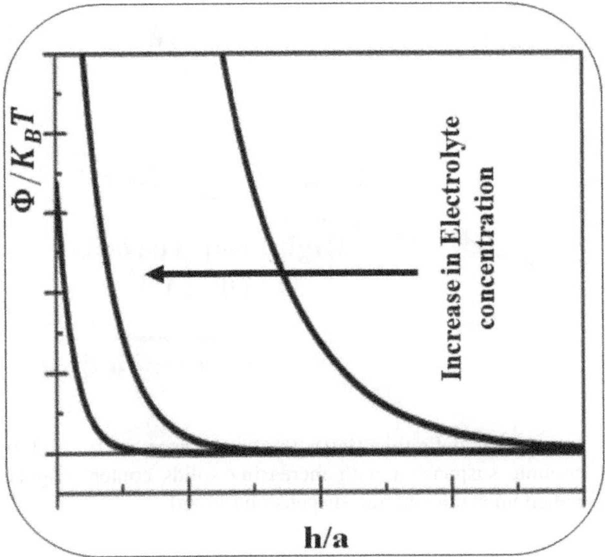

FIGURE 4.12 The comparison of electrostatic potential with distance radius ratio by gradual increase in electrolyte concertation in a typical nanosized oxide ceramic suspension.

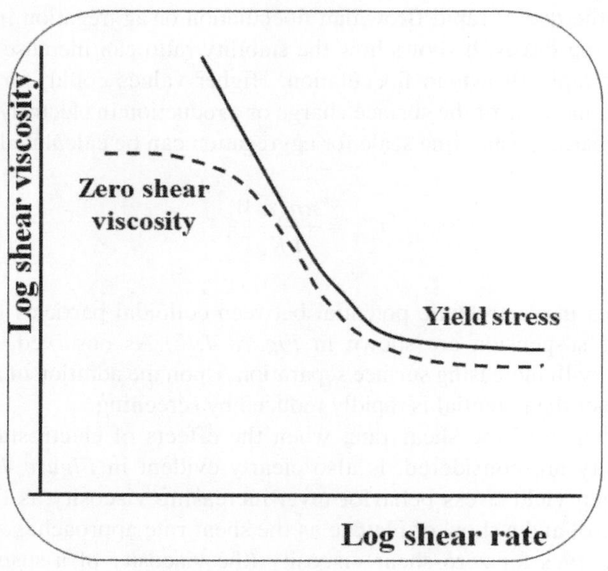

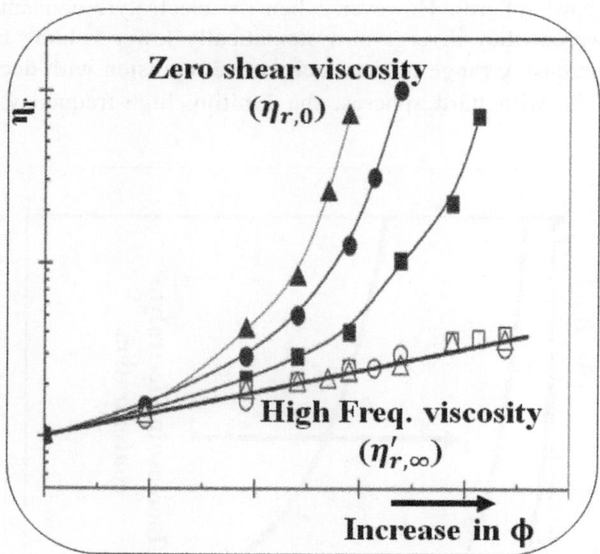

FIGURE 4.13 A graphical definition (left), the comparison of η'_∞ and η_r for a typical nanosized oxide ceramic suspension with increasing solids content (right), with a visual presentation of the phenomena in ceramic suspension (down).

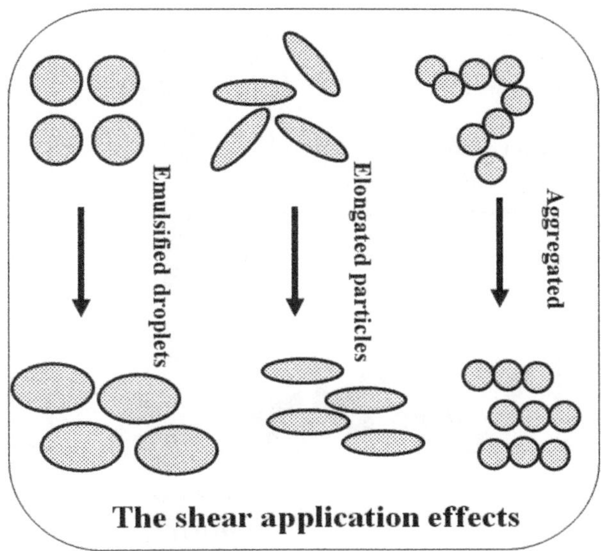

FIGURE 4.13 (Continued)

(a decrease in molecular/particle interaction and an increase in free space between dispersed components, which both contribute to the large drop in viscosity with the maximum degree of orientation achievable and hence the minimum attainable viscosity and are influenced largely by the solvent viscosity and related hydrodynamic forces) reflects the purely hydrodynamic component of the viscosity and is therefore not directly influenced by electrostatic repulsion. At relatively high frequencies, electrostatic forces contribute to the elastic modulus but not to the viscosity. Hence, adding electrolyte has little effect on the high-frequency viscosity; the contribution of electrostatic repulsion to the viscosity depends on both pH and the added electrolyte concentration.

The divergence of the low shear viscosity curves (*Figure 4.14*) indicates a liquid-solid (or sol-gel) transition with the gradual emergence of elasticity at higher volume fractions. Similar trends can be seen in *Figure 4.14 (right)*. It is noteworthy that electrostatically stabilized ceramic suspensions can form weak crystalline solids or repulsive driven glasses at much lower volume fractions than Brownian spheres.

The flocculation of the ceramic high solids content suspensions (slurry) is also affected by the concentration of the electrolytes. According to the *Schulze-Hardy* rule, in an aqueous suspension at 25°C:

$$n_c = 8\times10^{-22}\left(\frac{\gamma^4}{A^2 z^6}\right) \text{ for } \gamma = \left[\exp\left(\frac{z}{2}\right)\right] / \left[\exp\left(\frac{z}{2}\right)+1\right]\left(Z = \frac{ze\phi_0}{KT}\right) \quad (4.6)$$

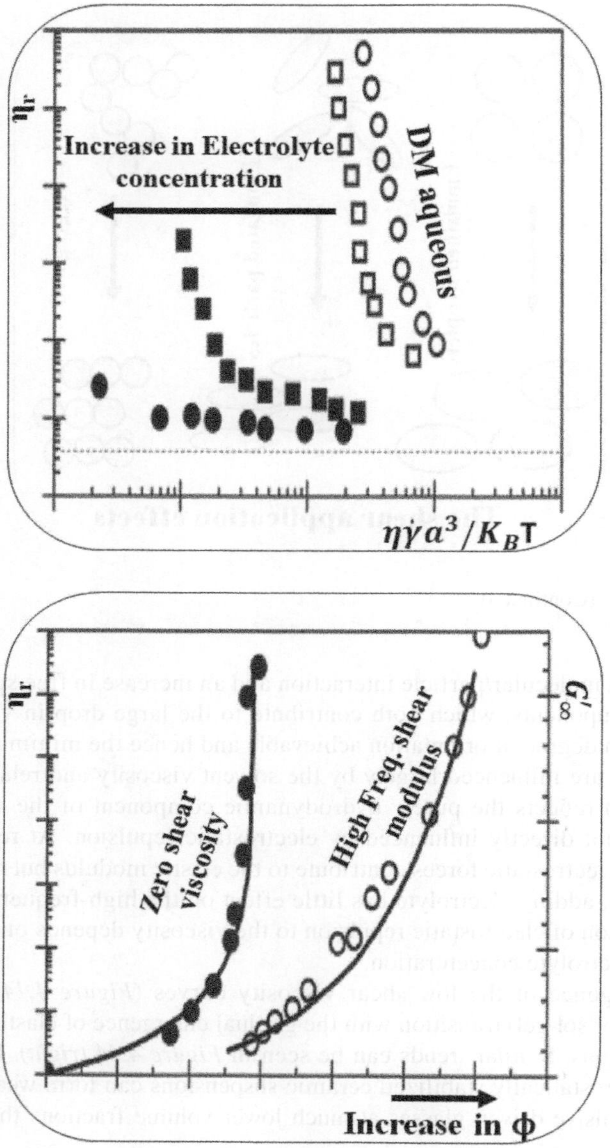

FIGURE 4.14 The comparison of media concentration (left) and the solids content dependency of shear modulus (right) for a typical suspension.

Where n_c is the critical flocculation concentration, z is the valency of the counterion, e is the electron charge, k is the Boltzmann constant, T is the absolute temperature, and A is the Hamaker constant. The critical flocculation concentration n_c is inversely proportional to the sixth power of the counterion valency. Furthermore, if the counterion valency is higher, then the critical flocculation concentration becomes lower. When compared with the monovalence counterions, high-valence counterions are observed to lead to coagulation at lower molar concentrations.

Among stable suspensions, stable suspensions of Brownian hard spheres are the most convenient. As noted, such dispersions are difficult to realize in practice as the ubiquitous van der Waals attractive forces necessitate some explicit method of imparting stability. These stabilizing forces can be of electrostatic and/or steric origin, and when the interparticle interaction is repulsive at all but short separations and the barrier to aggregation is sufficiently large, the suspension is kinetically (sometimes referred to as colloidally) stable. Under these conditions, the microstructure and rheology have many similarities to the case of Brownian hard spheres. The strength of the Brownian force scales with $k_B T / a$, which sets the scale for the elasticity of Brownian hard sphere suspensions. Imparting a significant electrostatic (charge), steric, or electrosteric stabilizing force can lead to much greater repulsive forces and, hence, larger elastic moduli. Furthermore, these forces can act over a significant range and, as a result, can drive crystallization and glass formation at much lower particle concentrations than those required for Brownian hard spheres.

The steric stabilization typically results from the presence of an adsorbed or grafted polymeric layer on the particle surface. Adsorbed surfactants, nanoparticles, or macroions provide a similar effect. Repulsion occurs when the adsorbed or grafted layers on two particles start to overlap, and then, typically, increases rapidly as the layers are compressed. The presence of an electric double layer surrounding charged particles can affect suspension viscosity in various ways. Often a distinction is made between three electroviscous effects [6].

In a dilute system, interactions between particles do not directly contribute to the viscosity. Mechanically, the suspending medium could be expected to flow around a charged particle exactly as it would around a hard sphere. The flow transfers the ions of the electric double layer. This tends to distort the double layer, which in turn will be counteracted by diffusive and electrostatic forces. The net result will be an increase in energy dissipation and viscosity. The effect occurs in dilute systems and will increase the intrinsic viscosity above the Einstein value. This first or primary electroviscous effect is small. In slightly more concentrated systems, electrostatic particle interactions contribute to the energy dissipation and suspension elasticity. The repulsion between like-charged particles will keep them farther apart and push them across the streamlines of the fluid. The corresponding increase in viscosity due to this secondary electroviscous effect can be substantial. The electrostatic repulsion will also contribute to the suspension elasticity. A tertiary electroviscous effect occurs in systems where the molecular configuration depends on the solution ionic strength. Polyelectrolytes grafted or adsorbed on the particle surface swell and/or collapse, depending on the medium's pH and ionic strength. Therefore, the stabilization is a combination of coupled steric and electrostatic effects, known as *electrosteric stabilization*. The dilute viscosity of electrostatistically stable particles systems can be studies via φ^2 coefficient. The decrease in ionic strength of the medium and increase in electrostatic repulsion causes higher viscosity. The coefficient of the φ^2term can be significantly larger in electrostatically stabilized suspensions than, for instance, in Brownian systems (see *Figure 4.15*).

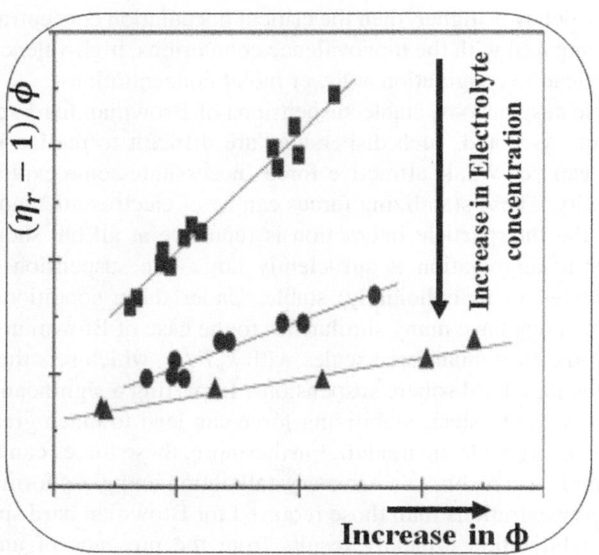

FIGURE 4.15 The comparison of viscosity as a function of ϕ with increasing electrolyte concentration in a typical suspension.

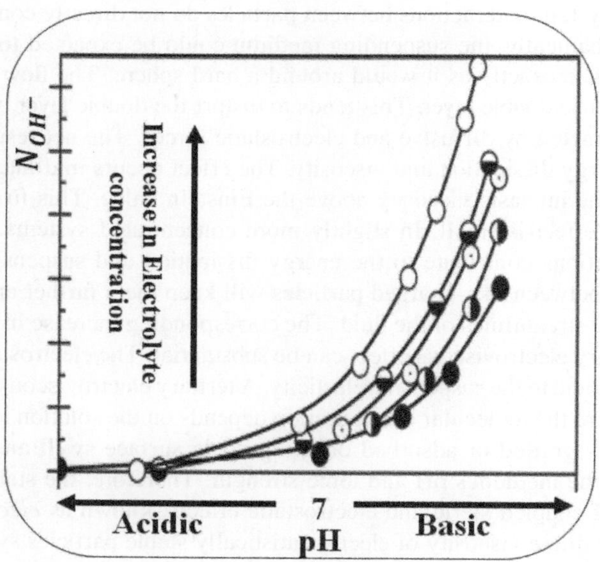

FIGURE 4.16 The pH dependence of dispersion estimation by surface charge in an oxide ceramic suspension with increasing electrolyte concentration.

In more concentrated systems, electrostatic contributions are more pronounced. This can be studied by the *Krieger and Eguiluz* model [7]. At low ionic strength, the zero-shear viscosity appears to diverge at low stress levels, giving rise to an apparent yield stress, i.e., the dispersion flows only when sheared above a minimum stress level. Adding electrolyte screens the interactions and greatly reduces the viscosity. The electrostatic contribution to the viscosity also decreases with increasing shear stress, leading to significant shear thinning. The acidification of the medium will also reduce the dissociation of the surface acid groups, leading to a reduction in the surface potential and, therefore, in the strength of the electrostatic repulsion (see *Figure 4.16*).

4.3 THE BASIS FOR CERAMIC SUSPENSION STABILITY

This criterion is expressed as molar critical coagulation concentration (*CCC*) of an indifferent electrolyte, above which the interaction between the particles is purely attractive. The DLVO theory exclusively considers van der Waals forces and double-layer interactions. Its predictive power is, therefore, lost when other types of interactions prevail or when the effect of (interfacial) additives is not properly accounted for. This might be the case when polymeric additives are used to induce or to prevent coagulation. Processing additives are used in ceramic suspension systems to obtain homogeneously dispersed, stable, and concentrated suspensions with reduced viscosities. They may be organic (polymers, surfactants, or other small molecules) or inorganic (such as pyrophosphates) and aim to enhance and/or create the repulsive forces or restrict the attractive interactions by different mechanisms.

For oxides and/or particles with an oxide surface layer, the surface charge can be changed by adjusting the suspension pH, as going away from the IEP, the repulsive interactions are enhanced. By adding charged species that adsorb to the particles, such as surfactants, the effective surface charge can be changed, and thus the repulsive interactions can be adjusted.

The ceramic particles, in which the surfaces are filled with several hydroxyl groups, have a strong tendency to agglomerate, thus increasing the suspension viscosity. The van der Waals forces promote an attractive potential between particles, especially in finer (nanosized) powders, causing the particles to aggregate in a strong flocculated state. The van der Waals potential can also be reduced by suspending. Dispersants can improve the formation of highly loaded ceramic suspensions.

The other type of dispersion mechanism is steric stabilization. In ceramic high solids content suspensions and slurries, their type and quantity significantly affect what occurs through the adsorption of long organic molecules at the surface. The rheological behavior and stability of suspensions is directly affected by these absorptions. The surface of the particles generates repulsion forces, and the polymeric effectiveness can be evaluated by the viscosity and stability of the suspension.

The use of organic molecules is also helpful in stabilizing the suspension via steric repulsion mechanisms. The soluble polymers, capable of adsorbing on the particle surface, introduce a polymeric layer between the particles and prevent the agglomeration of the particles. Because of the adsorption, the electrostatic properties of the particles may also change; in this case, the combined mechanism is called electrosteric stabilization. The mechanism is based on adsorption, and therefore temperature, pH, and the concentration of the ionic species affect stabilization. For aqueous alumina suspensions, acrylic acid groups, including sodium or ammonium polyacrylates and methacrylates, are commonly used for neutral pH. Oligo- or polysaccharides (e.g., dextrins, dextrans, maltodextrins) are also used as a dispersant for micron-sized ceramic particles in aqueous suspensions.

The organic polymers used as dispersants in conventional ceramic powder systems barely work well with nanoscale powders. The polysaccharides, for instance, significantly reduce the viscosity of micron-sized alumina suspensions while resulting in an increase in viscosity for alumina nanopowder suspensions, whereas saccharides with low molecular weight significantly reduce the viscosity of the same system. To understand the viscosity-reduction mechanism, smaller but similar organic dispersants and the changes in the system parameters, such as water mobility and the available water content, were also studied.

As an example, the effects of electrolyte concentration, particle size, and volume fraction on zeta potential of nanopowder suspensions by using γ and α-alumina powders with average particle diameters $< 100nm$ show that the zeta potential decreases with particle size and the volume fraction of the suspensions. The increase in electrolyte concentration resulting from dissociation of ionizable surface sites on the alumina and soluble species resulting from the dissolution of alumina can be presented as a possible reason.

Most conventional additives used to decrease the viscosities of submicron-sized powder suspensions either increase the viscosity of the suspensions rather than decrease it or limit the maximum achievable solids content because they may induce bridging or depletion flocculation in concentrated suspensions of nanopowders.

The degree of flocculation depends on the strength of the interparticle attraction. Weakly flocculated systems show reversible flocculation and strong shear-thinning behavior because the shear during the flow breaks the weak links between the structures. In flocculated systems where particles are held strongly, on the other hand, flocculation (or aggregation) is irreversible, and these systems exhibit substantial yield stress as well as shear-thinning behavior. In this case, the aggregates are considered as the primary flow units, and with an increase in shear rate, the viscous forces reduce the size of the aggregates and release the immobilized water within the aggregates. This, in turn, leads to a decrease in the effective solids content and hence in the viscosity of the suspension. Even though this explanation is widely accepted, the cause of the shear-thinning behavior is still controversial. Recently, it was reported that a decrease in interparticle forces at high shear rates caused by strong hydrodynamic forces leads to shear-thinning behavior.

The flocculation of sterically stabilized dispersions should be considered in nanosized ceramic suspensions. Two main types of possible flocculation may be distinguished as:

(1) Weak flocculation. This occurs when the thickness of the adsorbed layer is neglectable ($< 5nm$), especially when the particle radius and Hamaker constant are considerable.

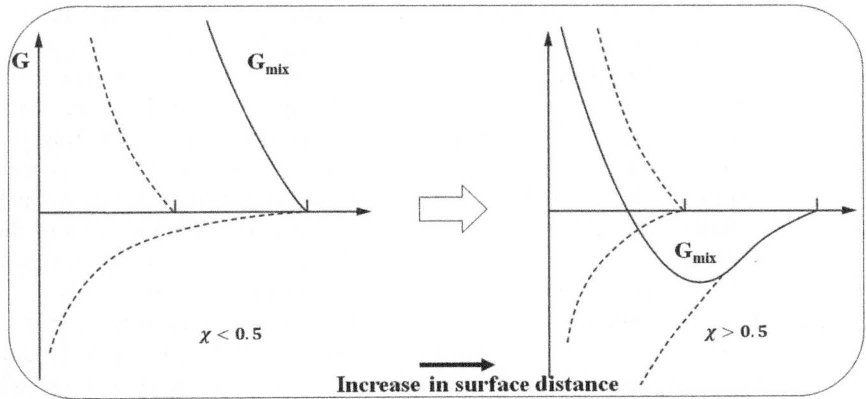

FIGURE 4.17 The decrease in particle dispersion in the media by reduction in layer formation in a typical suspension.

(2) Developing flocculation. This occurs when the dispersibility of the medium is reduced to toward worse (i.e., $\chi > 0.5$). *Figure 4.17* represents this, where χ was changed from < 0.5 (good dispersing media) to > 0.5 (poor dispersing media).

The osmotic pressure in the overlap region is a result of the unfavorable mixing of the surfactant chains. This is referred to as *osmotic repulsion* or mixing interaction, and it is described by a free energy of interaction G_{mix}.

The other non-DLVO interactions are the strong short-range repulsive interactions, frequently called *hydration interaction*, usually observed for high electrolyte concentrations in suspension. This type of interaction is commonly used to explain the surprisingly high stability of titania suspensions.

Through these systems responses, the *stability ratio* can be introduced. The examination of the free energy of two interacting particles reveals the main factors that govern the suspension stability and provides criteria for classifying a colloidal system as stable or not. However, in order to understand how the particle-particle interactions quantitatively affect the suspension stability, it is necessary to consider their influence on the particle aggregation rate. A simple approach to this ratio is:

$$W = 2\int_{2}^{\infty}\left[\exp\left(V / k_B T\right) / S^2\right] ds,\ S = R/a \tag{4.7}$$

Where V and S are the total potential of interaction between the particles and the scaled distance of approach between the particle's centers, respectively. The R and a are related to interparticle distance and particle radius, respectively. The factor W describes the retardation of aggregation due to particle interactions; it has values below 1 when attractive interactions prevail ($V<0$), whereas for repulsive interactions ($V>0$), W amounts to values above 1. For this reason, it is called the stability ratio. A colloidal suspension can be considered as to be long-term stabilized when the stability ratio exceeds these values. This ratio can be redesigned to include the effects of viscous interactions in the calculation of the stability ratio.

The similar affecting the minimum collision distance, the exact composition of a given suspension, and the numerous factors affecting stability (e.g., zeta potential) are necessary for evaluation of suspension stability. To achieve an acceptable stability ratio, it is necessary to monitor the initial stages of particle aggregation, which are mainly measurable as slight changes of the particle size distributions. For other parameters, the focus lies on changes in the macroscopic behavior (e.g., rheology, sedimentation), which become more evident at later stages of aggregation. Such macroscopic changes are caused by the apparent increase of volume fraction of the dispersed phase (since liquid is "immobilized" within the aggregates) and may even result in phase transition, e.g., to gelation or to the formation of colloidal glass.

The evaluation of suspension stability via its rheology gives a clear indication of suspension stability via customary macroscopic parameters. The rheology of a colloidal suspension reflects, similar to sedimentation, its microscopic properties and, in particular, the particle concentration, the particle interactions, and the resulting suspension structure. Therefore, it is responsive to dispersion-depreciating phenomenon, such as aggregation. This reaction primarily results from the increase in the effective volume of the dispersed phase (the main factor determining the viscosity of suspensions) with ongoing aggregation because liquid is "immobilized" in the pores of the evolving aggregates or flocs.

Since the aggregates usually grow in a fractal-like manner, the effective volume fraction of the dispersed phase as well as the suspension viscosity increase steadily. Therefore, the temporal evolution of viscosity can be used to quantify the aggregation kinetics. If the aggregation is not affected by sedimentation or hydrodynamic agitation, the suspension eventually starts to gel and the viscosity separates. Likewise, the gelling suspensions are characterized by the yield stress and by the viscoelastic properties, i.e., the storage modulus and loss modulus for the linear viscoelastic regime.

According to this concept, the stability of dense colloidal suspensions is evaluated via the yield stress; the absence of a measurable yield stress is then interpreted as the absence of significant aggregation. Such an approach is mainly used in parameter studies, when the pH value, the salt concentration, or the dose of some additive is diversified.

As a result, the stability analysis should consist of exploring the influence of particle concentration, the influence of time, and the significance of homoaggregation and heteroaggregation, i.e., aggregation between similar particles and aggregation between dissimilar particles, respectively. The typical stability ratio curves are mostly U-shaped, and suspensions can be stable at low surfactant concentrations due to strong electrostatic repulsion that is induced by the high repulsive charge of the adjacent particles. The stability ratios decrease with increasing concentration before reaching a minimum close to the IEP. Near this point, the stability ratios are close to unity, and the aggregation is close to diffusion controlled. When the surfactant concentration is increased further, the particles progressively accumulate attractive charge. This accumulation of charge again induces repulsive double-layer forces, which result in the restabilization of the suspension and an increase of the stability ratio (see *Figure 4.18*).

The evaluation of ceramic suspension stability by monitoring sedimentation behavior is the other applicable route. The stability of a colloidal suspension can be

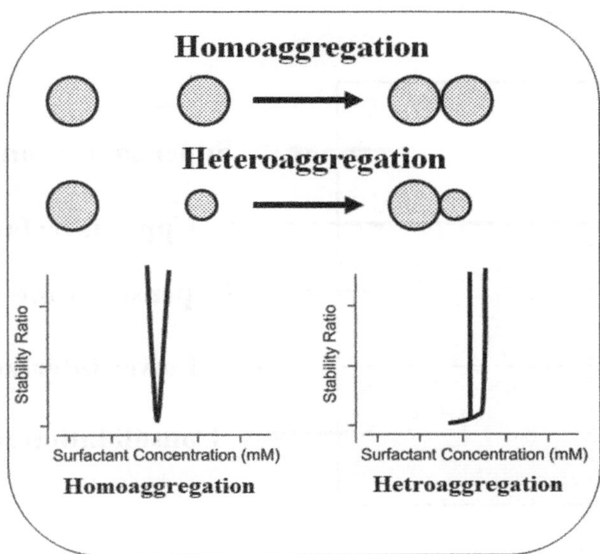

FIGURE 4.18 The decrease in particle dispersion in the media by reduction in layer formation in a typical suspension.

frequently evaluated by its (visible) sedimentation behavior. The settling velocity of particles grows with size. Large, micrometer-sized aggregates ("*flocs*") settle quickly, which results in a separation of the dispersed from the liquid phase within a few minutes up to hours. In contrast, suspensions of very fine nanoparticles ($< 100\,nm$) reach a diffusion-sedimentation equilibrium, in which the dispersed phase remains perpetually yet not uniformly distributed throughout the whole sample volume. The sedimentation behavior of ceramic suspensions is affected not only by the size of the particles but also by their concentration—or, more precisely, by the interparticle distance. Only for very dilute suspensions can this influence be ignored (free settling).

The viscous interaction between the settling particles and the backflow of the displaced liquid retard the sedimentation process even for concentrations far below the maximum packing density and lead to a hindered settling of individual particles. In this regime, the coarse particles still settle faster than the fine ones. Therefore, the interface between the original suspension phase and the supernatant is gradually broadening, and the concentration-height profile reflects the particle size distribution. If the interparticle distances are reduced in such a way that coarse particles cannot pass finer ones due to steric effects, then all particles are forced to settle with the same velocity and the sedimentation becomes equivalent to the flow through a fixed bed (zone sedimentation). As a result, the dispersed phase subsides with a sharp interface between the suspension phase and the supernatant. Zone sedimentation occurs for high particle concentrations or is a result of particle aggregation because aggregates "immobilize" liquid and thus increase the apparent volume of the dispersed phase (see *Figure 4.19*).

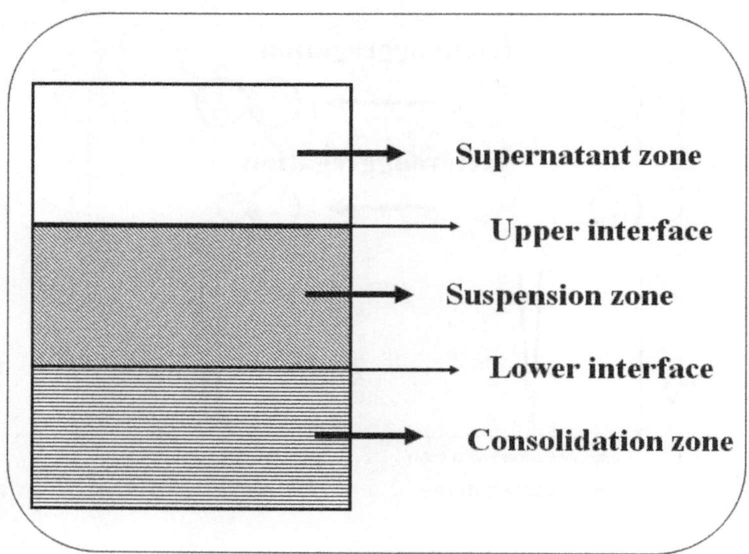

FIGURE 4.19 Supernatant, suspension, and consolidation zones during sedimentation and self-weight consolidation of fine-grained typical suspension.

The zone sedimentation of aggregating ceramic suspensions coincides with the formation of a particle network (gel) in the lower part of the sedimentation vessel. This gel is gradually consolidated under its own weight, which reduces its porosity and the sedimentation velocity. The gel is, therefore, called compression or compaction zone; it eventually comprises all particles of the suspension. The compression zone has a much higher porosity than the sediment of a well-stabilized suspension. Due to these principal sedimentation effects, one can employ the following relationships for the evaluation of suspension stability: aggregation leads to aggregates which settle faster than the primary particles, advanced aggregation in dense suspensions results in zone sedimentation with a sharp interface between the supernatant and the suspension phase whereas this interface is diffuse for the hindered settling in stable suspensions, zone sedimentation eventually leads to compression, and the sediment porosity of a stable suspension is considerably lower than that of a flocculated one.

Then, apparently stable suspension without visible separation of the dispersed phase may indicate gelation, aggregation of the primary particles, or a diffusion sedimentation equilibrium, which occurs for nanosized particles in the absence of aggregation. For instance, the pH dependence ($pH=2-10$) of these parameters for three essential ceramic oxide materials (SiO_2, TiO_2, and Al_2O_3) indicates that the corresponding IEPs of these suspensions were 4.0 (SiO_2), 7.2 (TiO_2), and 8.8 (Al_2O_3). Around the IEPs, an increased initial settling velocity happens, which is related to the increased size due to the coagulation of the pyrogenic aggregates.

The electrostatic stabilization solely does not fully explain high viscosity observed in concentrated nanoparticle ceramic suspensions. The increase in available surface leads to an increase in ionizable alumina surface sites resulting in high concentration of dispersed particles in the suspensions. The van der Waals attractive forces are dominant in the alumina nanoparticle suspensions, and the magnitude of the electrostatic repulsion is not sufficient to overcome the attractive forces. Therefore, steric stabilization for nanoparticle suspensions is dominant. The use of a more robust steric repulsion for stabilization of nano oxide suspensions and the suspensions with chemically adsorbed molecules with higher stability during processing explain the nature of alumina nanoparticle suspensions.

The silica suspensions can virtually stay stable over the 2–10 pH range, whereas the interface sharpness and the sediment porosity values are significantly increased around the IEP and indicate the beginning of coagulation.

The ceramic suspension stability can also be considered as a function of solvent properties. The suspension properties (e.g., solids content, liquid phase, concentration of ionic, or polymeric additives) are the controlling parameters in inducing stability in these systems.

4.4 STABILITY OF BINARY AND TERNARY CERAMIC SUSPENSIONS

The binary ceramic suspension, a suspension consisting of two particle components, can experience more complex stability behavior than unary systems. A very common example is the preparation of mullite ceramic suspension made of silica and alumina particles. The extensively dominant phenomenon in these systems can include homoaggregation of each component and/or heteroaggregation between both components. The heteroaggregation may be further even used to produce core-shell particles.

The binary ceramic suspension can be considered microscopically stable only in the absence of both types of aggregation. The microprocesses in binary ceramic suspensions are affected not only by the particle interactions but also by the size ratio and the mixing ratio of the particulate components. Evaluation by three partial stability ratios, namely W_{11}, W_{22}, and W_{12}:

$$W_{ij} = r_{ij,0} \int_{r_{ij,0}}^{\infty} \exp(\frac{V_t^{ij}}{k_B T}).\frac{dr}{r^2} \tag{4.8}$$

The formation of aggregates from particles of the components i and j obeys the following kinetic when dominated by Brownian collisions:

$$\frac{dc_{N,i+j}}{dt} = 2\left(2-\delta_{ij}\right)\pi.c_{N,i}c_{N,j}.r_{ij,0}.\left(D_i + D_j\right) / W_{ij} \tag{4.9}$$

Whether the particles of both components are approximately equal in size or slightly different, an effective stability ratio for the whole suspension can be defined as:

$$\frac{1}{W_{eff}} = \frac{\varphi_{N,1}^2}{W_{11}} + \frac{\varphi_{N,2}^2}{W_{22}} + \frac{2\varphi_{N,1}\varphi_{N,2}}{W_{12}} \tag{4.10}$$

Where $\varphi_{N,i}$ is the relative number frequency of particles of the component i in the whole particle population. The effective stability ratio is dominated by the least stable combination. It thus indicates the total stability behavior of the binary suspension.

The stability of aqueous suspensions of binary oxidic components is an appropriate practice for these theories. In the case of aqueous suspensions of oxide materials, the stability is mainly determined by the van der Waals and double-layer interactions (DLI) between the particles (for particle collision). Whether the DLI is attractive or repulsive depends on the signs of the surface charges, the absolute values of the zeta potential, and the regulation capacity of the double layer. When all particles are identically charged, the DLI is repulsive. The most influential factors on the stability are the pH value, the total electrolyte concentration, and the valency of the ions. The pH range in which a binary suspension remains stable is typically different to the respective stability ranges of the two particle components. The absolute microscopic stability of the binary oxide suspension occurs when the zeta potentials of the two particle components are equal in sign and large in absolute value.

This theory is based on the fact that for the description of particle aggregation, the interfacial properties of the two components do not affect each other. The theory does not account the consequences of charge regulation between interacting double layers. As a result, heteroaggregation is expected when both components are oppositely charged, while homoaggregation occurs when the attractive interaction forces between similar particles dominate due to low surface potentials. In this condition, when both components are weakly charged and if their charges are of opposite sign, they experience homoaggregation as well as heteroaggregation. Simultaneous occurrence of homoaggregation of only one particle component as well as of heteroaggregation results from oppositely charged surfaces with one strongly charged and one weakly charged component. Nevertheless, if the sign of the surface charge is equal for the two particle components and one is weakly charged and the other is strongly charged, then the homoaggregation of the weakly charged component is predominantly possible.

Simultaneously, it is probable that double-layer overlap induces a charge reversal on the surface of the weakly charged component and thus promotes heteroaggregation, where double-layer repulsion is expected. The specific aggregation in binary suspensions is the morphology of heteroaggregates. While in suspensions with only one dispersed phase, a significant rate of aggregation coincides with low (microscopic) suspension stability. For binary suspensions, the situation can be more complex. During heteroaggregation between a huge number of small particles and few coarse particles, the system can experience agglomerates that resemble core-shell particles and that do not agglomerate with each other. A suspension of such aggregates and excess fine particles may be microscopic. In a binary suspension of titania and alumina, the influence of pH on the stability is considered the key factor. The freshly prepared suspensions rapidly coagulate in the pH range between the IEPs of the two components, i.e., the pH range where the two oxides are oppositely charged. The rapid aggregation is detectable for a small pH range around the IEP of alumina. This effect is explained as precipitation of $Al(OH)_3$ on the surface of the titania particles with the $Al(III)$ ions originating from the dissolution of the alumina surface.

Other examples include heteroaggregation of alumina with other oxides (zirconia, silica) without interfacial harmonization and subsequent stabilization.

The suspension stability in similar system like rutile/anatase titania-amorphous silica involves dispersion of fractal aggregates of nanoparticles with high specific surface areas, highly dense suspensions event at medium solids contents of about 5 $wt\%$. This situation with the high specific surface area leads to a strong impact of the double-layer formation (i.e., [de-]protonation of hydroxyl groups, ion adsorption, etc.). The suspension system exists between the zeta potential curves of pure silica ($pH_{IEP} \approx 4$) and pure titania (($pH_{IEP} \approx 7.2$). The addition of silica particles to a titania suspension has a stabilizing effect for basic pH values. Even small amounts of silica can considerably reduce the settling velocity. The different dissociation of silanol ($Si-OH$) and titanol ($Ti-OH$) groups indicates the zeta potential curves of mixtures. In the acid pH range, the reduced value of the (positive) zeta potential promotes agglomeration, which results in an increased settling velocity and higher sediment porosity. Different to this, the dissolved silica increases the negative surface charge under basic conditions and, as a result, decelerates agglomeration. Correspondingly, the settling velocity and sediment porosity are reduced.

The effect of sample age on the quantifiable suspension's stability can be controversial. The response to resting suspensions results from the fact that coagulation affects the measurable suspension properties. Only after having reached a certain stage and, additionally, from the fact that the interface of suspended ceramic particles reaches their chemical equilibrium (e.g., with regard to dissociation of surface groups or the adsorption of ions and polymers) are its responses meaningful. After finite periods of time, the suspension system can be considered at an indicating state. This means that the ceramic suspension stability is not only a function of the present suspension parameters (e.g., pH, or ionic strength) but can also be considerably affected by previous processing steps (including dispersing procedure, temperature, or pH). In this respect, there is an influence of the "suspension history." As an example, in previously mentioned system, the dissolution of silica and the subsequent adsorption and precipitation of silicic acid on the titania surface are chemical reactions with a characteristic kinetic. The mechanism evolves the stabilizing effect of silica on the binary suspensions not immediately after the suspension preparation but after resting periods about one day later.

The stability of viscoelasticity of formed stabilized suspensions is qualitatively similar to that of suspensions of Brownian hard spheres. With increasing volume fraction, the oscillatory response shifts from that of a viscoelastic fluid to that of a weak viscoelastic solid. At low volume fractions, there is a liquid-like relaxation at low frequencies, whereas at high frequencies, the storage moduli tend to be at a plateau value. In electrostatically stabilized suspensions, the maximum in loss modulus is evident. This effect can be seen with dynamic moduli G' and G'' versus stirring in *Figure 4.20*.

In contrast to electrostatic stabilization, not only does steric stabilization increase the zero-shear viscosity, but the adsorbed or grafted surface layer also increases the hydrodynamic size of the colloidal particle. Hence, the hydrodynamic contribution to the viscosity will increase accordingly. Unlike electrostatic stabilization, steric stabilization can impart true thermodynamic stability, and so such dispersions can often

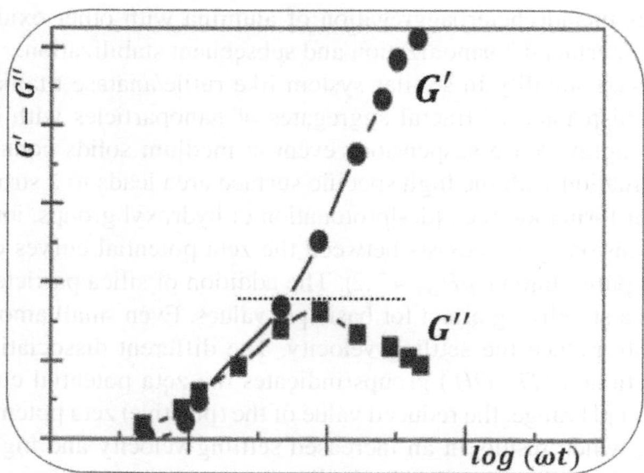

FIGURE 4.20 A typical stabilized suspension relation on dynamic storage moduli to stirring.

be highly concentrated and still stable, leading to high amounts of viscoelasticity. Represented is the viscosity and first normal stress difference for a sterically stabilized dispersion. At higher shear rates, the first normal stress difference is of comparable magnitude to the shear stresses for the concentrated plastisol dispersions.

As with the electrostatically stabilized dispersions, the limiting low shear viscosities for sterically stabilized dispersions can often be compared to those of Brownian hard spheres via an effective volume fraction. The trends in *Figure 4.22* show the zero-shear viscosities of a series of dispersions with stabilized particles of different core particle sizes but with the same stabilizer layer thickness. At a given core volume fraction, the smaller particles have a higher viscosity because the additional volume of the layer coat is proportionally greater for smaller core particle size. However, when plotted versus an effective hard sphere volume fraction, the data for the various core sizes come much closer together (indeed, the lower viscosities for the medium particle sizes can be attributed to greater dispersity).

The suitable effective volume fraction can be used to reduce the zero-shear viscosity of these sterically stabilized dispersions. In the plot, the effective hard sphere diameter is determined from measurements of the elastic moduli and is determined from data taken at high concentrations rather than in dilute limiting conditions. The analysis shows a reasonable reduction of the data to hard-sphere-like behavior, which is useful for designing steric stabilizing layers in dispersion formulation.

In electrostatically stabilized systems, the dilute and semidilute nature of suspensions directly attribute to stabilization condition. As noted, very dilute dispersions follow the linear viscosity dependency to solids content for noninteracting hard spheres with an added contribution due to the primary electroviscous effect. The particle collisions can be ignored; as a result, the viscosity is linear in a different volume fraction. This also requires that the average interparticle distance be much larger than the range of the electrostatic interactions. For charged particles, the flow of the suspending medium around a charged particle will transport the ions in the

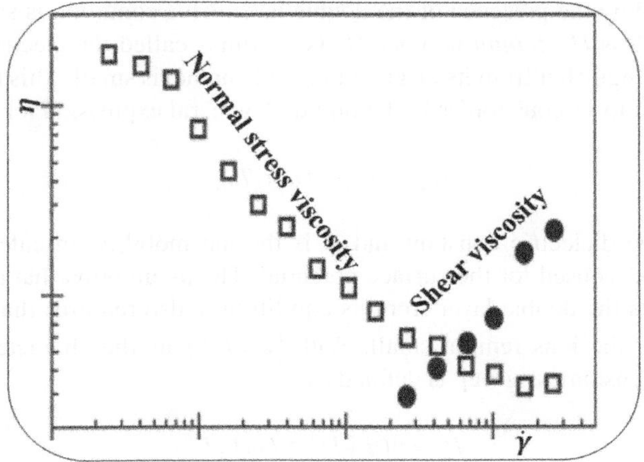

FIGURE 4.21 A typical sterically stabilized suspension relation to shear rate.

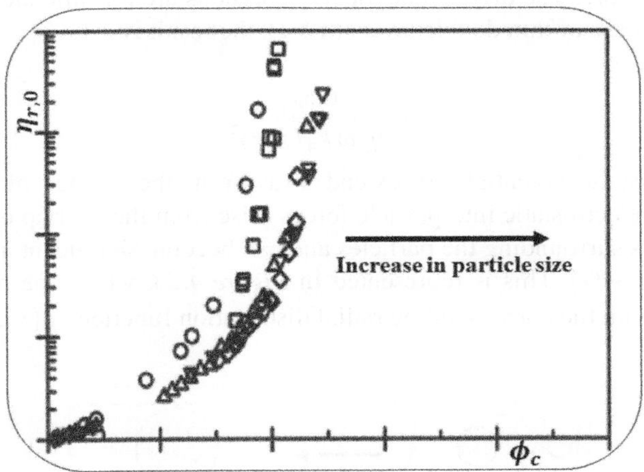

FIGURE 4.22 A typical sterically stabilized suspension zero-viscosity core solids content dependency for different particle sizes.

diffuse electric double layer and distort their distribution. The distortion of this layer will cause electric stresses that will distort the flow lines and increase the dissipated energy. The resulting viscosity can be described by:

$$\eta = \eta_m \left[1 + 2.5(1 + \rho)\right] \tag{4.11}$$

An expression for the **viscosity** increase p caused by electrostatic effects in dilute suspensions in the case of a small surface potential $\psi_S = e\psi_s / k_B T \ll 1$ and arbitrary thickness of the electric double layer k^{-1}. The flow around the particle is only slightly altered by the presence of the double layer. This condition is satisfied when the dimensionless *Hartmann* number H_a (sometimes called the electric Hartmann number to distinguish it from its original magnetic name) is small. This measures the ratio of electric to viscous forces in the liquid. A general expression for $H_a$is:

$$H_a = \varepsilon\varepsilon_0\psi_s^2 / \omega_i k_B T\eta_m \tag{4.12}$$

Where ε is the dielectric constant and ω_i is the ion mobility. In calculations, the zeta potential ζ is used for the surface potential. The assumption that the flow only slightly distorts the double layer from its equilibrium also requires that the *P´eclet* number P_{ei} for the ions remain small. With $\left(a+k^{-1}\right)$ as the characteristic length scale, this dimensionless group is defined as:

$$P_{ei} = \dot{\gamma}(a+k^{-1})^2 / \omega_i k_B T \tag{4.13}$$

It should be noted that ion mobility is much larger than particle mobility. Hence, much higher shear rates are required to achieve high Pei than for the particle *P´eclet* number used before. The driven analytical expressions are for thin and thick double layers. For the case of thin double layers ($a \gg 1$), the result is:

$$p = \frac{6\varepsilon\psi_s^2}{\eta_m\omega_i k_B T(ak)^2} \tag{4.14}$$

The electrostatic potential can extend away from the surface of the particle. Therefore, the electrostatic interparticle forces arise from the overlap of the electrostatic potentials surrounding the particles and can become significant at low volume fractions (even ~1%). This is represented in *Figure 4.23*, where the potentials are plotted along with the corresponding radial distribution functions $g\left(r\right)$.

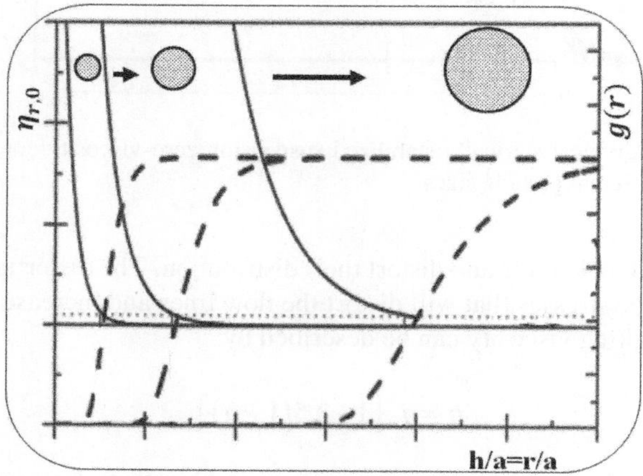

FIGURE 4.23 The electrostatic repulsion (solid) and radial distribution function (dashed) trends versus gradual particle distance increase for a typical stabilized suspension with schematic depiction of effective particle size drawn above the curve.

In dilute systems, the radial distribution function is simply given by the Boltzmann factor for radial distribution function $g(r)$ (depends on the particle situation angle and interparticle distance) as:

$$\lim_{\phi \to 0} g(r) = \exp\left[-\frac{\phi(r)}{k_B T}\right] \tag{4.15}$$

The probability of finding a neighboring particle nearby is vanishingly small when the potential is very large. The schematic illustrations can show the effective hard sphere size relative to the core particle size, where one can define, as a first approximation, this effective size as extending to where the repulsive potential is of the order of the thermal energy $k_B T$. Brownian motion cannot drive particles close together when the repulsive potential is significantly greater than the thermal energy.

In semidilute systems with higher volume fractions, contributions from pairwise interactions should be included in the calculation of the stresses. The basic procedure for this follows the fact that the extra stress arising from these interactions is calculated as the product of the interparticle force and the probability of finding a neighboring particle. The viscosity of semidilute dispersions of charged spherical Brownian particles occurs in certain conditions such as that the electrostatic forces were strong enough to prevent close encounters between particles. This permits hydrodynamic interactions to be neglected. In addition, the double layer is assumed to remain at equilibrium, i.e., the Hartmann and ionic Péclet numbers should remain small, which is a good assumption for highly charged colloidal dispersions at low ionic strengths ($ka<1$). The force acting between particles is given by a linear superposition of the electrostatic potentials. For charged particles, a characteristic separation length L can then be defined such that the Brownian and electrostatic forces balance at that distance $\phi(L)/k_B T \approx 1$ and for $\alpha = \varepsilon\varepsilon_0 \Phi_0^2 a^2 k exp(2ak)/k_B T$ and L:

$$L \sim \frac{1}{k} ln\left\{\alpha / \ln\left[\frac{\alpha}{\ln\left(\frac{\alpha}{\dots}\right)}\right]\right\} = \frac{1}{k} ln \frac{\alpha}{\ln\left(\frac{\alpha}{ln\alpha}\right)} \tag{4.16}$$

This distance determines the location of the nearest-neighbor particles at equilibrium, which is then used in an approximate calculation of the distribution of neighboring particles under shear flow. For sufficiently large separations ($L \gg 2a$), the analytical expression valid equation is:

$$\eta_r = 1 + 2.5\phi + \left[2.5 + \frac{3}{40}\left(\frac{L}{a}\right)^5\right]\phi^2 + \dots \tag{4.17}$$

The coefficient of the ϕ^2term is very sensitive to L or, according to Eq. (4.6), to α and k, with trend shown in *Figure 4.24*. At low ionic strengths, the ϕ^2 coefficient clearly can be very large. The same phenomena can be expected to cause a strong

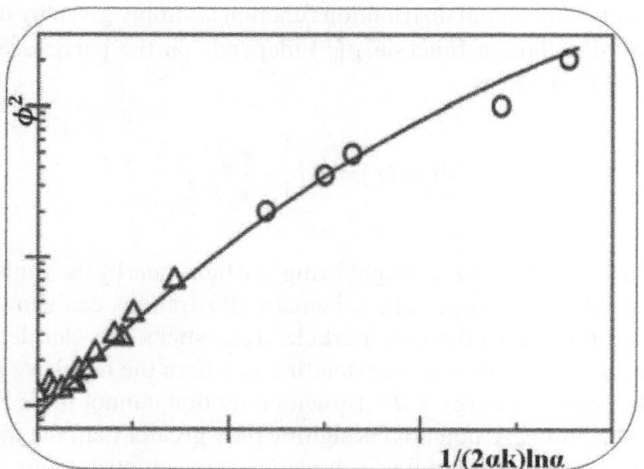

FIGURE 4.24 The ϕ^2dependency to L, α, K in stabilized suspensions.

increase in viscosity in more concentrated suspensions. The ceramic suspensions are occasionally treated with ion exchange resin to remove all ions other than the counterions required to satisfy electroneutrality (deionized suspensions). Results for the high-frequency limiting viscosity can be calculated by dynamics methods, which are described to third order in volume fraction:

$$\eta'_{r,\infty} = 1 + 2.5\phi(1+\phi) + 7.9\phi^3 \tag{4.18}$$

The predicted high-frequency viscosity for deionized suspensions is below that of hard spheres. This is again due to weaker hydrodynamic interactions because of the excluded volume arising from electrostatic repulsion, as represented in *Figure 4.25.*

The direct result of the electrostatic repulsion preventing particles from close approach where the hydrodynamic interactions are greatest in magnitude can be indicated with 2.5 as the coefficient of the ϕ^2term, which is substantially below the value of 5.0, calculated for a hard sphere microstructure. The schematic drawing of electric double layers and its relation to surface electrical charge (*Figure 4.26*) shows that the specific adsorbed cations exist in the Stern layer, and when the concentration of OH^- in the diffusion electronic double layer increases along with the increase in the pH value, this results in pH_{iep} when the potential at the slide-plane zeta potential becomes zero. When the density of the cations in the Stern layer decreases, the OH^- concentration at which the zeta potential is zero, or the pH_{iep}, also decreases.

In concentrated suspensions, the zero-shear viscosity of electrostatically stabilized dispersions can diverge at very low volume fractions relative to that for Brownian hard sphere dispersions. Indeed, the calculation of a characteristic excluded shell permits the definition of an effective volume fraction:

$$\phi_{eff}^{hs} \approx n\frac{\pi}{6}\left(\frac{L}{a}\right)^3 = \phi\left(\frac{L}{2a}\right)^3 \tag{4.19}$$

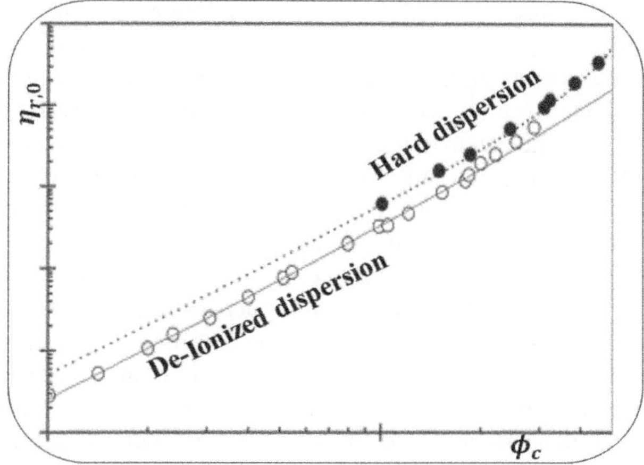

FIGURE 4.25 The high-frequency relative viscosity of a typical suspension in deionized media and calculated hard sphere model versus solids content of suspension.

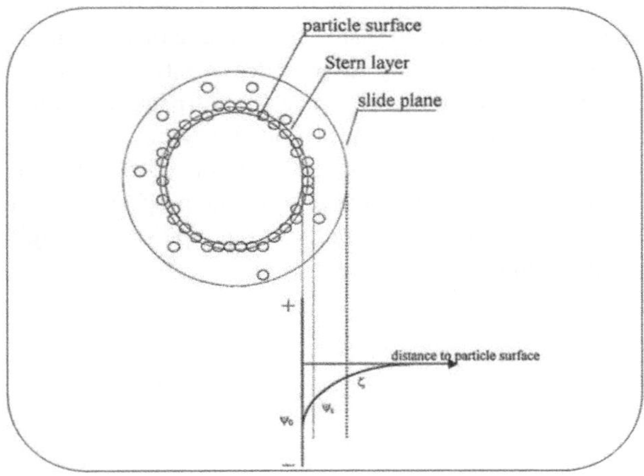

FIGURE 4.26 The schematic layer formation around a well-dispersed particle and the developed surface charge and electrical potential trend in a typical stabilized semidilute suspension.

If accounting for the electrostatic repulsion in this manner provides an effective hard sphere particle size and volume fraction, the zero-shear viscosities should collapse onto a master curve when plotted as a function of this effective volume fraction. The previous approach is useful as a concept and to identify the possible influences of particle size, surface potential, and added electrolyte on the effective size of the electrostatically stabilized particles. In electrostatically stabilized dispersions of highly concentrated suspension, the ionic strength, and therefore the range of the potential, varies with particle volume fraction because of the simultaneous changes in counterion concentration (a consequence of electroneutrality). Therefore, it is important (and very convenient) that a single parameter like L can reduce the data to a master curve for the entire range of volume fractions. Indeed, the possible complexities of electrostatic interactions include variations in surface charge density and/or potential with particle concentration due to the accompanying counterions. Therefore, the strength and range of the electrostatic repulsion can vary from low to high particle concentrations. Such effects are most evident at low added electrolyte concentrations and high particle surface charge. This can be complicated further by the presence of attractive interactions (such as the ubiquitous dispersion forces).

At low packing fractions, particles repel each other at a distance close to where the potential is $\sim k_B T$. The equilibrium dispersion microstructure changes with increasing particle concentration and increasing salt concentration (see *Figure 4.27*).

According to trends, the decreasing salt concentration amplifies the relative particle separation and heightens the nearest-neighbor peak, in direct correspondence with the pair potentials. As observed for hard spheres, increasing the particle concentration leads to an improved order in the liquid and more nearest neighbors, as expressed by the magnitude of $g(r)$. Meanwhile, the particles come into increasingly closer approach at higher concentrations. As the electrostatic potential increases more steeply with decreasing average separation, the repulsive force acting between particles, the derivative of the potential, increases as well. Therefore, the normalized viscosity rises slowly as compared to that of hard sphere dispersions at lower relative packing fractions, where the neighboring particles experience a softer potential. On the other hand, they increase more rapidly at higher packing fractions, where the neighboring particles experience a stronger repulsive force.

Another empirical approach for treating the effects of electrostatic repulsion on the zero-shear viscosity and on other, thermodynamic properties is to use the idea of an excluded volume that depends on the range of the repulsion through ka. A scaling parameter k can be introduced, such that:

$$\phi_{eff} = \phi(1 + \frac{\alpha}{ka})^3 \tag{4.20}$$

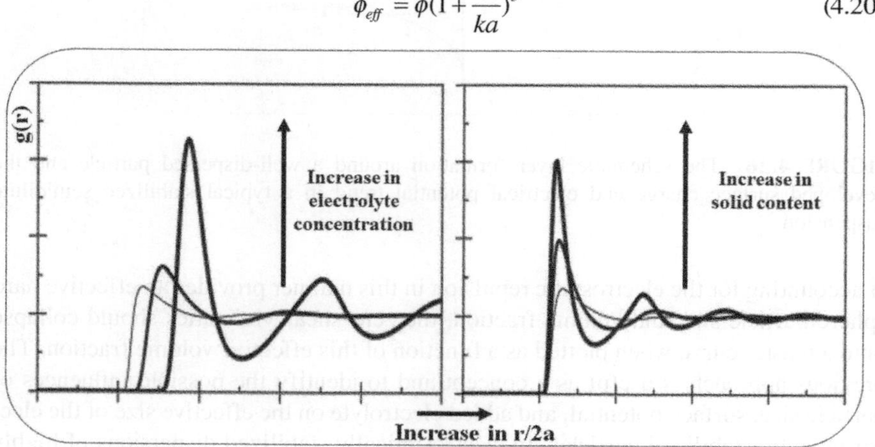

FIGURE 4.27 The trends in g(r) function against increasing r/2a ration for constant solids content (left) and constant electrolyte concentration (right).

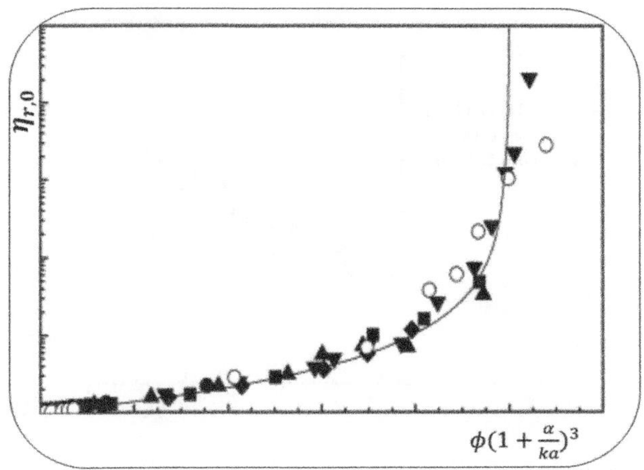

FIGURE 4.28 The trend zero-shear viscosity of a typical oxide suspension in organic media with different particle dimensions.

Mapping the viscosities of electrostatically stabilized dispersions onto hard sphere behavior, the value of α varies from ~1 for highly charged latices to ~0.5 for nanoparticles in nonaqueous solvents. This factor can also be as large as ~4 for micro-sized stabilized silica particles in an alcohol media. The parameter α is not predictable from the particle and medium properties. A further example of this type of scaling for the zero-shear viscosity of nonaqueous dispersions and acts as an effective hard sphere dispersion. This correlative approach works reasonably well for a very broad range of particle sizes (see *Figure 4.28*). Deviations become apparent at higher particle volume fractions, a consequence of the softness of the potential.

The dispersions of charged stabilized particles as the system of hard spheres is derived from a similar formula for the contribution of the interparticle forces to viscosity:

$$" \eta^{I}_{r,0} = \frac{12}{5}\frac{a}{b}\phi_b^2 \frac{g(2;\phi_b)}{D_0^2(\phi)} \tag{4.21}$$

Where b is the effective hard sphere radius due to a strong repulsive force acting between particles, and ϕ_b is the effective volume fraction based on this size. This formula is nearly the same as that derived for hard spheres. However, as $\phi_b > \phi$, and considering that the radial distribution at the nearest neighbor peak increases greatly with increasing electrostatic stabilization, theory predicts that additional electrostatic repulsion leads to an increase in the zero-shear viscosity.

$$\eta^{B}_{r,0} \sim \frac{12}{5}\phi^2 \frac{g(2a;\phi)}{D^{ss}(\phi)} \Rightarrow " \eta^{I}_{r,0} = \frac{12}{5}\frac{a}{b}\phi_b^2 \frac{g(2;\phi_b)}{D_0^2(\phi)} \tag{4.22}$$

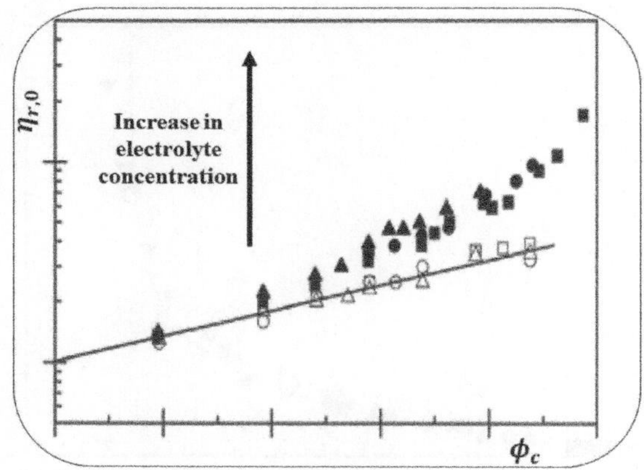

FIGURE 4.29 The trend zero-shear relative viscosity of a typical oxide suspension in organic media with increasing solids content and electrolyte concentrations.

Where D^{ss} is the short-time self-diffusion coefficient. The high-frequency limiting viscosity is independent of the interparticle repulsion (like changes in the salt concentration) and depends largely on long-range hydrodynamic interactions and is less sensitive to the details of the microstructure for stable dispersions, similar to high-shear limiting viscosity $\eta_{r,\infty}$ without shear-thickening effects. The shear limiting viscosities, compared in *Figure 4.29* for the high-frequency limiting viscosity, show that the high shear viscosity is greater than the high-frequency viscosity, even though both are due to hydrodynamic interactions.

The shear thinning is typically more extreme in electrostatically stabilized dispersions because the zero-shear viscosity is greatly enhanced by the interparticle repulsion, whereas the high shear is relatively insensitive to the potential.

As indicated in the following equation, the shear stress decreases to zero as the shear rate tends toward zero. Note that even in a shear-thinning power law fluid, the shear rate has to be nonzero to produce a nonzero stress. This condition defines a fluid, which by definition cannot be in equilibrium under a non-zero-shear stress. For the shear viscosity of charged nanoparticle dispersions and $k_{\dot{\gamma}}^{'}$ replaced by σ/σ_c :

$$\eta - \eta_\infty = \frac{\eta_0 - \eta_\infty}{1 + (k_{\dot{\gamma}}^{'})^m} \Rightarrow \eta_r = \eta_{r,\infty} + \frac{\eta_{r,0} - \eta_{r,\infty}}{1 + (\sigma/\sigma_c)^m} \tag{4.23}$$

Where the parameter σ_c, which is typically of the order of 1 in units of $k_B T / a^3$ for hard spheres, now ranges from 0.10 to 0.025. The exponent m ranges from 1 to ~1.8 with increasing volume fraction. As the high shear and low shear viscosities depend on different effective hydrodynamic radii, it is not possible to simply map the shear-thinning behavior onto that of a hard sphere dispersion.

The ceramic suspensions can show sterically stabilized mechanisms, with interfaces on dispersion particles that can induce interparticle repulsion and colloidal stability when grafted or adsorbed onto the particles. The created repulsion depends on ϕ increases with $r/2a$. For terminally anchored particles with the media quality of the suspending medium, the graft density and the molar mass of the particle are determinative. Systems tends to stretch out the surface layer and make the stabilizer layer thicker. In order to induce a repulsive force between particles, sufficiently thick and dense polymer layers should overlap (see *Figure 4.30*). The trends show that with sufficient graft density and molecular weight, the steric repulsion can prevent particles from aggregating.

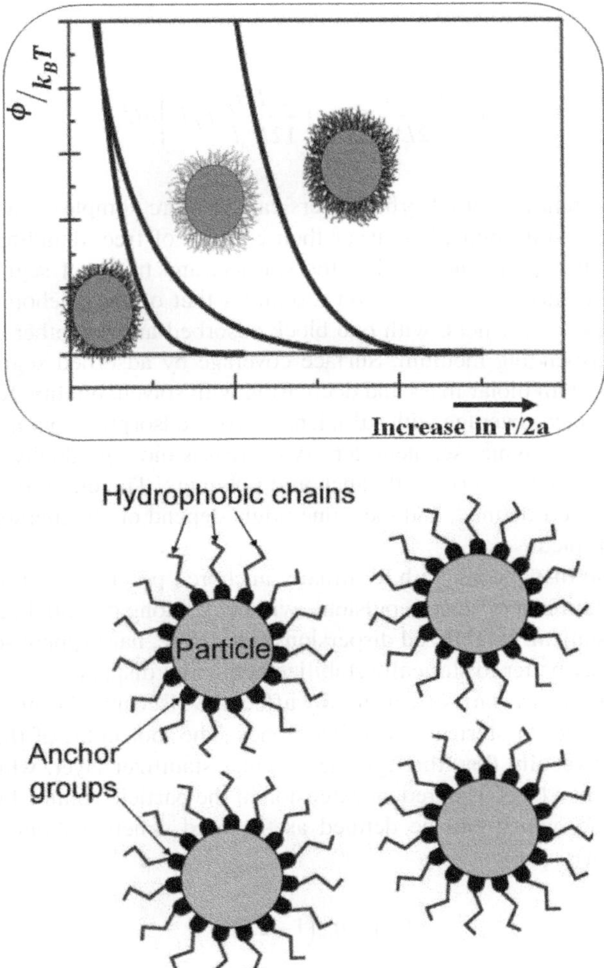

FIGURE 4.30 The trend in steric repulsion and the potential of interaction in increasing r/2a. Notice the formation of thick stabilizing layer on particle surfaces with a closer view of the anchored ceramic particles.

The particle's response to possible overlap can be considered the main source of the steric repulsion. The potential as the product of the osmotic pressure in the overlap region (Π) and the volume of overlap (Vo), which is essentially the work required to bring the particles into overlap, is $\Phi^{pol} = \Pi V_o$ which:

$$\Phi^{pol}(r) = \begin{cases} r < 2a \Rightarrow \infty \\ 2a < r < 2(a+L) \Rightarrow \Phi_0[-\ln(y) - \frac{9}{5}(1-y) + \frac{1}{3}(1-y^3) - \frac{1}{30}(1-y^6) \\ r > 2(a+L) \Rightarrow 0 \end{cases} \quad (4.24)$$

Where,

$$y = \frac{r-2a}{2L}, \Phi_0 = \left(\frac{\pi^3 L\sigma_P}{12N_p l^2} k_B T \right) aL^2 \quad (4.25)$$

This typical behavior of adsorbed layers can be more complex. Large molecules can adsorb at several points. The layer then consists of free, dangling ends (tails), loops between two points attached at the surface, and trains of segments that lay adsorbed on the surface. The situation resembles that of end-anchored layers similar to AB-block copolymers, with one block adsorbed and the other dangling as a "tail" in the suspending medium. Surface coverage by adsorbed segments is variable, increasing with molar mass and decreasing with solvent quality. Repulsion now requires full surface coverage with sufficiently strong adsorption, but in a sufficiently good solvent. The polymer segment density decreases more gradually with distance from the surface than in terminally anchored polymers. Therefore, the thickness of the layer is less well-defined, and the value might depend on the method by which it has been determined.

The colloidal dispersions with terminally anchored polymers of low polydispersity typically exhibit very steep repulsions, without the long potential tails characteristic of electrostatically stabilized dispersions. Therefore, hard sphere scaling can be expected to apply better to sterically stabilized colloidal dispersions.

Although the electrostatic effects hardly affect the viscosity of dilute suspensions, this is not the case for sterically stabilized ones. The molecules of the suspending medium do not readily flow through the polymer stabilizer layer, which, to a first approximation, can be considered an extension of the particle volume hydrodynamic effective volume, which can be defined as the hard sphere volume fraction that reduces the $\eta(\varphi)$:

$$\eta(\varphi) = \eta_m \left(1 + 2.5\phi_{eff}^h\right) \quad (4.26)$$

From ϕ_{eff}^h, hydrodynamic effective particle radius $a_{eff}^h = a + \delta_h$ can be derived, where δ_h is the hydrodynamic effective layer thickness of the stabilizer:

$$\phi_{eff}^h = \phi(1 + {}^{\delta_h}\!/_{a})^3 \quad (4.27)$$

Due to the similarities in the interparticle potentials, sterically stabilized systems display qualitatively similar structures and similar phase behavior to electrostatically stabilized dispersions, both at equilibrium and under flow. Differences arise from dense brushes, which have a steeper, short-range repulsion. For instance, higher concentrations are typically required for crystallization of sterically stabilized dispersions because repulsion only occurs when the polymer layers overlap, and as the polymer brush provides resistance to the penetration and flow of the suspending medium, hydrodynamic interactions are also substantially affected, unlike for the case of electrostatic stability.

The hard sphere mapping of these systems is probable if the $\eta(\phi)$ curves have the same intrinsic shape and can be superimposed using a single scaling factor. The scaling factor should reflect the thickness and softness of the polymer layer. A possible scaling factor of this kind is the maximum packing.

The scaling factor should reflect the thickness and softness of the polymer layer. A possible scaling factor of this kind is the maximum packing. Plotting the viscosities as a function of either ϕ / ϕ_{max} or $\phi^h_{eff} / \phi^h_{eff,max}$ approximately coincides with that for hard spheres. (See *Figure 4.31.*)

As sterically stabilized dispersions often have steep interparticle repulsions, they are expected to be very suitable for mapping onto the rheology of hard sphere dispersions. (See *Figure 4.32.*)

The curves for the largest particles, with the lowest δ_h / a ratios, cluster together. When this ratio tends to zero, logically these curves should asymptotically approach the hard sphere case; however, here the maximum packing at high shear rates is below the Brownian value and actually is close to the low shear limit for a hard sphere glass. The softer systems have systematically greater maximum packing fractions, for which compression of the soft polymer layers provides a logical explanation.

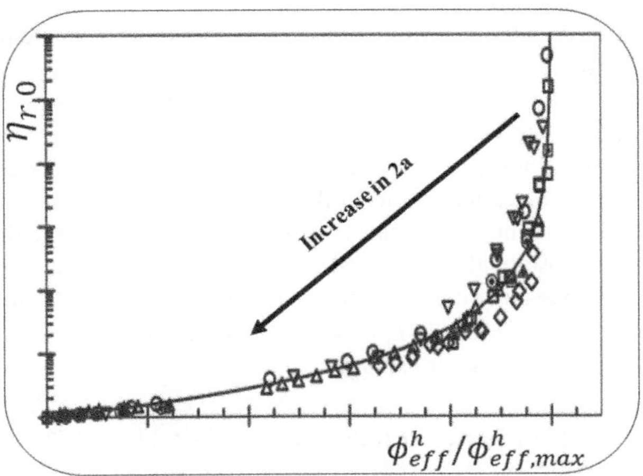

FIGURE 4.31 The trend in relative zero-shear viscosity with volume fraction ϕ^h_{eff} for a typical sterically stabilized suspension with various particle sizes.

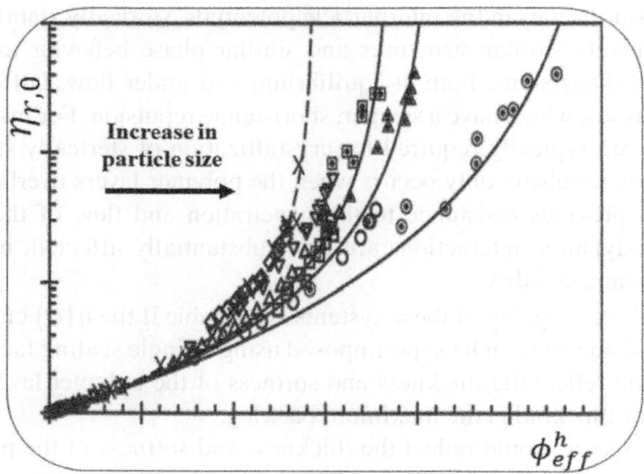

FIGURE 4.32 The trend in relative zero-shear viscosity with volume fraction ϕ_{eff}^{h} for a typical sterically stabilized oxide suspension with various particle sizes.

As indicated, the volume fraction at maximum packing is actually larger than unity. The data in this figure are for the limiting high shear rate viscosity and clearly do not reduce to the curve for Brownian hard sphere dispersions, in contrast to electrostatic systems. Similar to other sterically stabilized systems, the limiting value of the maximum packing is somewhat smaller than the corresponding value of 0.58 for Brownian hard sphere glasses or 0.638 for the random close packed limit for spheres. The reasons for this are not well understood, but residual attractive forces or electrostatic contributions are possible reasons for the discrepancy. Interestingly, the softness afforded by the steric repulsion enables the probing of dispersions at much higher effective volume fractions and zero-shear viscosities than typically achievable in hard sphere dispersions.

In order to describe the viscosity curves, it is necessary to supplement the limiting viscosities with the intermediate shear-thinning behavior expression to link the viscosity to the reduced shear like equation 4.23. The shear stress decreases to zero as the shear rate tends to zero. Note that even in a shear-thinning power law fluid, the shear rate has to be nonzero to produce a nonzero stress. This condition defines a fluid, which by definition cannot be in equilibrium under a non-zero-shear stress. The exponents **m** can be obtained by fitting. This behavior seems closer to the hard sphere values than for electrostatically stabilized systems. Unlike in the case of electrostatically stabilized dispersions, the high shear viscosity is affected by the steric layer. The critical values of $\sigma_{rc} = \sigma_c a^e / k_B T$ for sterically stabilized systems are closely related to those for Brownian hard spheres.

4.5 COLLOIDAL ATTRACTIONS AND FLOCCULATED DISPERSIONS IN CERAMIC SUSPENSIONS

Although the main goal of colloidal science is focused toward dispersion of particles into their core radius, many stable ceramic suspensions may experience flocculation.

The *flocs* are held together by relatively weak interparticle forces. Hence, their structure will be readily affected by flow, while the floc structure will in turn have an effect on the flow. The result is a complex interplay between microstructure and flow, which often results in a strongly non-Newtonian behavior. A small quantity of flocculated fine particles, just a few percent, might be enough to induce pronounced rheological changes. The low shear viscosity in particular can increase enormously and can even evolve into an apparent yield stress below which the suspension does not flow. The lack of flow at low stress levels can cause problems during processing. However, it can also be a desirable feature, for instance, if one does not want particles to settle or the suspension to flow under gravity. Flow-induced changes in microstructure might take a significant amount of time; accordingly, the viscosity will then also evolve in time. Thixotropy, proper adjustment of attractive forces, the shear and time dependence of the viscosity of suspensions, can be optimally adapted to London–van der Waals or dispersion forces, which are an omnipresent source of attraction, although they seldom provide a practical route to modifying the properties of a system. Dispersion forces depend on the nature of the particles and the suspending medium, which can rarely be changed freely.

Meanwhile, in electrostatically stabilized systems, the interparticle repulsion can be reduced by adding salts or components that reduce or neutralize the surface charges. For example, changing pH toward the point of zero charge can lead to colloidal aggregation. The addition of ions gives a similar effect by screening the surface charges. Nevertheless, these methods typically lead to irreversible aggregation in the primary minimum and often do not provide the desired rheological control that depends on achieving a shear dependent structure. Interestingly, not all dispersions can be so easily destabilized; salting out of aqueous oxide dispersions is often limited by the presence of hydroxide surface layers that impart substantial stability to aggregation under the right conditions. The DLVO potential can include a second, weak attractive region that is separated from the primary minimum by a potential barrier. Such dispersions, however, are only kinetically stable against irreversible primary minimum aggregation. Examples of secondary minimum flocculation in the literature are rare and include studies of shear effects on model colloidal flocs.

The direct electrostatic attraction can also be used to induce aggregation, such as in clay dispersions, where the faces and edges can be oppositely charged over a range of pH values. Such systems readily form colloidal gels at low concentrations and are often discussed in terms of a "house of cards" structure. Mixing particles of unlike charge often leads to precipitation if they are in nearly equal proportion, or to the formation of electrostatically stable aggregates.

Due to the presence of some additional steric repulsion, neutralization or reduction of electrostatic stability can lead to weakly flocculated suspensions. For example, adsorbing surfactants of opposite charge on the colloid can both neutralize surface charges and prevent overly strong attractions. Similarly, adsorbed water-soluble particles can lead to weakly flocculated suspensions at high added electrolyte concentration, by providing steric stability. These flocculation methods are, however, generally of limited use for rheology control. They mostly rely on dispersion forces as the source of interparticle attraction, and on reducing the relevant mechanism of colloidal stability to generate the required strength of the aggregates.

Several other techniques are available to induce weak flocculation in otherwise-stable suspensions. For example, sterically stabilized suspensions can be flocculated by reducing the solubility of the suspending medium for the polymer coat. This can be achieved by changing the solvent or by adjusting the temperature (thermal gelation).

The weak interparticle attractions can also be induced and accurately controlled by adding polymers to an otherwise-stable dispersion. Adding a nonadsorbing polymer causes depletion flocculation. The range of attraction is similar to the radius of gyration of the polymer in the solution, and the strength is regulated by the polymer concentration. Similar effects have been noted in the presence of micelles and nanoparticles. It should be noted that adsorbing polymers can also lead to depletion flocculation when the adsorption is essentially irreversible and there is excess polymer in solution. The polymer adsorbed onto particles at low surface coverage can lead to flocculation. The simultaneous adsorption of a polymer molecule on two particles then results in a molecular bridge between the particles.

The amphiphilic and cationic polyelectrolytes are commonly used to flocculate colloidal particles by a bridging mechanism in water clarification. Block copolymers constitute a special case in this respect. Simple AB-block copolymers that contain a single adsorbing group and a single nonadsorbing group provide steric stability. With several such groups in or attached to the backbone chain, reversible bridging flocculation can be achieved. An example is provided by the associative polymers, which consist of a water-soluble chain to which two or more hydrophobic groups are attached. Depending on the nature of the surface, one of these chain elements can adsorb on the particle, while the hydrophobic groups can associate together in the water phase. Therefore, these materials are effective thickeners for waterborne suspensions

Some small particles (e.g., fumed silica or clay) can gel at very low concentration. They can be effective thickeners without strongly affecting the stability and dispersion of other, often larger, particles in a mixture. Still, other phenomena can contribute to floc formation. Solvation forces can be used, in which case changes in temperature and composition can induce aggregation.

Among ceramic particles, silica particles have complex and variable surfaces, where hydrogen bonding has been reported to contribute to aggregation. It is important to recognize that particles generally have heterogeneous surfaces, which can lead to "*patchiness*." The presence of patched surfaces will further complicate the aggregation behavior as the bonding will depend on the particle orientation (such as proteins).

In a dilute suspension of Brownian hard spheres, an increase of the interparticle attraction will result in a higher viscosity. This effect was used as a method to track particle aggregation over one hundred years ago in investigations of the atomic theory of matter.

Alumina particles can be the source of coagulation in aluminum hydroxide sols. The increase in suspension viscosity for various suspension concentrations is used to calculate the effective volume fraction of fractal-like aggregates "similar to *snowflakes*" for different initial particle concentrations.

The reduced effective volume fraction versus time accounts for the concentration dependence of the characteristic aggregation time. The deviations from hard sphere behavior increase the viscosity of more concentrated dispersions. The attractive forces can be manipulated by means of temperature changes, and so aggregation is reversible. It is obvious that attractive forces can have a very significant effect on the rheology of colloidal dispersions, a phenomenon that is used industrially for practical control of product properties. The viscosity increase is caused by the tendency of the particles to cluster together and to form flocs. The flocculation induced by interparticle attraction is reduced or eliminated, and the particles peptized or redispersed, by shearing at high shear rates, a phenomenon that gives rise to shear-thinning behavior. The zero-shear viscosity can be increased substantially, and a yield stress can even be induced, by means of attractive interparticle forces. With decreasing temperature, i.e., increasing 1/B, the logarithmic viscosity curve tends to a slope of -1, meaning that the stress becomes a constant, and therefore there is a yield stress.

The presence of a yield stress indicates that the behavior at lower stress levels will be solid-like. It suggests that attractive forces can lead to the formation of a particulate network of flocs that spans the whole sample. (See *Figure 4.33*.)

The linear relationship between viscosity and effective volume fraction, inspired by the Einstein viscosity formula to deduce an effective hydrodynamic volume of the aggregates. Attractive interparticle forces can generate substantial storage moduli much larger than those caused by repulsive forces. With increasing interparticle attraction or volume fraction, the dynamic moduli often evolve in a characteristic fashion where interparticle attraction was controlled by means of the temperature. The low frequency response at low volume fractions follows the general pattern for viscoelastic fluids, the so-called terminal zone. (See *Figure 4.34*.)

With increasing interparticle attraction or volume fraction, the moduli increase and the slopes gradually decrease until a plateau region is approached for *G'*. The curves for *G* and the slopes gradually decrease until a plateau region is approached. The curves

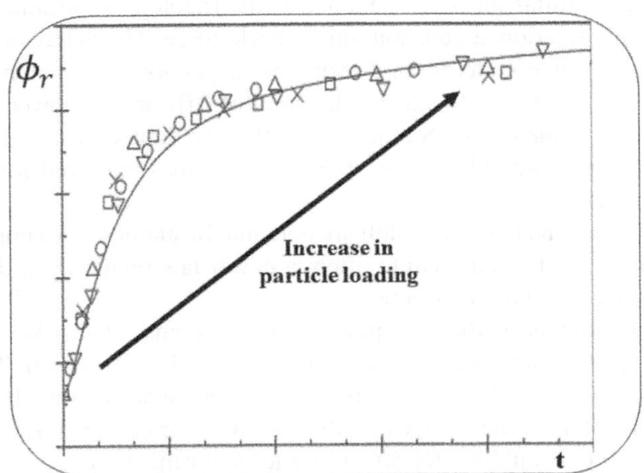

FIGURE 4.33 The trend in time dependency of ϕ_r in aluminum oxide suspensions with increasing solid load.

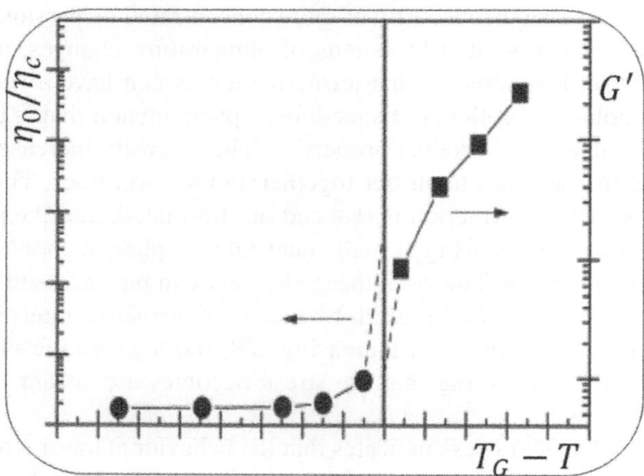

FIGURE 4.34 The trend in zero-shear viscosity and dynamic moduli near the gel point in a typical suspension.

for G'' become quite flat, often showing a shallow minimum, suggesting the existence of a maximum at still lower frequencies. In the plateau region, G'' is larger than G'.

The behavior is qualitatively similar to that of stable suspensions. The evolution of the curves is typical for a transition from a liquid to a solid and is consistent with the divergence of the steady-state zero shear. In the present case, the "solid" can be very soft and weak and is normally called a gel. In the vicinity of the gel point, the zero-shear viscosity and the low-frequency plateau modulus are evolved.

The viscosity diverges near the transition point, where a plateau modulus emerges instead. Together with the plateau modulus, a dynamic yield stress appears in the viscosity curves. Similar plots are obtained if the rheological parameters are plotted versus volume fraction at constant interparticle force. The behavior qualitatively resembles that of stable systems. In flocculated dispersions, the volume fraction at the gel point can be very small, e.g., on the order of 0.01 or even lower. This is quite different from the value of ~0.58 required for the divergence of the zero-shear viscosity for Brownian hard spheres, or ~0.3, as sometimes observed for electrostatically stabilized particles.

The yield stresses and plateau moduli are both manifestations of a sample-spanning, solid-like particulate structure. Quite often, a power law relation can describe their dependence on particle volume fraction.

The flocs and fractals are the main parts of these systems. At very low volume fractions, increasing the interparticle attraction results in floc formation. This involves dispersions of spherical colloids with strong, short-range attractions. If the particles are not density-matched, the flocs will either sediment or cream. The energy barrier required to prevent flocculation for 50 nm particles, for this barrier, was estimated to be approximately 20–30 kT.

The percolation of flocs may lead to very weak solids at very low particle concentrations as the floc network spans the sample. Such systems can exhibit delayed settling and are often very shear sensitive. The percolation is defined as the formation of a sample-spanning connectivity between the structural elements. Percolation depends on the definition of "connectedness." The connectivity of structure is clearly a requirement for some colloidal structures. During gel formation and as gelation requires a network structure with permanent (at least on the time scale of observation) stress-bearing capacity, gels may be equilibrium structures or nonequilibrium states, where the particles become trapped during phase separation.

The percolation theory has often been used to describe the formation of colloidal networks (such as gels) and is achieved when there is at least one continuous path of "linked" particles throughout the sample. The percolation concept has been used to describe gelation in suspensions comprised of fractal aggregates. A space-filling network develops when the effective volume fraction $\phi^h_{eff,floc}$ of flocs in the system becomes equal to ~1. The flocs with more open structures can develop a space-filling structure at lower volume fractions. (See *Figure 4.35.*)

For intermediate volume fractions, increasing the interparticle attraction changes the rheological behavior quite suddenly from that of a liquid to that of a weak solid or gel. Whether flocculation leading to fractal structures results in gel formation can also be limited by gravitational settling. For particles not closely density-matched, the growing clusters will either sediment or cream as they increase in size.

The time scale for sedimentation versus diffusion-limited growth gives an estimate of the gelation limit in terms of the density difference ($\Delta\rho$) between the particles and suspending medium:

$$\phi_{gel} \sim \left(\frac{9k_BT}{2\pi\text{''}\ \rho a a^4}\right)^{\frac{D_f-3}{D_f+1}} \tag{4.28}$$

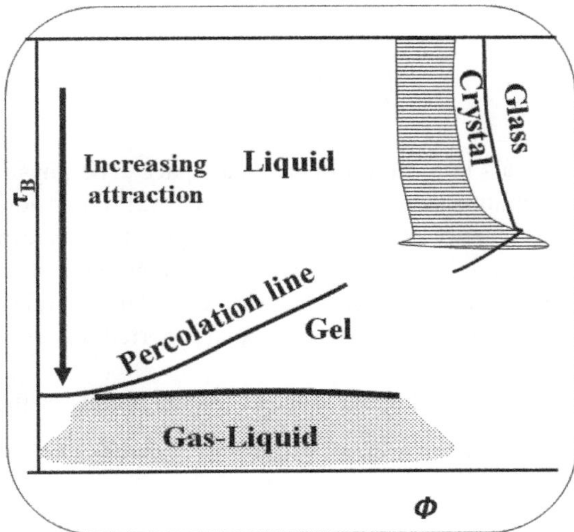

FIGURE 4.35 The schematic phase and state diagrams of typical suspension with sort-range attractions.

For sufficiently strong interparticle attractions, structural arrest can now occur at much lower nominal volume fractions, because it is due to particle bonding rather than to jamming. This approach assumes that local interparticle attractions bring particles into contact with each other and that these particles then become arrested (nonergodic) at a volume fraction dependent upon the strength and range of the interparticle attraction. Increasing the strength of attraction (decreasing τ_B) actually increases the volume fraction required to form an ideal glass. In simplified terms, weak attractions stabilize the liquid phase, as particles will tend to weakly cluster, leaving more free volume to diffuse by one another. Once τ_B <~ 2<, a new type of glass can be formed, an attractive driven glass (ADG). The floc size, shape, and compactness are straightforward structure parameters, but they do not provide a complete description of the internal structure of an aggregate. The internal organization of aggregates often can be described by fractals as self-similar structure, at least on length scales between that of a few particles and the overall floc size. In other words, the object looks similar when viewed at different length scales or magnifications. More specifically, in fractal structures, the total number N of elementary particles or the total mass M of particles scales with the distance R from the center of the aggregate, according to the relation for where D_f is the mass fractal dimension:

$$N \propto M \propto F^{D_f} \propto (\frac{R_g}{a})^{D_f} \tag{4.29}$$

The relation between N and the radius of gyration R_g, which is commonly used to characterize the size of a floc. This also means that the average internal density of the aggregate varies with R. The volume fraction $\phi_{i,floc}$, floc of particles in a floc of radius R_{floc}, and fractal dimension D_f:

$$\phi_{i,floc}\left(R_{floc}, D_f\right) \propto (\frac{R_{floc}}{a})^{D_f-3} \tag{4.30}$$

Which applies to fractals in a three-dimensional space. The physical limits for D_f are $1 \leq D_f \ll 3$. The chain-like aggregates have a fractal dimension of 1; branching leads to intermediate values of D_f, whereas values closer to 3 correspond to compact, solid objects, such as a solid sphere. For example, coalescence of emulsion droplets leads to a larger droplet with a fractal dimension of 3. The fractal structures are not limited to spherical particles; they have also been reported for rodlike and platelike particles.

The power law behavior over some range of the magnitude of the scattering vector q is:

$$I(q) \sim q^{-D_f} \tag{4.31}$$

The form of aggregation involves collisions between clusters (cluster-cluster aggregation). This seems to be the best way to describe the properties of many real systems where aggregation is driven by Brownian motion (perikinetic aggregation). When it is assumed that each collision results in the formation of a permanent interparticle bond, the fractal dimension D_f would be 1.7–1.9. This is known as diffusion-limited cluster aggregation (DLCA). It is representative of very strong interparticle forces that result in irreversible bonds.

The reduced sticking probability provides the possibility for the particles to explore different positions of attachment. This results in a denser packing, with $D_f \simeq 2 - 2.1$ for so-called reaction-limited cluster aggregation (RLCA). The range of fractal dimensions reported experimentally for dispersions aggregated at rest is roughly within the limits determined by DLCA and RLCA. On occasion, the degree of compactness of an aggregate is expressed as a coordination number, i.e., the number of neighbors in contact with a particle. This is not uniquely related to the fractal dimension and can be applied even when the floc is not fractal. A distribution of coordination numbers might be a more suitable indicator of floc structure for some properties. It should also be mentioned that neither fractal dimension nor coordination number describes anisometric structures, which can occur in flowing flocculated dispersions.

The effect of flow on floc structure is the main indicator of flowability in these systems. The flow affects floc structure by accelerating the rate of aggregation above that by pure Brownian motion as it brings particles together more quickly (orthokinetic aggregation). For high shear rates (high Pe), the rate of doublet formation is enhanced in direct proportion to the rate of shear. Flow, however, also pulls particles apart and can be used to redisperse aggregates. These two processes can sometimes lead to a balance and a steady state floc size distribution.

The combination of these theories can be manifested in the stability diagram of fluctuated suspension systems (see *Figure 4.36*). The shear-induced coagulation and breakup of colloidal doublets creates the "stability plane" for shear flow, which, as has been determined, is cast in terms of two dimensionless groups: the strength of electrostatic stabilization relative to the strength of attraction N_r, and a dimensionless flow rate N_f. Interestingly, there is a regime ($N_r \sim 5$) where secondary minimum flocculated dispersions can be driven into primary minimum flocculation with increasing shear rate, which ultimately leads to dispersion at even higher shear rates. (See *Figure 4.37*.)

The decrease in floc size with shear rate can be explained by a balance reached between the rate of aggregation and the rate of breakage due to shear. The force acting on a floc in a flow field is estimated to be proportional to the stress acting across the cluster by the flow field. This is given by the medium viscosity times the shear rate, $\sigma \approx \eta_m \dot{\gamma}$. The shear rate dependence on the number of particles per floc and of its radius of gyration:

$$N_{floc} \propto (\eta_m \dot{\gamma})^{-0.9} \tag{4.31}$$

$$(R_g / a)^3 \propto (\eta_m \dot{\gamma})^{-1.06} \tag{4.32}$$

The average floc size versus wall shear stress for aggregates is represented in *Figure 4.38*.

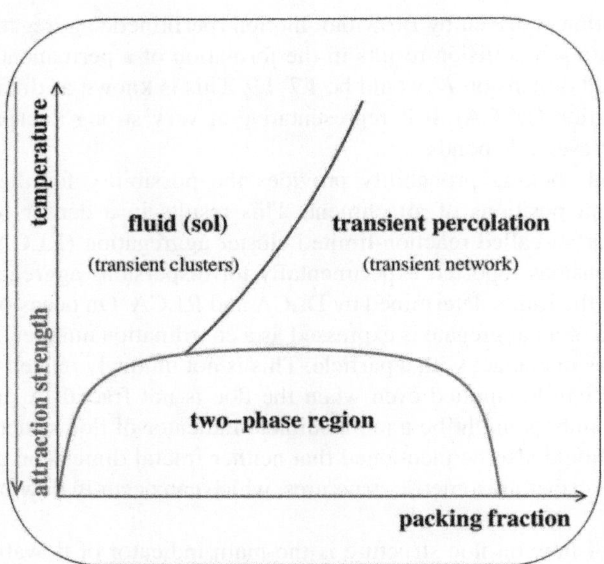

FIGURE 4.36 The schematic phase and state diagrams of typical fluctuated suspension system with regards to temperature and attraction strength.

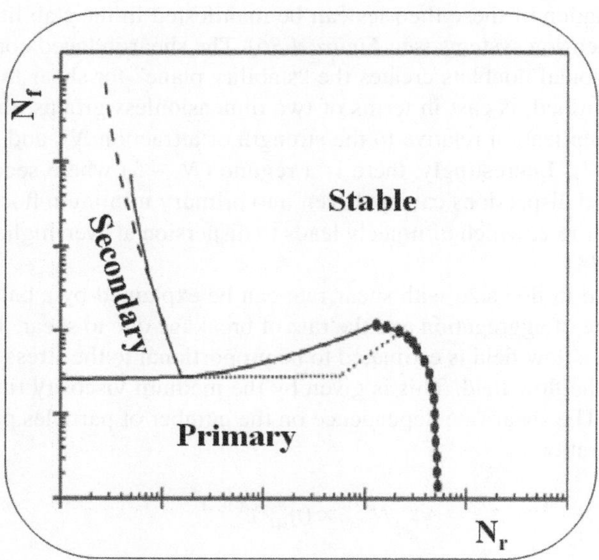

FIGURE 4.37 The schematic shear stability diagram with indicating state of particles in contact.

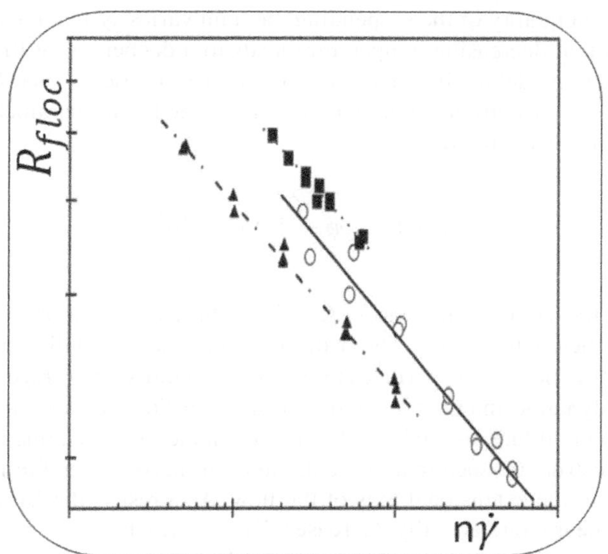

FIGURE 4.38 The schematic relation between floc radius and the applied shear stress.

The floc strength is given by the floc elasticity, which is assumed to be of the form of an elastic constant $K(R)$ divided by size of the cluster R_c, or$G \sim K(R_c)/R_c$. The breaking up of aggregates under shear does not result in fragments of a uniform size. The breakup is a somewhat-random process that results in a distribution of fragment sizes. The applied shear can also change the internal structure of the flocs. A common phenomenon is that shear densifies the flocs. This should be reflected in an increase in the fractal dimension. In addition to size and density, the shape of flocs can also be affected by flow. It is often assumed that flocs are spherical. Under normal conditions, this is, on average, roughly the case, as demonstrated by the circular symmetry of scattering patterns. At high shear rates, however, scattering experiments produce "butterfly" patterns, indicating structural anisotropy. These have been reported for various types of suspensions and different scattering techniques.

The suspension structure would involve stable clusters; electrostatically stabilized dispersions can be destabilized by the addition of salts to yield fractal-like aggregates. The presence of a weak, long-range repulsion can lead to the formation of stable clusters of particles of finite size. This is in contrast to fractals, which grow until there are no additional particles or neighboring clusters to aggregate with. The formulation and microstructure of these stable clusters remain an active area of research.

The rheology at low volume fractions of these systems is directly related to cluster amount and radius. Both interparticle repulsion and attraction increase energy dissipation during flow, and hence also the viscosity. The viscosity increases systematically with this increasing well depth (decreasing temperature). These effects accumulate to interparticle interactions and start to affect shear viscosity at the φ^2

level. The solvent quality of the suspending medium varies systematically with temperature, such that decreasing temperature leads to a deepening of the square-well attraction used to model the interactions between these particles. (See *Figure 4.39.*)

The viscosity-concentration equation in terms of the Baxter parameter τ_B for systems with short-range attractions:

$$\eta_r = 1 + 2.5\phi + \left(5.9 + \frac{1.9}{\tau_B}\right)\phi^2 \tag{4.33}$$

Physically, the viscosity increases stem from three contributions to the stresses due to interparticle interactions. The hydrodynamic viscosity will increase because interparticle attractions bring particles into closer proximity on average, thus increasing the hydrodynamic interactions and resistance to flow. This is, in general, the dominant effect in dilute dispersions. Attractive interactions fundamentally alter the colloidal microstructure such that in the dilute limit, particles tend to form doublets that orient along the extensional axis of the flow. As a result, the Brownian contribution to the shear stress actually decreases. During which, the viscosity due to the direct interparticle force, which is attractive and of opposite sign to the Brownian force, greatly increases. It is essential to recognize that the stress is the product of the force and the neighbor distribution so that strong interparticle interactions lead to a microstructure fundamentally different from that of hard spheres or electrostatic, steric, or electrosterically stabilized dispersions. Indeed, detailed calculations for the

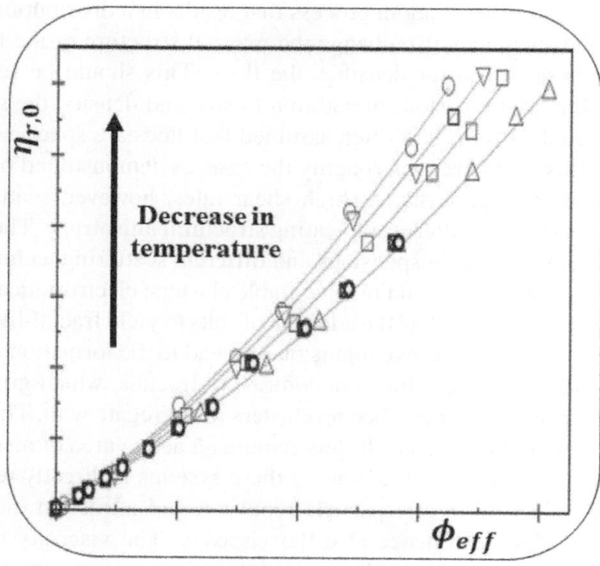

FIGURE 4.39 The schematic relation between viscosity, solids content, and temperature in a semidilute, sterically stabilized typical suspension.

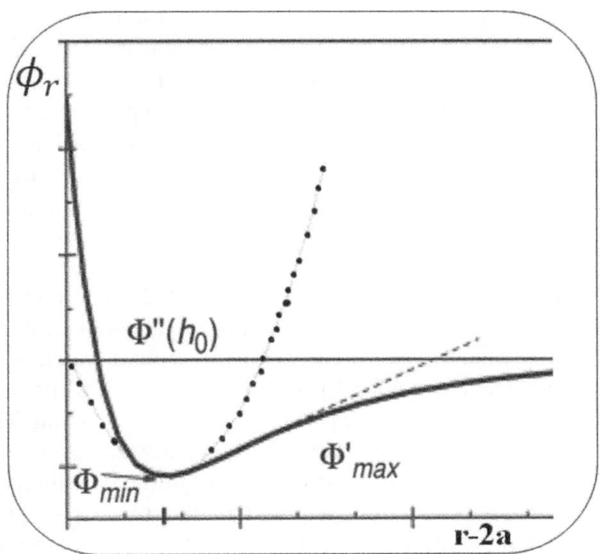

FIGURE 4.40 The schematic interparticle potential with short-range attractive interactions with the potential min and max in a typical separated particle suspension.

square-well fluid show that, for very weak attractive interactions, the viscosity can in fact be slightly lower than that for a hard sphere suspension.

In concentrated dispersions, weak attractive forces, semiconcentrated dispersions still display a zero-shear viscosity, although a higher one than that of Brownian systems. For nondilute depletion flocculated system by connecting the viscosity to the well depth of the attractive interactions between particles for hard sphere viscosity η_{r0}^{hs} and the interparticle potentialΦ_{min}:

$$\eta_{r0} = \eta_{r0}^{hs} \exp\left(-\frac{\alpha(\phi.a)\Phi_{min}}{k_B T}\right) \tag{4.34}$$

The interparticle potential for a dispersion with short-range attractive interactions, defining the potential minimum at a dimensionless, surface-to-surface separation distance *ho*. Motion of the particles is by activated hopping, so the activation energy for flow is proportional to the work required for particles to separate, which is proportional to the energy at the potential minimum. Hence, the viscosity is expected to depend exponentially on the strength of attraction. (See *Figure 4.40*.)

4.6 NANOPOWDER OXIDE CERAMIC SUSPENSION

Studies of oxide suspensions have revealed that there are certain deviations from the behavior exhibited by larger particle suspensions while using nanosized particles. The key parameters include those which control the effects of electrolyte concentration.

For instance, the increase in electrolyte concentration resulting from dissociation of ionizable surface sites on the oxide particles and soluble species resulting from the dissolution of oxide was offered as a possible reason regarding the large reaction surface area sites. The particle size directly controls the zeta potential, which decreases with particle size and the volume fraction of the ceramic suspensions.

The other aspect of nanosized ceramic suspension indicates the nanopowder suspensions and the effect of bound media, such as water. Upon heating of oxide nanopowder suspensions, water shows two separate events in contrast to a single event observed for pure water. This interesting behavior was observed and recognized as the effect which differs from the bulk water and is called "bound water." Three distinctive types of water are present in oxide nanopowder aqueous suspension systems: free water, physically bound water, and chemically bound water. The chemically bound water was found to account for 1.1 wt.% of the powder, while the thickness of physically bound water varies with solids content from 3 nm to 6 nm. The increase in solids content (> 0.30) leads to an overlapping of bound water layers and results in a dramatic increase in viscosity. Overall, the thickness of the bound layer is estimated as 2–3 nm in oxide nanopowder suspensions. The thickness of the gel layer at IEP is approximately 15 nm, whereas at low pH it is approximately 3–5 nm. The higher packing densities can be obtained in an acidic medium compared to basic conditions, which indicates the thinner water layer (1–5 nm). Therefore, the bound water layers exhibit lower molecular mobility. (See *Figure 4.41.*)

The introduction of surfactant additive (fructose) to the oxide suspension system increases the average mobility of the water molecules. It is suggested that fructose

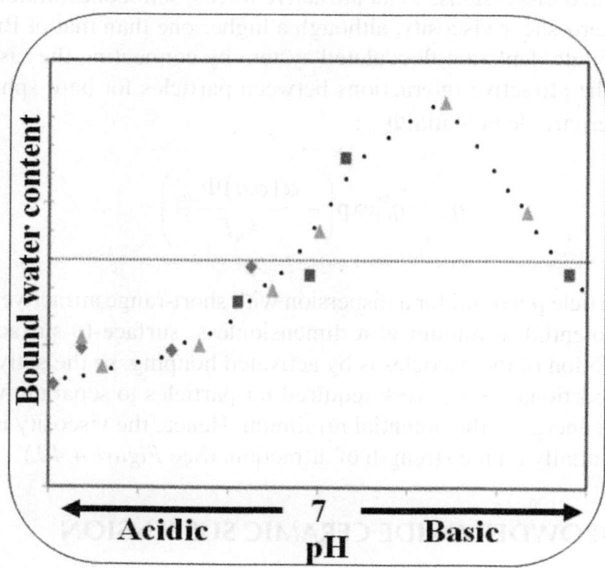

FIGURE 4.41 The schematic pH dependence of bound water content in an aqueous oxide ceramic suspension.

molecules release some of the bound water from the surface, increasing the availability of liquid for flow, and thus decrease the viscosity of oxide nanopowder suspensions. The compatibility of the additive molecule with three-dimensional hydrogen structure of water affects the suspension viscosity. The zeta potential of oxide nanoparticle suspensions is related to electrolyte concentration, particle size, and volume fraction. The electrostatic stabilization alone does not fully explain the high viscosity observed in concentrated nanoparticle suspensions. The increase in available surface leads to an increase in ionizable oxide surface sites, resulting in high concentration of soluble species in the suspensions.

The van der Waals attractive forces are dominant in the oxide nanoparticle suspensions, and the magnitude of the electrostatic repulsion is not sufficient to overcome the attractive forces. The steric stabilization for nanoparticle suspensions is most likely. The use of a more robust steric repulsion for stabilization of nano-oxide suspensions is essential, and the suspensions with chemically adsorbed molecules have higher stability during processing.

As the available surface is much higher in a nanoparticle suspension, a larger quantity of additive is required to cover the particle surfaces. As this adlayer polymer associates with the particle surface, the effective solids content increases in proportion to the thickness of the layer and limits the maximum achievable solids content. Although decrease in viscosity of the suspensions is evident, the maximum achievable solids content is limited to 30–40 vol%. When dispersants like oligo- or polysaccharides (e.g., dextrins, dextrans, maltodextrins) are used, the viscosity might increase. The small separation distances between particles in submicrometer particle systems can even increase viscosity after particle size reduction (high molecular weight polysaccharides).

As a result, many attempts have been made to explain the mechanism of viscosity reduction in ceramic suspensions by various additives. The electrostatic interactions have the most dominant effect due to a change in pH of the suspensions with the addition of these additives. However, for the nondissociating macromolecules, adsorption of these molecules on the particle surface was suggested as the viscosity reduction mechanism (different low molecular weight additives like tetraalkylammonium hydroxides, tetraalkylammonium chlorides, phenols, and carboxylic acids). Also, the adsorption of surfactant like monosaccharides on the alumina surface is important. Neither shift in the isoelectric point nor significant change in zeta potential of oxide is evident with the addition of surfactant such as fructose, glucose, and their derivatives with carboxylic group attachments. Therefore, although the low molecular weight saccharides adsorb on the oxide surface, the viscosity reduction in oxide nanosuspensions cannot be explained entirely by electrostatic or steric stabilization mechanisms. The water availability has shown a significant increase in the average water mobility with the addition of fructose to the oxide suspensions. Therefore, finding an optimum is necessary.

The increase in surfactant (i.e., ascorbic acid concentration) first reduces the viscosity of the suspensions. After reaching a critical concentration, the viscosity starts to increase. A similar behavior is conceivable when additives are adsorbed on the surface. Mixing with polymeric additives shows minima when the monolayer coverage of the surface is obtained. Excess polymeric additives result in bridging and depletion flocculation. (See *Figure 4.42.*)

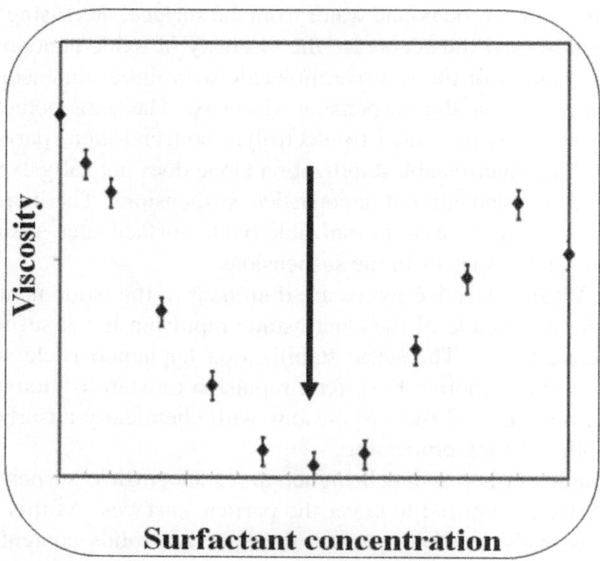

FIGURE 4.42 The schematic behavior of a typical nanoparticulate oxide suspension with surfactant concentration.

Once the monolayer surface coverage is completed, excessive amounts of ascorbic acid would probably accumulate in solution. Because the separation distance is very short in concentrated nanoparticle suspensions, accumulated ascorbic acid will act as a bridge between particles, resulting in higher viscosities.

The comparison of the surfactant's effect on the oxide nanopowder suspensions with and without addition is shown in *Figure 4.43*. For all solids content levels, addition of surfactant leads to a reduction in suspension viscosity. However, the reduction at high solids content is not as significant as that at low solids content.

The effect of surfactant on zeta potential of a nanoparticulate oxide suspension should be considered in suspension preparation. The adsorption of the charged molecules on the oxide nanopowders surface may cause changes in surface charge. A strong adsorption results in shifts in IEP of the powders. Increase in surfactant concentration leads to shifts in IEP to lower pH values, which indicate the more acidic alumina surface. Adsorption of ascorbate ions on the surface also leads to a decrease in the IEP of oxide from basic to approximately neutral range by the monolayer coverage of oxide surface. (See *Figure 4.44*.)

The bound water behavior in suspension directly controls rheological properties, such as viscosity. The release of water, as a result of applying shear stresses, and increase in mobility decrease viscosity. As expected, when the solids content increases, the viscosity increases, and the suspension becomes more shear thinning. (See *Figure 4.45*.)

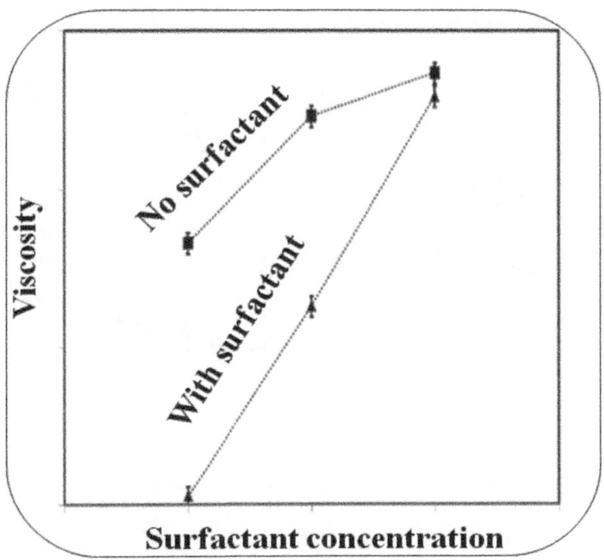

FIGURE 4.43 The behavior comparison of a typical nanoparticulate oxide suspension with and without surfactant.

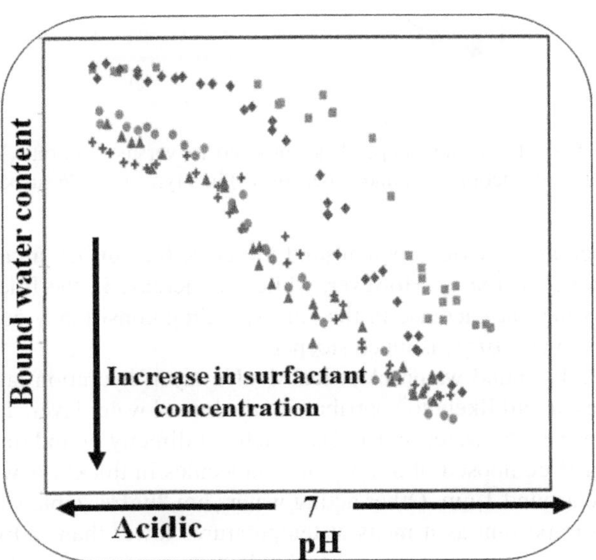

FIGURE 4.44 The effect of surfactant on a typical nanoparticulate oxide suspension.

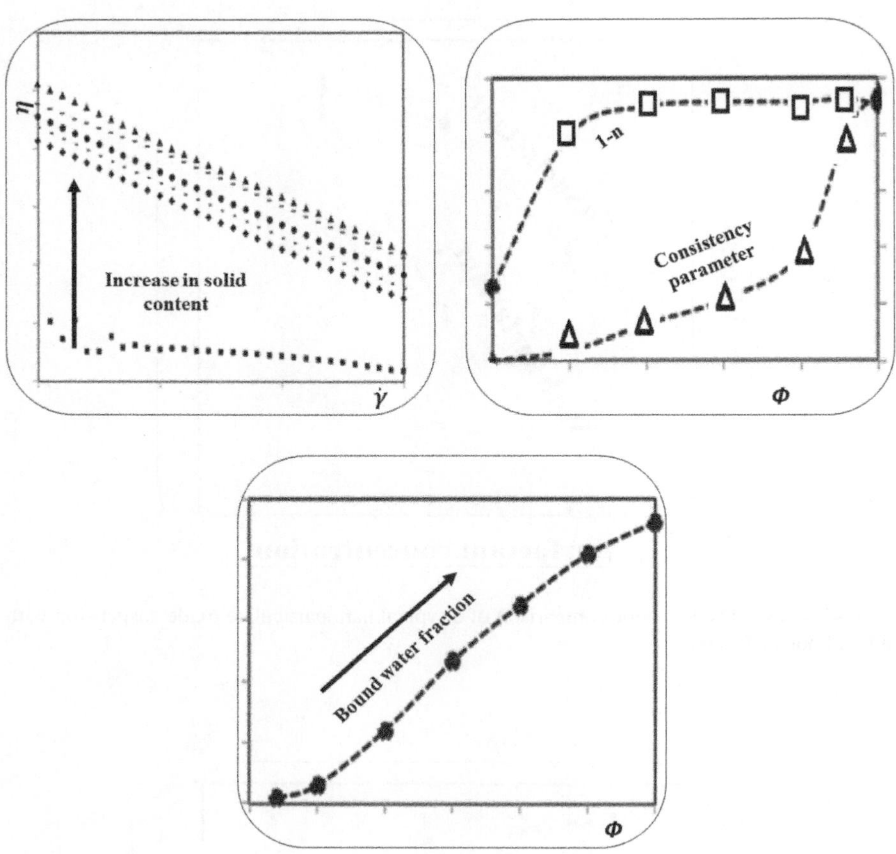

FIGURE 4.45 The solids content dependency of viscosity versus sheer rate (left) and K-(1-n) parameters (right) and fraction of bound water (down) in a typical oxide suspension.

With the increase in oxide content (and therefore the surface area), the fraction of the bound water increases. However, while the increase in the fraction of bound water is monotonic, the increase in the corresponding consistency coefficient, K (a term analogous to *viscosity*), is much steeper.

The chemically bound water, electrical double layer, hydration layer, and water trapped in pores are all likely to contribute to the bound water layer. The chemically bound water layer is the water molecules which are directly bound to the oxide surface. This layer is composed of a few water molecules in thickness which has been measured to be around 1 nm. Other acting waters are "unfreezable water," which is not considered important as it melts at temperatures lower than -20°C. The bound water is primarily due to the electrical double layer around the particles.

The profound effect of potential determining ions (i.e., pH) and indifferent ions on the stability of colloids and the rheological behavior of suspensions is well-established. Fructose has a very low acid dissociation constant ($pKa \approx 12.2$). Neither the change in pH nor that in ion concentration of the suspension results in significant change in zeta potential of the nano alumina particles.

Controlling physical parameters leads to electrostatic stabilization of oxide nano-powder suspensions. The basic control mechanisms and parameters for the rheological behavior and electrostatic stabilization of oxide nanopowder suspensions beside bound water (or swelling) involve surface charge and the compression of the electrical double layer. The concentrated oxide nanopowder suspensions (average particle size of about 50 nm, stable suspensions, which were obtained with $0.020 \leq [NaCl] \leq 0.040\, M$ or in the range $4 \leq pH \leq 7$) can be stabilized at specific ranges of ionic strength and pH.

The role of polymeric surfactants at the solid/liquid interface in ceramic with aqueous/nonaqueous media are especially important in nanopowder suspensions. The use of surfactants (ionic, nonionic, and zwitterionic) and polymers to control the stability behavior of ceramic suspensions is of considerable technological importance. They facilitate a particularly robust form of stabilization, which is useful at high disperse volume fractions and high electrolyte concentrations, as well as under extreme conditions of high temperature, pressure, and flow. Their important rule involves surfactants and polymers which are essential for stabilizing suspensions in nonaqueous media, where electrostatic stabilization is less successful.

The surfactant adsorption occurs through hydrophobic bonding, solvation forces, and chemisorption. The adsorption of ionic surfactants involves electrostatic forces, particularly with polar surfaces containing ionogenic groups. The adsorption of ionic surfactants on hydrophobic surfaces happens by hydrophobic surfaces such as carbon black, polymer surfaces, and ceramics (silicon carbide or silicon nitride), i.e., sodiumdodecyl sulphate (SDS) on carbon black surfaces. By hydrophobic interaction between the alkyl chain of the surfactant and the hydrophobic surface instead of electrostatic interaction. If the surfactant head group is of the same sign of charge as that on the substrate surface, electrostatic repulsion may oppose adsorption. On the other hand, if the head groups are of opposite sign to the surface, adsorption may be enhanced. Since adsorption depends on the magnitude of the hydrophobic bonding free energy, the amount of surfactant adsorbed increases directly with increasing alkyl chain length in accordance with *Traube's* rule.

The Stern-Langmuir isotherm states the details of adsorption of ionic surfactants on hydrophobic surfaces. A substrate containing Ns sites ($mol.m^{-2}$), on which $\Gamma\ mol.m^{-2}$ of surfactant ions are adsorbed. The surface coverage θ is Γ / Ns, and the fraction of uncovered surface is $1-\theta$. The rate of adsorption is proportional to the surfactant concentration expressed in mole fraction C/55.5 and the fraction of free surface $1-\theta$:

$$Rate\ of\ adsorption = k_{ads}\left(\frac{c}{55.5}\right)(1-\theta) \tag{4.35}$$

Where k_{ads} is the rate constant for adsorption. The rate of desorption is proportional to the fraction of surface covered θ, with rate of desorption $n = k_{des}\,\theta$. At equilibrium, the rate of adsorption is equal to the rate of desorption, and the ratio of $k_{des} = k_{ads}$ is the equilibrium constant K:

$$\frac{\theta}{1-\theta} = \frac{C}{55.5} K \tag{4.36}$$

The equilibrium constant K is related to the standard free energy of adsorption by:

$$-\Delta G^{o}_{ads} = RTlnK \tag{4.37}$$

R is the gas constant, and T is the absolute temperature at low surface coverage $\theta < 0.1$:

$$\frac{\theta}{1-\theta} = \frac{C}{55.5} \exp\left(-\frac{\Delta G^{o}_{ads}}{RT}\right) \tag{4.38}$$

At low surface coverage $\theta > 0.1$:

$$\frac{\theta}{1-\theta} \exp(A\theta) = \frac{C}{55.5} \exp\left(-\frac{\Delta G^{o}_{ads}}{RT}\right) \tag{4.39}$$

Where A is Frumkin-Fowler-Guggenheim constant. The Stern-Langmuir equation in a simple form to describe the adsorption of surfactant ions on mineral surfaces:

$$\Gamma = 2rC\exp\left(\frac{\Delta G^{o}_{ads}}{RT}\right) \tag{4.40}$$

ΔG^{o}_{ads} may be taken to consist of two main contributions, $\Delta G^{o}_{ads} = \Delta G^{o}_{elec} + \Delta G^{o}_{spec}$, where ΔG^{o}_{elec} accounts for any electrical interactions and ΔG^{o}_{spec} is a specific adsorption term that contains all contributions to the adsorption free energy that depend on the "specific" (nonelectrical) nature of the system:

$$\Delta G^{o}_{spec} = \Delta G_{cc} + \Delta G_{cs} + \Delta G_{hs} + \ldots \tag{4.41}$$

The amount of adsorption Γ:

$$\Gamma = \frac{(C_1 - C_2)}{mA_s} \tag{4.42}$$

Where mass $m(g)$ of the particles (substrate) and specific surface area $A_s\left(m^2S^{-1}\right)$ are taken into account. The adsorption of ionic surfactants on polar surfaces that contain ionizable groups may show characteristic features due to additional interaction between the head group and substrate and/or possible chain-chain interaction. For instance, in sodium dodecyl sulphonate (SDSe) adsorption on oxides, the oxides surface (at pH 7.2) is positively charged (the isoelectric point of oxides is at pH ~9) and the counterions are Cl^{-1} from the added supporting electrolyte. The saturation adsorption Γ is plotted versus equilibrium surfactant concentration C on logarithmic scales. The changes in adsorption isotherm (circle) and corresponding zeta potential (square) in a typical oxide nanopowder suspension show the increasing trend in both for higher equilibrium concentrations. (See *Figure 4.46*.)

The adsorption and zeta potential results show three distinct regions:

I. Showing a gradual increase of adsorption with increasing concentration, with virtually no change in the zeta potential, corresponding to an ion-exchange process. In other words, the surfactant ions simply exchange with the counterions of the supporting electrolyte in the electrical double layer.

II. At a critical surfactant concentration, the desorption increases dramatically with further increase in surfactant concentration (region II). The positive zeta potential gradually decreases to zero (charge neutralization), after which a negative value is obtained, which increases rapidly with increasing surfactant concentration. At a critical surfactant concentration (the critical

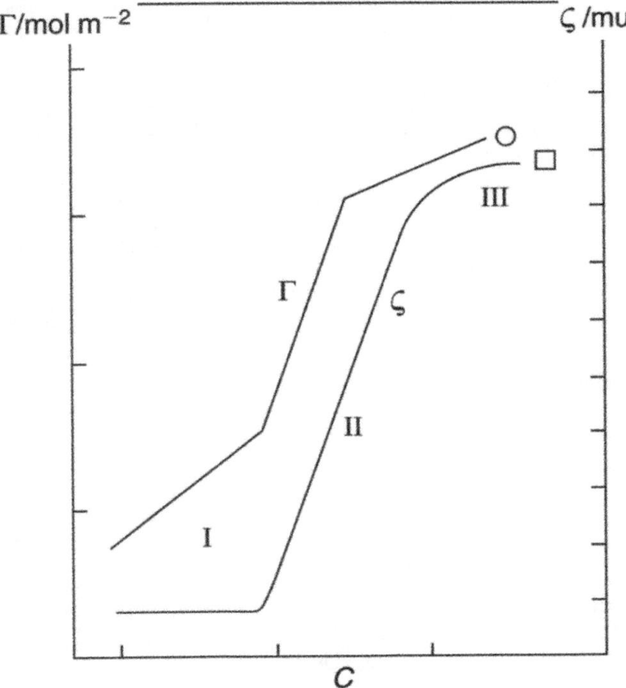

FIGURE 4.46 The trends in zeta potential and adsorption isotherms versus surfactant concentration in a typical oxide nanopowder suspension.

aggregation concentration CAC), the hydrophobic moieties of the adsorbed surfactant chains are "squeezed out" from the aqueous solution by forming two-dimensional aggregates on the adsorbent surface.

III. At a certain surfactant concentration in the hemi-micellisation process, the isoelectric point is exceeded, and thereafter, adsorption is hindered by the electrostatic repulsion between the hemi-micelles, and hence, the slope of the adsorption isotherm is reduced.

Similarly, the adsorption of nonionic surfactants involves adsorption, which is generally reversible, and obeys Langmuirian adsorption isotherm. The are many interactions, such as various adsorbate-adsorbate, adsorbate-adsorbent, and adsorbate-solvent interactions, where most surfactants produce a vertically oriented monolayer just below the *c.m.c* (see *Figure 4.47*). In the first stage of adsorption (denoted by I),

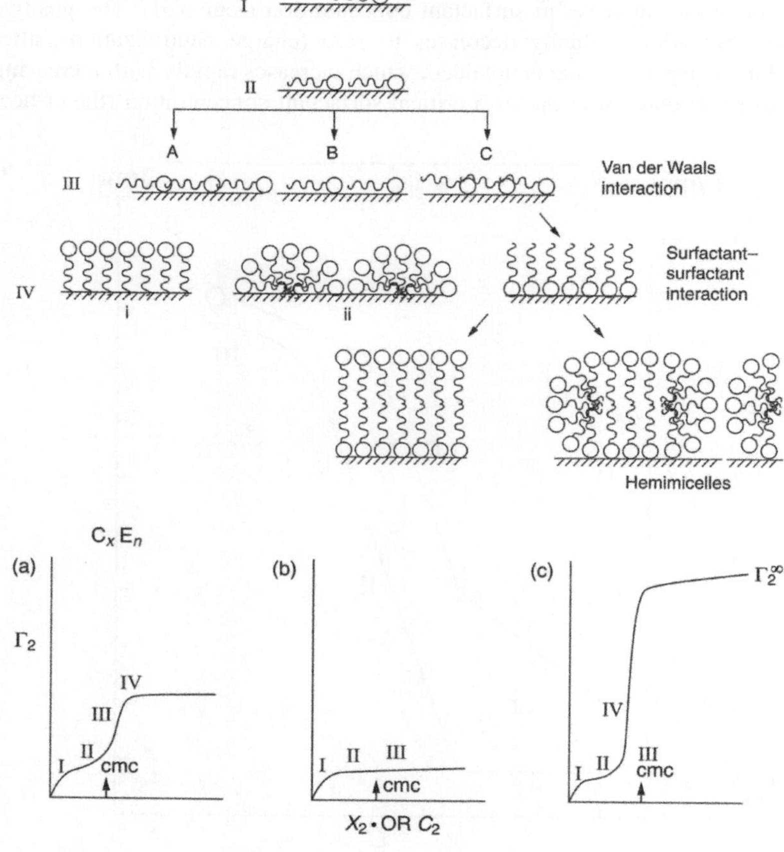

FIGURE 4.47 The I to V adsorption steps and orientations with adsorption isotherms corresponding to the steps.

surfactant-surfactant interaction is negligible (low coverage) and adsorption occurs mainly by van der Waals interaction. On a hydrophobic surface, the interaction is dominated by the hydrophobic portion of the surfactant molecule. If the chemical is hydrophilic, the interaction will be dominated by the EO chain. The subsequent stages of adsorption (regions III and IV) are determined by surfactant-surfactant interaction, although surfactant-surface interaction initially determines adsorption beyond stage II. This interaction depends on the nature of the surface and the hydrophilic-lipophilic balance of the surfactant molecule (HLB). For a hydrophobic surface, adsorption occurs via the alkyl group of the surfactant.

As an example, polyethylene glycol group, i.e., $(CH_2CH_2O)_nOH$ (where n can vary from as little as 2 to as high as 100 or more units), linked either to an alkyl $(CxH2x+1)$ or alkyl phenyl $(CxH2x+1-C6H4$ a) group. These ethoxylated surfactants are characterized by a relatively large head group compared to the alkyl chain (when n>4), with small head group such as amine oxides $(-N \rightarrow O)$ head group, phosphate oxide $(-P \rightarrow O)$, or sulphinyl-alkanol $(-SO-(CH2)n\text{–}OH)$.

Therefore, the surfactants have shown potential to be used as dispersants and stabilization of ceramic oxide nanopowder suspensions. The surfactant accumulation at the solid (ceramic)/liquid interface in aqueous or nonaqueous. The dispersion by aggregates and agglomerates of powders is into "individual" units, and stabilization of the resulting dispersion against aggregation and sedimentation. In solid/liquid dispersions (suspensions), the stabilization of suspensions by surfactants happens both electrostatically and sterically. The role of surfactant can be studied in either suspension preparation methods: condensation method and dispersion method.

The role of surfactants in condensation methods includes nucleation and growth. The major processes involve, namely, nucleation and growth similar to solid materials. The addition of surfactants can be used to control the process of nucleation and the size of the resulting nucleus. As the free energy of formation of a nucleus and the critical radius r^*, above which the cluster formation grows spontaneously, depend on two main parameters, g and $S = S0$, both of which are influenced by surfactants; g is influenced directly by adsorption of surfactant on the surface of the nucleus, which lowers g, and this reduces r^* and $" G^*$.

The spontaneous formation of clusters occurs at smaller critical radii. In addition, surfactant adsorption stabilizes the nuclei against any flocculation. The presence of micelles in solution also affects the process of nucleation and growth, both directly and indirectly. The micelles can act as "nuclei" on which growth may occur. In addition, they may solubilize the molecules of the material, thus affecting the relative supersaturation, which can affect both nucleation and growth.

The role of surfactants in dispersion methods involves three stages which should have been considered: wetting of the powder by the liquid, breaking of the aggregates and agglomerates, and comminution (milling) of the resulting particles into smaller units. These three stages are considered in detail in the passages that follow.

The powder wetting, as the wetting, is a fundamental process in which one fluid phase is displaced completely or partially by another fluid phase from the surface of a solid. A useful parameter to describe wetting is the contact angle θ of a liquid drop on a solid substrate. If the liquid makes no contact with the solid, i.e., $\theta : 180°$, the solid is referred to as nonwettable by the liquid in question. This may be the case for a perfectly hydrophobic surface with a polar liquid such as water. However, when $180° > \theta > 90°$, one may refer to a case of poor wetting. When $0° < \theta < 90°$, partial (incomplete) wetting is the case, whereas when $\theta : 0°$, complete wetting occurs and the liquid spreads on the solid substrate, forming a uniform liquid film. The value depends on (1) the history of the system and (2) whether the liquid is tending to advance across or recede from the solid surface (advancing angle θ_A, receding angle θ_R, usually $\theta_A > \theta_R$). The shape that minimizes the free energy of the system relates to interfacial tensions. Three interfacial tensions can be identified: γ_{SV}, solid/vapor area A_{SV}; γ_{SL}, solid/liquid area A_{SL}; γ_{VL}, liquid/vapor area A_{VL}. A minimum at equilibrium of $\gamma_{SV}A_{SV} + \gamma_{SL}A_{SL} + \gamma_{VL}A_{VL}$, leading to the well-known Young's equation:

$$\gamma_{SV} = \gamma_{SL} + \gamma_{VL}cos\theta \Rightarrow cos\theta = \frac{\gamma_{SV} - \gamma_{SL}}{\gamma_{VL}} \Leftrightarrow$$

$$\gamma_{VL}cos\theta = \gamma_{SV} - \gamma_{SL} = Adhesion\,tension \tag{4.43}$$

The result of the balance between the adhesion force between solid and liquid and the cohesive force in the liquid. When the surface of the solid is in equilibrium with the liquid vapor, we can consider the spreading pressure π:

$$\pi = \gamma_S - \gamma_{SV} \Rightarrow \gamma_{LV}cos\theta = \gamma_S - \gamma_{SL} - \pi \tag{4.44}$$

The adhesion tension t as the difference between the surface pressure of the solid/liquid and that between the solid/vapor interface:

$$\tau = \pi_{SL} - \pi_{SV} \Rightarrow \tag{4.45}$$

$$\pi_{SV} = \gamma_S - \gamma_{SV} \Rightarrow$$

$$\pi_{SL} = \gamma_S - \gamma_{SL} \Rightarrow$$

$$\tau = \gamma_{SV} - \gamma_{SL} = \gamma_{LV}cos\theta$$

The work of adhesion (W_a) is a direct measure of the free energy of interaction between solid and liquid:

$$W_a = (\gamma_{LV} + \gamma_{SV}) - \gamma_{SL} \Rightarrow W_a = \gamma_{LV} + \gamma_{SV} - \gamma_{LV}cos\theta = \gamma_{LV}(cos\theta + 1) \tag{4.46}$$

The spreading coefficient S defines the spreading coefficient as the work required to destroy a unit area of SL and LV and leaves a unit area of bare solid SV, where the spreading coefficient S = surface energy of final state—surface energy of the initial state.

$$S = \gamma_{SV} - (\gamma_{LS} + \gamma_{LV}) \tag{4.47}$$

$$\gamma_{SV} = \gamma_{SL} + \gamma_{VL} cos\theta$$

$$S = \gamma_{LV}\left(cos\theta - 1\right)$$

If S is zero (or positive), i.e., $\theta = 0°$, the liquid will spread until it completely wets the solid. If S is negative, i.e., $\theta > 0°$, only partial wetting occurs. Alternatively, one can use the equilibrium (final) spreading coefficient. For dispersion of powders into liquids, one usually requires complete spreading, i.e., θ should be zero.

The contact angle hysteresis for a liquid spreading on a uniform, nondeformable solid (idealized case) shows that there is only one contact angle—the equilibrium value. With real systems (practical solids), several stable contact angles can be measured. Two relatively reproducible angles can be measured: largest—advancing angle θ_A; smallest—receding angle θ_R. The θ_A is measured by advancing the periphery of a drop over a surface (e.g., by adding more liquid to the drop). The θ_R is measured by pulling the liquid back. The difference $\theta_A - \theta_R$ is referred to as contact angle hysteresis. (See *Figure 4.48*.)

The causes for hysteresis behavior are:

(1) Penetration of wetting liquid into pores during advancing contact angle measurements.
(2) Surface roughness. The first and rear edges both meet the liquid with some intrinsic angle θ_0 (microscopic contact angle). The macroscopic angles θ_A and θ_R vary significantly. This is best illustrated for a surface inclined at an angle a from the horizontal.

The effect of surfactant adsorption might involve combinations of these phenomenon. The surfactants lower the surface tension of water, γ, and they adsorb at the solid/liquid interface. A plot of γ_{LV} versus $log\,C$ (where C is the surfactant concentration) results in a gradual reduction in γ_{LV}, followed by a linear decrease of γ_{LV} with $log\,C$ (just below the critical micelle concentration, c.m.c.), and when the c.m.c. is reached, γ_{LV} remains virtually constant. (See *Figure 4.49*.)

The Gibbs adsorption isotherm for a typical surfactant in a suspension is:

$$\frac{d_\gamma}{dlogC} = -2.303RT\text{``} \qquad (4.48)$$

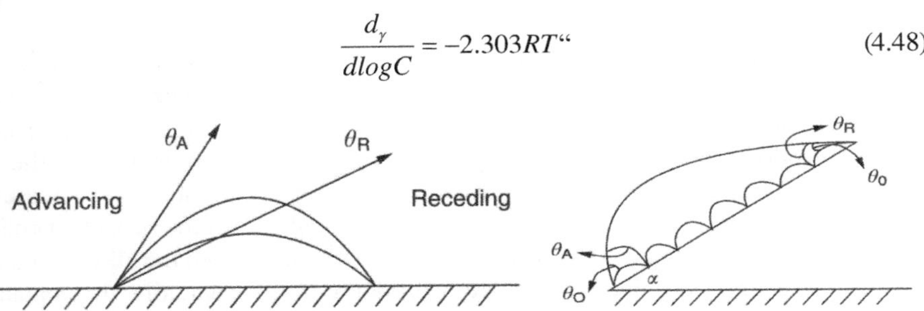

FIGURE 4.48 The schematic drawing of advancing and receding contract angles (left) and contact angle hysteresis (right).

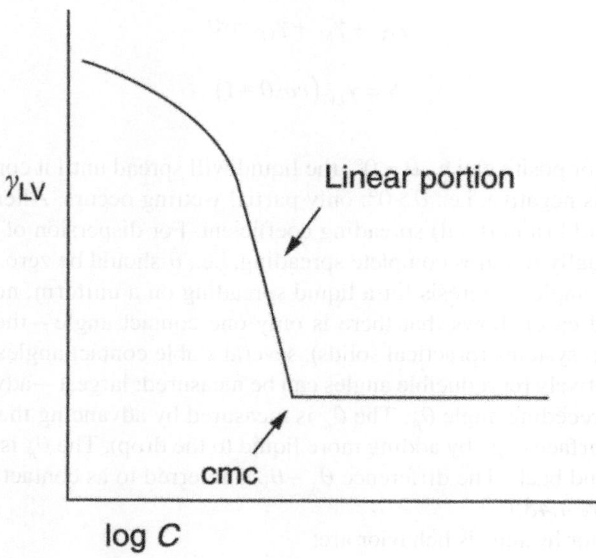

FIGURE 4.49 The schematic drawing of concentration dependency of γ_{LV} for a typical surfactant.

Where T is temperature, R is the gas constant, and " is the surface excess ($mol.m^{-2}$).

From " one can obtain the area per molecule:

$$A_{molecular} = \frac{1}{"N_{av}}\left(m^2\right) = \frac{10^{18}}{"N_{av}}\left(nm^2\right) \tag{4.49}$$

For instance, the area per molecule is usually determined by the cross-sectional area of the head group. For ionic surfactants containing, say, $-OSO_3-$ or $-SO_3-$head groups, the area per molecule is in the region of 0.4 nm. For nonionic surfactants containing several moles of ethylene oxide (8–10), the area per molecule can be much larger (1–2 nm^2).

The surfactants will also adsorb at the solid/liquid interface. For hydrophobic surfaces, the main driving force for adsorption is by hydrophobic bonding. This results in lowering of the contact angle of water on the solid surface. For hydrophilic surfaces, adsorption occurs via the hydrophilic group, e.g., cationic surfactants on silica. Initially, the surface becomes more hydrophobic and the contact angle θ increases with increasing surfactant concentration. However, at higher cationic surfactant concentrations, a bilayer is formed by hydrophobic interaction between the alkyl groups, and the surface becomes more and more hydrophilic until, eventually, the contact angle reaches zero at high surfactant concentrations. The relationship for change of θ with C is:

$$\frac{d\gamma_{LV}}{dlnC}=\frac{d\gamma_{SV}}{dlnC}-\frac{d\gamma_{LS}}{dlnC}\Rightarrow sin\theta\left(\frac{d\gamma}{dlnc}\right)=RT\left(\Gamma_{SV}-\Gamma_{SL}-\gamma_{LV}cos\theta\right) \quad (4.50)$$

Since γ_{LV} sin y is always positive, then $d\theta / dlnC$ will always have the same sign as the right-hand side mentioned prior, so some cases may be distinguished, like:

Addition of surfactant improves wetting: $(d\theta / dlnc) < 0, \Gamma_{SV} < \Gamma_{SL} + \Gamma_{LV}cos\theta$

Surfactant has no effect on wetting: $\left(\frac{d\theta}{dlnc}\right) = 0, \Gamma_{SV} = \Gamma_{SL} + \Gamma_{LV}cos\theta$

Surfactant causes dewetting: $\left(\frac{d\theta}{dlnc}\right) > 0, \Gamma_{SV} > \Gamma_{SL} + \Gamma_{LV}cos\theta$

As a result, one can predict the wetting of powders by liquids. The wetting of powders by liquids is very important in their dispersion, e.g., in the preparation of concentrated suspensions. The particles in a dry powder form either aggregate or agglomerates. (See *Figure 4.50.*)

It is essential in the dispersion process to wet both external and internal surfaces and displace the air entrapped between the particles. Wetting is achieved by the use of surface-active agents (wetting agents) of the ionic or nonionic type that can diffuse quickly (i.e., lower the dynamic surface tension) to the solid/liquid interface and displace the air entrapped by rapid penetration through the channels between the particles and inside any "capillaries." For wetting of hydrophobic powders into water, anionic surfactants, e.g., alkyl sulphates or sulphonates or nonionic surfactants of the alcohol or alkyl phenol ethoxylates, are usually used. A useful concept for choosing wetting agents of the ethoxylated surfactants is the hydrophilic-lipophilic balance (HLB) concept, $HLB = Percent\ of\ hydrophobic\ groups / 5$.

Most wetting agents of this class have an HLB number in the range 7–9. The process of wetting of a solid by a liquid involves three types of wetting: adhesion wetting, Wa; immersion wetting, Wi; and spreading wetting, Ws. (See *Figure 4.51.*)

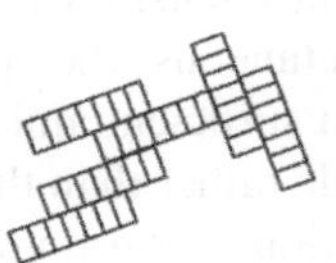

• Aggregates (particles joined by edges or corners) Loose structures
• Low power bulk density (air entrapped in the pores)

FIGURE 4.50 The schematic drawing of aggregates and agglomerates.

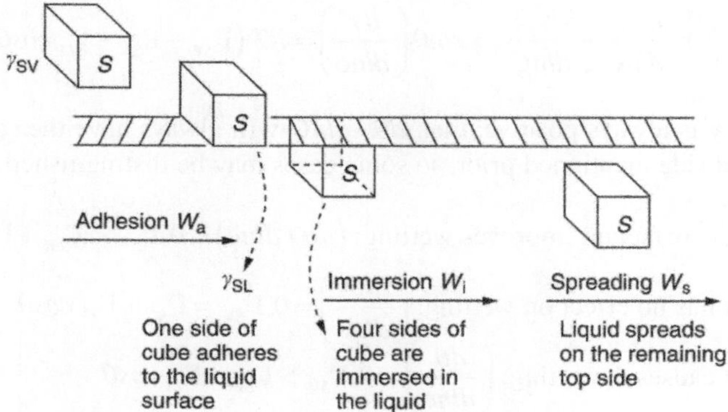

FIGURE 4.51 The schematic drawing of wetting steps.

The calculations involve:

$$\gamma_{SV} = \gamma_{SL} + \gamma_{VL} cos\theta \tag{4.51}$$

$$W_a = \gamma_{SL} - \left(\gamma_{SV} + \gamma_{VL}\right) = -\gamma_{VL}\left(cos\theta + 1\right)$$

$$W_i = 4\gamma_{SL} - 4\gamma_{SV} = -4\gamma_{VL} cos\theta$$

$$W_S = (\gamma_{SL} + \gamma_{VL}) - \gamma_{SV} = -\gamma_{VL}\left(cos\theta - 1\right)$$

The work of W_d dispersion is the sum of W_a, W_i, and W_S:

$$W_d = W_a + W_i + W_S = 6\gamma_{VL} cos\theta$$

The wetting and dispersion depend on γ_{VL} (liquid surface tension) and the contact angle y between liquid and solid. W_a, W_i, and W_S are spontaneous when $\theta < 90°$. W_d is spontaneous when $\theta : 0°$. Since surfactants are added in sufficient amounts ($g_{dynamic}$ is lowered sufficiently), spontaneous dispersion is the rule rather than the exception.

The wetting of the internal surface requires penetration of the liquid into channels between and inside the agglomerates. The process is similar to forcing a liquid through fine capillaries. To force a liquid through a capillary with radius r, a pressure p is required that is given by:

$$p = -\frac{2\gamma_{VL} cos\theta}{r} = \left[\frac{-2\left(\gamma_{SV} - \gamma_{SL}\right)}{r\gamma_{VL}}\right] \tag{4.52}$$

Where γ_{SL} has to be made as small as possible; rapid surfactant adsorption to the solid surface, low θ.

According to the surfactant and particle mutual behavior and interactions, the dispersion/adsorption mechanisms can be established. The dispersion mechanisms are classified into main types. The first is electrostatic stabilization. In aqueous suspensions, a repulsive electrostatic potential can be generated if the surfaces of the particles are electrically charged. Opposed-sign ions on the liquid are attracted to the particle surface forming a repulsive electric potential, which exponentially decreases with the distance and is proportional to the dielectric constant of the suspension medium. Water has a high dielectric constant, while those of nonpolar solvents are much lower. Therefore, effective electrostatic stabilization is exclusive of aqueous suspensions and can be neglected in nonpolar polymeric solvents. Furthermore, electrostatic stabilization alone might be insufficient for the achievement of a stable suspension. According to the DLVO theory, if the particles have sufficient energy to overcome the electrostatic repulsive barrier, they might fall on the primary minimum forming a very stable aggregate.

The second type of dispersion mechanism is steric stabilization, which occurs through the adsorption of long organic molecules at the surface of the particles, generating repulsion force when the polymeric layer of two neighboring particles overlaps, avoiding their aggregation.

Unlike electrostatic stabilization, steric stabilization can disperse nonaqueous suspensions. Due to the hydrophilic surface of ceramic particles, the stabilizing materials must have a hydrophilic anchor group that interacts with the particle surface and a hydrophobic chain soluble in the liquid medium. The steric barrier that stabilizes the dispersed particles in the suspension is created at this stage.

The repulsive steric potential is dependent on Flory-Huggins parameter χ (wettability of the dispersant in the medium), adsorbed layer volume fraction ϕa, layer thickness δ, molar volume of the solvent $v1$, and temperature T. The following conditions are required for a good steric dispersion. The complete coverage of particle surface by the steric layer (high ϕa), sufficient thickness of the adsorbed layer ($\delta > 5$ nm), medium as a good solvent for the stabilizing chain ($\chi < 0.5$), and strong attachment of dispersant with the surface particle are achieved with anchor groups of strong affinity with the surface. On the other hand, if the polymer is not attached to the surface but freely moves dissolved in the solvent, it can produce depletion agglomeration instead of dispersion. Therefore, the high affinity of the dispersant with both ceramic particle and solvent medium is fundamental for good dispersion, which can be evaluated with the use of contact angle measurements.

The steric stabilization offers advantages when compared to electrostatic stabilization. Its steric repulsive potential is steeper than electrostatic potential; therefore, the total potential has no primary minimum, enabling an easy redispersion of agglomeration. Furthermore, steric stabilization is independent of the ionic concentration of the liquid medium and promotes good dispersion at intermediate pH values, whereas in electrostatic stabilization, a better dispersion is achieved at extreme pH values.

The third type of dispersion mechanism is electrosteric stabilization, which combines both types of stabilization. Polyelectrolytes, which are polymers charged along the length of the polymeric chain (in contrast to those with charged species only at the molecule terminations), promote a steric barrier and an electrostatic potential.

The superficial charge of the particles, altered by the pH, can either increase or decrease polymer adsorption on the surface. The electrosteric mechanism yields good results for aqueous suspensions. Researchers have tested some of the most used dispersants for aqueous suspensions, namely ammonium polyacrylate, sodium polyacrylate, polyvinylpyrrolidone, and polyacrylic acid, and reported ammonium polyacrylate, which is a polyelectrolyte, resulted in the lowest viscosity suspension.

The adsorption of steric dispersants can be classified as either chemical or physical. In chemical adsorption, molecules chemically react with the surface, forming covalent bonds. This process requires a considerable amount of energy since it involves disruption and formation of covalent bonds, which can have the magnitude of hundreds of kilojoules.

The chemisorption can also be the adsorption/dispersion mechanism. The ceramic particles are previously mixed with the dispersant and a solvent (such as ethanol or acetone) and then stirred in an ultrasonic mixer, ball mill, or planetary mill. Finally, the powder is dried for the elimination of the solvent and eventually heated for enhancing chemical reactions and then sieved. This two-step process is employed for suspensions that use carboxylic acids (monocarboxylic and dicarboxylic acids) and silane couplings as dispersants.

On the other hand, the physical adsorption involves van der Waals bonds, which occur between the adsorbent and the adsorbent surface and do not require a large amount of energy, reaching a maximum of a few kilojoules. Examples of such interactions are hydrogen bonds between hydroxyl groups present on the surfaces of oxides and the oxygen of carbonyl groups of esters or ethoxy segments. Mixing requires no preliminary step, enabling ceramic particles and dispersants to be mixed at once. Several commercial dispersants with long polymeric chains were mixed in this single-step process (e.g., KOS110 [Guangzhou Kangoushuang], BYK w969, disperBYK-180 [BYK-Chemie], and Variquac CC 42 [Evonik]). In all these systems, the addition of dispersants has an optimum point of minimum viscosity, resulting in a more viscous suspension for lower or higher amounts of dispersant.

For instance, adding varied amounts of dispersant in a suspension with 40 vol% zirconia particles at a constant shear rate demonstrates that the dosage of dispersant should be proportional to the superficial area of the powder rather than to its weight, since adsorption is a superficial effect. A low amount of dispersant can result in an incomplete coverage of the particle surface, flocculation. On the other hand, a large amount of dispersant above the adsorption limit increases the suspension viscosity, since the excess dispersant is diluted in the medium. The use of the optimum amount of dispersant, which is associated with the limit of adsorption of the Langmuir isothermal adsorption model, reduces suspension viscosity, shear-thinning behavior, and yield stress.

The thickness of the dispersive layer also displays an optimum intermediate size. If the layer is thin, the attractive potential is high, increasing viscosity. However, a thick dispersant layer might negatively affect the suspension viscosity. The adsorbed layer increases the apparent particle diameter, the apparent solid fraction, and the medium viscosity. Krieger-Dougherty's model can predict viscosity, replacing the solid fraction by the effective solid fraction, φ_{eff}, which includes adsorbed layer thickness (large adsorbed layer to particle radius ratio δ / a):

$$\varphi_{eff} = [1+\frac{\delta}{a}]^3 \tag{4.53}$$

The use of very high molecular polymers as dispersants can lead to bridging flocculation, hence, increasing viscosity. Li and Zhao compared the effect of oleic acid and ammonium polyacrylate used as dispersants for alumina in HDDA and observed a sharper increase in the viscosity of ammonium polyacrylate suspension when the solid load was increased from 35 vol% to 40 vol%. Such an increase was explained by the larger backbone chains of ammonium polyacrylate, which became entangled as the interparticle distance was reduced. Conclusively, the adsorbed layer of polymers of low molecular weight is excessively thin to prevent flocculation, whereas polymers of very high molecular weight considerably increase viscosity in highly loaded suspensions and might promote bridging flocculation. Therefore, the choice of the appropriate dispersant chain size must consider solid fraction and ceramic particle size.

For example, the α alumina-specific area of 10.2 $m^2\ g^{-1}$ (0.5–1 μm) and a refractive index of 1.70. The dispersant (azeotropic mixture of methylethylketone [MEK] and ethanol [60/40 vol]) acts both by electrostatic and steric repulsion and is efficient to disperse alumina particles in low polar organic media by compensating the van der Waals attractive forces between alumina particles and by preventing particles to approach 80 wt.%. For low amounts of dispersant (below 1.2 wt.%), the suspensions exhibit shear thickening. Above a concentration of dispersant of 1.2 wt.%, the slurries exhibit a shear-thinning behavior; at minimum of viscosity, the shear thinning behavior is evident. (See *Figure 4.52.*)

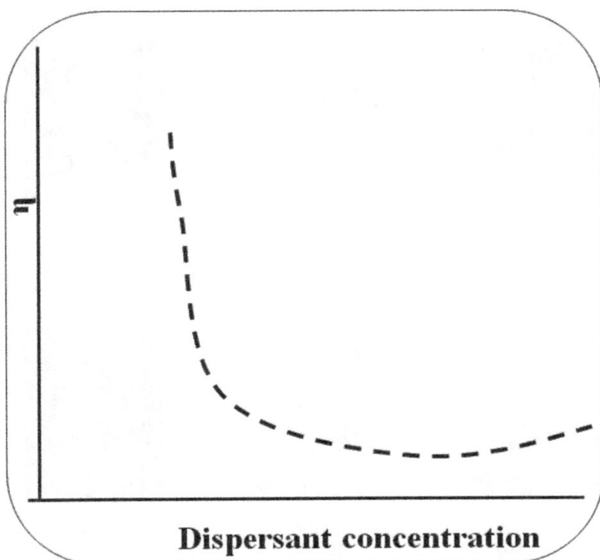

FIGURE 4.52 The schematic drawing of dispersant dependency in a typical oxide nanopowder suspension at constant solids content and shear rate.

In the concentrated ceramic suspensions, one of the main features is the formation of three-dimensional structure units, which determine their properties and, in particular, their rheology. The formation of these units is determined by the interparticle interactions. The particle number concentration and volume fraction, Φ, above which a suspension may be considered concentrated, is best defined in terms of the balance. Between the particle translational motion and interparticle interaction, the dispersion occurs. The standing, concentrated suspensions reach various states (structures) that are determined by (1) magnitude and balance of the various interaction forces, electrostatic repulsion, steric repulsion, and van der Waals attraction; (2) particle size and shape distribution; (3) density difference between disperse phase and medium, which determines the sedimentation characteristics; (4) conditions and prehistory of the suspension, e.g., agitation, which determines the structure of the flocs formed (chain aggregates, compact clusters, etc.); and (5) presence of additives, e.g., high molecular weight polymers that may cause bridging or depletion flocculation.

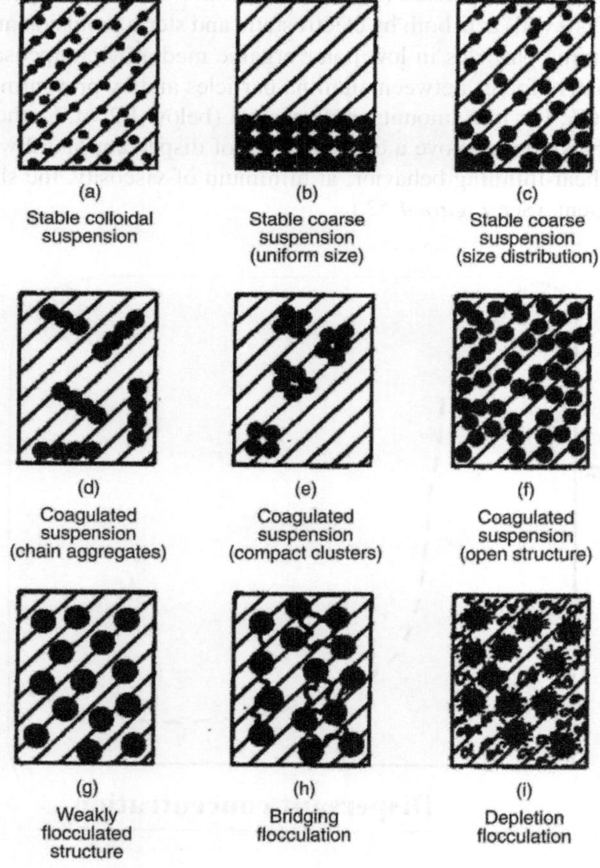

FIGURE 4.53 The schematic drawing of possible structures in concentrated suspensions.

TABLE 4.1
The rheological parameters of typical ceramic suspensions.

Ceramic	Particle size (nm)	Medium	Solids content	Zeta potential (mV)	surface potential (mV)	Viscosity (Pa.S)	Mechanism	Hamakar	Ref
alumina	55	DM water	1 M	43	162	0.09	van der Waals interactions	–	[8]
alumina	30	DM water	0.1 M	45	177	0.5	van der Waals interactions	–	[9]
alumina	50	DM water	200ppm	30–70	151	1.1	van der Waals interactions	–	[10]
alumina	50	DM water	25 wt.%	50	158	30	van der Waals interactions	3.67×10^{-20} J	[11]
Zirconia	30–60	DM water	–	–	–	0.011	van der Waals interactions	–	[12]
YSZ	30–60	DM water	–	–	–	0.013	van der Waals interactions	–	[13]
Titania	38	DM water	–	–	–	0.006	van der Waals interactions	–	[14]

These states may be described in terms of three different energy—distance curves may include (a) electrostatic, produced, for example, by the presence of ionogenic groups on the surface of the particles, or adsorption of ionic surfactants; (b) steric, produced, for example, by adsorption of nonionic surfactants or polymers; and (c) electrostatic and steric (electrosteric), as, for example, produced by polyelectrolytes, some of which are presented in *Figure 4.53*.

Table 4.2 summarizes some details on ceramic suspensions (mostly nanosized) in different oxide systems.

REFERENCES

[1] Lewis JA. Colloidal processing of ceramics. J Am Ceram Soc. 2000;83(10):2341–59. doi: 10.1111/j.1151-2916.2000.tb01560.x.

[2] Yaghtin M, Yaghtin A, Tang Z, Troczynski T. Improving the rheological and stability characteristics of highly concentrated aqueous yttria stabilized zirconia slurries. Ceram Int. 2020;46(17):26991–9. doi: 10.1016/j.ceramint.2020.07.176.

[3] Zhang C, Jiang Z, Zhao L, Guo W, Gao X. Stability, rheological behaviors, and curing properties of 3Y—ZrO2 and 3Y—ZrO2/GO ceramic suspensions in stereolithography applied for dental implants. Ceram Int. 2021;47(10A):13344–50.

[4] Kim IJ, Park JG, Han YH, Kim SY, Shackelford JF. Wet foam stability from colloidal suspension to porous ceramics: a review. J Korean Ceram Soc. 2019;56(3):211–32. doi: 10.4191/kcers.2019.56.3.02.

[5] Bai Y, Chi B-X, Ma W, Liu C-W. Suspension plasma-sprayed fluoridated hydroxyapatite coatings: effects of spraying power on microstructure, chemical stability and antibacterial activity. Surf Coat Technol. 2019;361:222–30. doi: 10.1016/j.surfcoat.2019.01.051.

[6] Bacciochini A, Ilavsky J, Montavon G, Denoirjean A, Ben-ettouil F, Valette S, Fauchais P, Wittmann-Teneze K. Quantification of void network architectures of suspension plasma-sprayed (SPS) yttria-stabilized zirconia (YSZ) coatings using ultra-small-angle X-ray scattering (USAXS). Mater Sci Eng A. 2010;528(1):91–102. doi: 10.1016/j.msea.2010.06.082.

[7] Krieger IM, Eguiluz M. The second electroviscous effect in polymer latices. Trans Soc Rheol. 1976;20(1):29–45. doi: 10.1122/1.549428.

[8] Tesar T, Musalek R, Lukac F, Medricky J, Cizek J, Rimal V, Joshi S, Chraska T. Increasing α-phase content of alumina-chromia coatings deposited by suspension plasma spraying using hybrid and intermixed concepts. Surf Coat Technol. 2019;371:298–311. doi: 10.1016/j.surfcoat.2019.04.091.

[9] Carpio P, Salvador MD, Borrell A, Sánchez E, Moreno R. Alumina–zirconia coatings obtained by suspension plasma spraying from highly concentrated aqueous suspensions. Surf Coat Technol. 2016;307:713–19. doi: 10.1016/j.surfcoat.2016.09.060.

[10] Carnicer V, Orts MJ, Moreno R, Sánchez E. Influence of solids concentration on the microstructure of suspension plasma sprayed Y-TZP/Al2O3/SiC composite coatings. Surf Coat Technol. 2019;371:143–50. doi: 10.1016/j.surfcoat.2019.01.078.

[11] Yaghtin M, Taghvaei AH, Hashemi B, Janghorban K. Effect of heat treatment on magnetic properties of iron-based soft magnetic composites with Al2O3 insulation coating produced by sol–gel method. J Alloys Compd. 2013;581:293–7. doi: 10.1016/j.jallcom.2013.07.008.

[12] Curry N, VanEvery K, Snyder T, Susnjar J, Bjorklund S. Performance testing of suspension plasma sprayed thermal barrier coatings produced with varied suspension parameters. Coatings. 2015;5(3):338–56. doi: 10.3390/coatings5030338.

[13] Mahade S, Curry N, Jonnalagadda KP, Peng RL, Markocsan N, Nylén P. Influence of YSZ layer thickness on the durability of gadolinium zirconate/YSZ double-layered thermal barrier coatings produced by suspension plasma spray. Surf Coat Technol. 2019;357:456–65. doi: 10.1016/j.surfcoat.2018.10.046.

[14] Vu P, Otto N, Vogel A, Kern F, Killinger A, Gadow R. Efficiently quantifying the anatase content and investigating its effect on the photocatalytic activity of titania coatings by suspension plasma spraying. Surf Coat Technol. 2019;371:117–23. doi: 10.1016/j.surfcoat.2018.07.064.

5 The Suspension Aspect of Suspension Plasma Spray Process (SPS)

Scope

Ensuing the physicomechanochemical aspects of general suspensions formation and further study of the ceramic (oxide/non-oxide) suspensions in aqueous/nonaqueous media, the specially formulated ceramic suspensions for suspension plasma spray process (SPS) must be considered. The most convenient route for preparation involves making a simple slurry with particles and media, particle sizes varying from a few tens of nanometers to micrometers, and applying appropriate surface agents. The mostly used medias are ethanol (organic) or water (nonorganic) or a mixture of both. After proper stirring, the suspension stability can be tested by sedimentation, turbidity, viscosity fluctuation, and chemical changes. Typical values of these suspensions' stability are a few tens of minutes to tens of hours, while the stability even increase with the solid load. The suspension with TiO_2, ZrO_2has been prepared that way, as well as with Al_2O_3 and ZrO_2-Al_2O_3mixtures. However, nanoparticles of oxides have the tendency to agglomerate or aggregate, even when stirring the suspension.

5.1 GENERAL OVERVIEW

This chapter focuses on the formation, development, inheritance, and control of the properties in colloidal ceramic suspensions. Generally, ceramic suspensions should have the following properties:

1. Low viscosity, which guarantees that the pumping machine and nozzles can transport suspensions in level, homogeneous layers
2. A high solids content to enable low shrinkage in the ceramic coating
3. A sufficient cure depth and width to generate a homogenous layer at the interface, which provides good cohesion with substrates

These properties are in direct relation with the considered process, the suspension plasma spraying (SPS), a process among the different possible routes to produce finely (i.e., submicrometer to nanometer scale) structured layers via accelerated ceramic suspension via plasma spraying, which appears as one of the most flexible processes for oxide, non-oxide, and functionally graded materials coatings [1]. Indeed, SPS is introduced as an alternative to conventional processes such as atmospheric plasma spraying (APS) to produce thinner layers (i.e., 10–100 μm) due to the specific size of the feedstock particles, from a few tens of nanometers to a few micrometers. It

DOI: 10.1201/9781003285014-5

consists of mechanically injecting within the plasma flow a suspension made of the feedstock particles, a liquid phase, and a dispersant. Upon penetration within the plasma flow, the liquid stream encounters two mechanisms, fragmentation and vaporization. In a first approximation, and whatever the suspension stream characteristic dimension (from a few micrometers to a few hundreds of micrometers in diameter), the fragmentation duration is at least two orders of magnitude shorter than the vaporization. Droplets encounter then liquid phase evaporation that leads to the formation of single particle or aggregates of a few particles, depending upon the size of the suspension droplets and the particles [2]. Then, these particles melt and form liquid drops that impact, spread, and solidify to form flattened lamellae of equivalent diameters between a few hundred nanometers to a few micrometers and of average flattening ratio varying from 1 to 2. The coatings, more or less cohesive, result from the stacking of such lamellae. Their architecture evolves from nearly fully dense to very porous, with a smooth (homogeneous) or irregular (heterogeneous) surface morphology, according to the spray parameters, among which plasma power parameters (plasma torch operating mode, plasma flow mass enthalpy, spray distance, etc.), suspension properties (particle size distribution, powder mass percentage, viscosity, surface tension, etc.), and substrate characteristics (topology, temperature, etc.) play relevant roles [3]. As can be seen in *Figure 5.1*, the process parameters can be categorized in plasma flow parameters, suspension parameters, and substrate conditions, which indicates the lateral importance of suspension condition on final coating product.

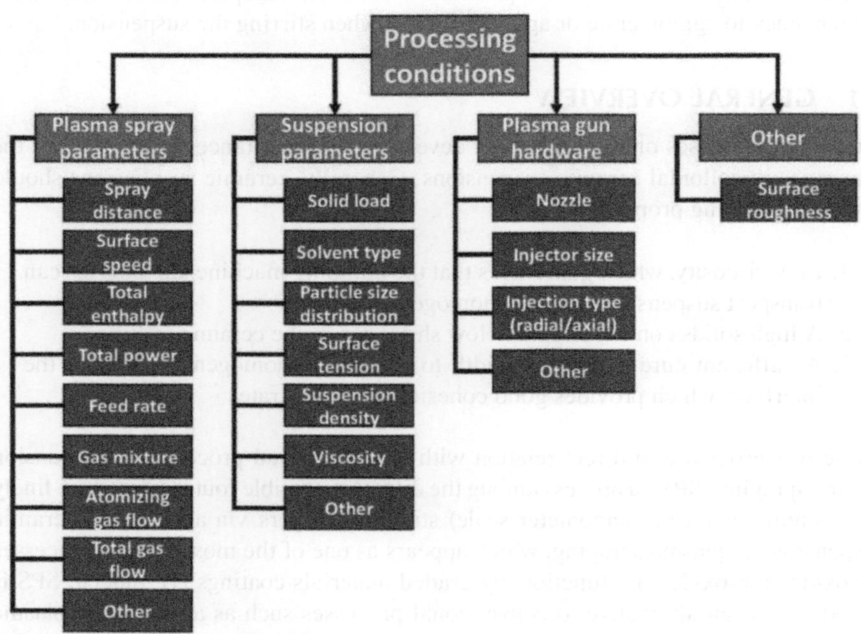

FIGURE 5.1 The precursor and operational parameters of SPS process.

It is worth mentioning that the rheological performance of the ceramic's suspension is principal for the colloidal ceramic suspensions; the ceramics suspensions processing requires the control of the initial suspension structure; SPS subsequent evolution happen through different fabrication stages, which can be summarized as (1) powder synthesis, (2) suspension preparation, (3) suspension stabilization, (4) removal of the liquid phase, and (5) densification to produce the desired final microstructure, for example, with the desired spatial distribution of phases and with controlled porosity; ceramic suspensions should be optimized via operating parameters, such as volumetric solid fraction, particle size, size distribution, monomer, diluent, dispersant, and temperature on such suspensions. It is reported that the best formulations involve those that achieved lower than 25 $mPa.s-(\text{cp})$ viscosity and higher than 40 wt.% solids content fractions [4].

The use of liquid mediums, to carry nanoparticles, has been developed in order to promote nanostructured coatings via plasma spraying. This approach ensures the *homogenous* structures, *vertically cracked structures*, and more *particularly columnar structures* [5]. The precursors include nanosized particles in suspension and/or ceramic precursors in solution. The stability problem can be overcome by using a suitable dispersant, which adsorbs on the particle surface and allows an effective dispersion of particles by electrostatic and/or steric repulsions. The percentage of dispersant must be adjusted in such a way that it displays the minimum viscosity of the suspension with a shear-thinning behavior. This behavior means that when the shear stress imposed by the plasma flow is low, the suspension viscosity is high, and it decreases drastically when the shear stress increases as the drop penetrates more deeply within the plasma flow. For instance, zirconia suspension with phosphate ester that can act as a combination of electrostatic and steric repulsion has been used. In a non-oxide system, the mixture of $WC-Co$, the stabilization is more complex due to the different acid/base properties of both components, as WC or, more precisely, WO_3, at its surface is a Lewis acid, while CoO is basic. Thus, a complex equilibrium between the dispersing agent and the suspension pH must be found; for example, the latter must be adjusted to less basic conditions, but avoiding the cobalt dissolution. It is also important to adapt the size distribution of particles within the suspension to the heat transfer of the plasma jet, limit the width of particle size distribution (as in conventional spraying) in order to reduce the dispersion of trajectories, and avoid powders that have the tendency to agglomerate or aggregate, which is often the case for nanosized particles, especially oxides.

The general behavior of these suspensions evolves from non-Newtonian behavior. The highly loaded ceramic suspensions for SPS applications are usually strongly nonlinear due to the dispersed solid phase, although the monomers and diluents used as a continuous phase generally display a Newtonian behavior. Because of the multiphase characteristic of ceramic suspensions, the rheological behavior is influenced by the properties of the continuous medium, dispersed phase, and remarkably, interactions between phases. In low to moderate solid loadings, ceramic suspensions exhibit a shear-thinning behavior characterized by a viscosity decrease with increasing shear rate. Shear thinning occurs through deflocculation of the particles and their increased arrangement in layers, caused by a shear rate increase (see *Figure 5.2*).

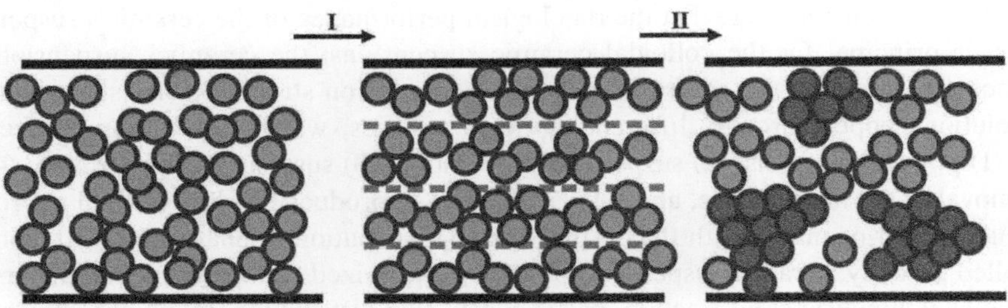

FIGURE 5.2 The particle arrangement in suspension during forced flow from at rest to semilayered ordering (I) and toward shear thickening (II).

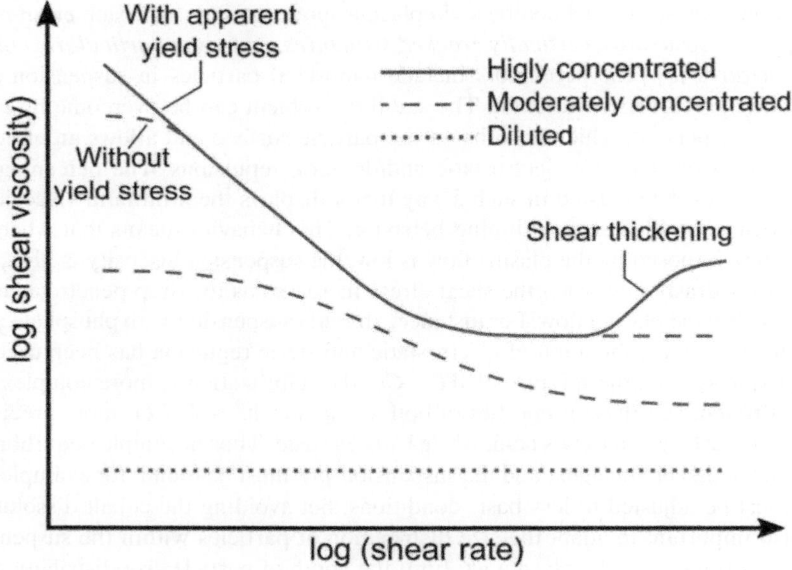

FIGURE 5.3 The particle shear viscosity–shear rate dependency in typical ceramic suspension according to concentration.

This behavior is desirable in ceramic suspension, since it prevents the sedimentation of the suspension at rest and promotes adequate flow when a shear rate is applied. On the other hand, highly loaded ceramic suspensions may display a shear thickening behavior, characterized by a viscosity increase with increasing shear rate, mainly in colloidal solutions (particles of at least one dimension between 1 nm and 1 μm) subjected to high shear rates. This phenomenon can be explained by a disarrangement in the layered flow of the particles, which follows the trend in *Figure 5.3* toward increasing viscosity [6].

The rheological behavior is strongly affected by solid fraction. As shown, some behaviors are characteristic of ceramic suspensions with different concentrations of solids. The very diluted suspensions tend to maintain a Newtonian behavior, moderately concentrated solutions display a shear-thinning behavior with Newtonian plateaus at low and high shear rates, and highly concentrated solutions may show shear thickening at high shear rates. The Brownian motion dominates the suspension behavior of colloidal suspensions for moderate and highly concentrated suspensions, maintaining the particles randomly distributed in a state of thermodynamic equilibrium, with some aggregation, thus resulting in higher viscosity at the first Newtonian plateau (low shear rate $\gamma^{\cdot} \rightarrow 0$). On the other hand, at the second Newtonian plateau (moderate to high shear rate), the flowing particles achieve an optimum layered arrangement, resulting in minimum viscosity [7]. The presence and intensity of this shear-thickening transition are related to factors such as choice of monomers, dispersants, dispersant quantity, particle size, and particle size distribution of ceramic powder. The critical shear rate, $\dot{\gamma}_{cr}$, characterizes the onset of shear-thickening behavior. The $\dot{\gamma}_{cr}$ decreases with an increasing volume percentage of solids, and higher solid loadings also result in a more intense shear-thickening behavior. The colloidal stability is governed by the total interparticle potential energy [7]:

$$V_{total} = V_{vdW} + V_{elect} + V_{steric} + V_{structural} \tag{5.1}$$

Where V_{vdW} is the attractive potential energy due to long-range van der Waals interactions between particles, V_{elect} the repulsive potential energy resulting from electrostatic interactions between like-charged particle surfaces, V_{steric} the repulsive potential energy resulting from steric interactions between particle surfaces coated with adsorbed polymeric species, and $V_{structural}$ the potential energy resulting from the presence of nonadsorbed species in a solution that may either increase or decrease suspension stability (see *Figure 5.4*). The long-range forces resulting from van der Waals (vdW) interactions are ubiquitous and always attractive between similar particles; V_{vdW} exhibits a power-law distance dependence whose strength depends on the dielectric properties of the interacting colloidal particles and intervening medium. For spherical particles of equal size, V_{vdW} is given by the Hamaker expression:

$$V_{vdW} = -\frac{A}{6}\left(\frac{2}{s^2-4} + \frac{2}{s^2} + ln\frac{s^2-4}{s^2}\right),\ s = \frac{2a+h}{a} \tag{5.2}$$

Where h is the minimum separation between the particle surfaces, a the particle radius, and A the Hamaker constant.

Some details of the constant for selected ceramics are presented in *Table 5.1*.

The electrostatic forces are determinative factors in aqueous ceramic fine-particle (submicrometer- and nanosized) suspensions. The stability of aqueous colloidal systems can be controlled by generating like charges of sufficient magnitude on the surfaces of suspended ceramic particles [8]. The resulting repulsive V_{elect} exhibits an exponential distance dependence whose strength depends on the surface potential induced on the interacting colloidal particles and the dielectric properties of the intervening medium.

$$For\, ka > 10 \tag{5.3}$$

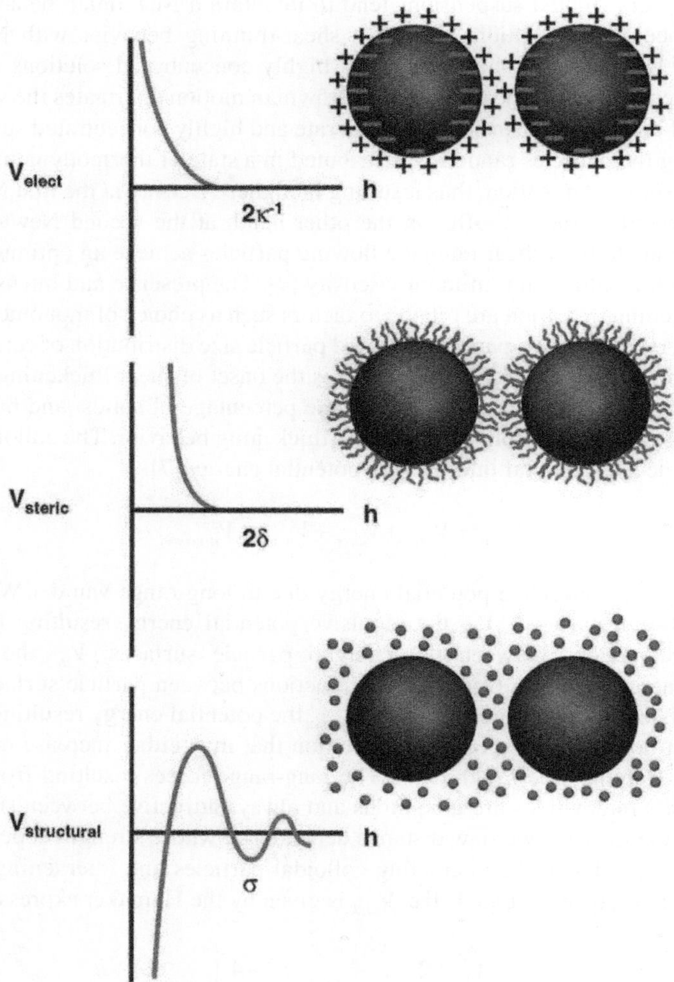

FIGURE 5.4 The schematic interparticle potential energy sources around a ceramic particle.

TABLE 5.1
Hamaker constants for ceramics.

Ceramic	Structure	Hamaker constant ($\times 10^{-20} J$)
$\alpha - Al_2O_3$	Hexagonal	3.67
$BaTiO_3$	Tetragonal	8
BeO	Hexagonal	3.35
$CaCO_3$	Trigonal	1.44

Ceramic	**Structure**	**Hamaker constant** $(\times 10^{-20} J)$
CdS	Hexagonal	3.4
$MgAl_2O_4$	Cubic	2.44
MgO	Cubic	2.21
$Mica$	Monoclinic	1.34
$6H-SiC$	Hexagonal	10.9
$\beta-SiC$	Cubic	10.7
$\beta-Si_3N_4$	Hexagonal	5.47
Si_3N_4	Amorphous	4.85
$q-SiO_2$	Trigonal	1.02
SiO_2	Amorphous	0.46
$SrTiO_3$	Cubic	4.77
TiO_2	Tetragonal	5.35
Y_2O_3	Hexagonal	3.03
ZnO	Hexagonal	1.89
ZnS	Hexagonal	5.74
$3Y-ZrO_2$	Tetragonal	7.23

$$V_{elect} = 2\pi\varepsilon_r\varepsilon_0 a\ddot{\ }^2 \ln\left[1+\exp(-kh)\right]$$

$$For\, ka < 5$$

$$V_{elect} = 2\pi\varepsilon_r\varepsilon_0 a\ddot{\ }^2 \exp(-kh)$$

$$k = \left(\frac{F^2\sum_i N_i z_i^2}{\varepsilon_r\varepsilon_0 kT}\right)^{1/2}$$

Where ε_r is the dielectric constant of the liquid, ε_0 the permittivity of vacuum, Ψ the surface potential, and $1/k$ the Debye-Huckel screening length, and where N_i and Z_i are the number density and valence of the counterions of type i, and F the Faraday constant. The Ψ originated from the dissociation of amphoteric hydroxyl groups on ceramic oxide surfaces and depends on pH and indifferent electrolyte concentration. It can be estimated from the zeta potential (ζ) [9]. The dispersions can be rendered unstable by either increasing ionic strength or adjusting pH toward the isoelectric point (IEP), as reported in *Table 5.2*.

For example, a weakly attractive, aqueous alumina suspensions at pH conditions below the IEP point, where they found that the yield strength increased with increased electrolyte concentration. These attractive networks were much weaker than those produced by flocculating the system at its *IEP* (pH:8.5) [10]. Because of the weak attraction between particles, such slurries could be consolidated under modest applied pressures to densities approaching those attainable in dispersed

TABLE 5.2
IEP values of ceramics.

Ceramic	Structure	IEP
$\alpha - Al_2O_3$	Hexagonal	8–9
$BaTi\,O_3$	Tetragonal	5–6
CeO_2	Cubic	6–7
Cr_2O_3	Hexagonal	7
CuO	Monoclinic	9.5
Fe_3O_4	Cubic	6.5
La_2O_3	Trigonal	10.4
MgO	Cubic	12.4
MnO_2	Tetragonal	4–4.5
NiO	Cubic	10–11
SiO_2	Amorphous	2–3
Si_3N_4	Amorphous	9
SnO_2	Tetragonal	7.3
TiO_2	Tetragonal	4–6
ZnO	Hexagonal	9
ZrO_2	Tetragonal	4–6

systems. The electrostatically stabilized suspensions are kinetically stable systems, where the rate of doublet formation is controlled by the stability ratio, W:

$$W = \frac{k_0}{k} = \exp\left(\frac{\frac{V_{max}}{k_B T}}{2ka}\right) \tag{5.4}$$

Where V_{max} is the maximum repulsive barrier height, $k0$ the rate constant for fast irreversible flocculation, and k the rate constant of flocculation for the system of interest. The stability ratio exhibits an exponential dependence on Vmax and a linear dependence on the normalized electrostatic double-layer thickness (K_a^{-1}). A system in which only Brownian motion acts to bring particles together. During colloidal processing, external forces can "push" particles over the repulsive barrier, further reducing suspension stability. In practice, it may be difficult to effectively design stable suspensions based only on electrostatic stabilization. Particle solubility concerns may limit the working pH range, whereas an extended double-layer thickness may lead to unacceptable drying shrinkage [11].

The steric stabilization provides an alternate route of controlling colloidal stability that can be used in aqueous and nonaqueous systems. In this approach, adsorbed organic molecules (often polymeric in nature) are utilized to induce steric repulsion.

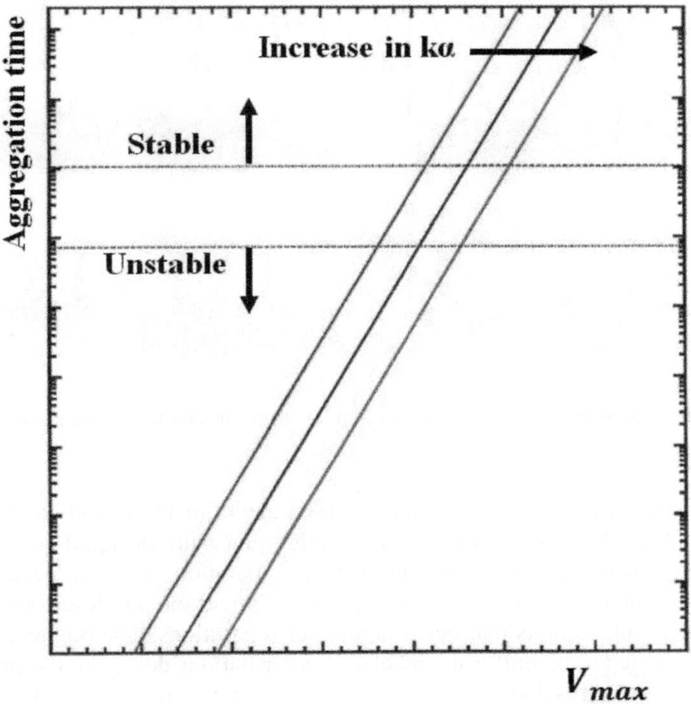

FIGURE 5.5 The schematic aggregation time versus repulsive potential in an aqueous ceramic suspension.

The adsorbed layers must be of sufficient thickness and density to overcome the *vdW* attraction between particles and to prevent bridging flocculation. The steric interactions occur when particles approach one another at a separation distance less than twice the adlayer thickness (d). The conformation of adsorbed layers can vary dramatically, depending on liquid quality, molecular architecture, number of anchoring groups, active surface site density, and colloid and organic concentrations in solution. The polymer surfactants can be adsorbed on ceramic particle surface according to *Figure 5.6*.

Their close approach can be divided into two domains, as interpenetrational domain for ($\delta < h < 2\delta$) a pseudohomopolymer model is used to describe the mixing interactions that occur in the region. The pseudohomopolymer model accounts for chain conformations other than tails (i.e., trains and loops) that are expected for such species.

$$V_{steric,mix} = \frac{32\pi a k_b T \bar{\phi}_2^a (0.5-\chi)}{5v_1\delta^4}(\delta - \frac{h}{2})^6 \tag{5.5}$$

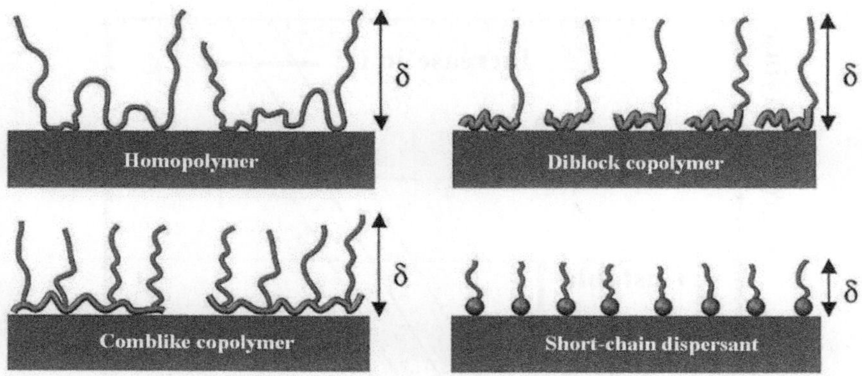

FIGURE 5.6 The schematic surface adsorption layer formation on ceramic particles.

Where $\bar{\phi}_2^a$ is the average volume fraction of segments in the adsorbed layer measured as 0.37, χ (the Flory—Huggins parameter) a measure of liquid quality, v_1 the molar volume of liquid, and h the interparticle separation. The interpenetrational-plus-compressional domain $\delta > h$, the uniform segment model, describes the mixing and elastic interactions that occur at smaller separations. The polymer segment density is assumed to be uniform, and elastic contributions dominate the interaction potential energy. In this domain, V_{steric} is given by the sum of the mixing ($V_{steric,mix}$) and elastic ($V_{steric,el}$) terms:

$$V_{steric,mix} = \frac{4\pi a\delta^2 k_b T\bar{\phi}_2^a (0.5-\chi)}{v_1}\left(\frac{h}{2\delta}-\frac{1}{4}-ln\frac{h}{\delta}\right)$$

$$V_{steric,el} = \frac{2\pi a\delta^2 k_b T\bar{\phi}_2^a}{M_2^a}\left\{\frac{h}{2\delta}\ln[\frac{h}{2\delta}(\frac{3-h/\delta}{2})^2 - 6\ln\left(\frac{3-\frac{h}{\delta}}{2}\right)+3\left(1-\frac{h}{\delta}\right)\right\}$$

Where δ^2 is the density and M_2^a the molecular weight of the adsorbed species. As predicted, such dispersions can be rendered unstable when liquid conditions become poor. The electrosteric forces that act in polyelectrolyte species are widely used additives that can impart electrostatic and steric stabilization to a given colloidal dispersion. Such systems are often referred to as electrosterically stabilized. Polyelectrolytes contain at least one type of ionizable group (e.g., carboxylic or sulfonic acid groups), with molecular architectures that range from homopolymers, such as poly (acrylic acid), to block copolymers with one or more ionizable segments. Polyelectrolyte adsorption is strongly influenced by the chemical and physical properties of the solid surfaces and liquid medium. At small adsorbed amounts, such species can promote flocculation either via surface charge neutralization or bridging mechanisms. At higher adsorbed amounts, particle stability increases because of long-range repulsive forces resulting from electrosteric interactions. For anionic polyelectrolytes, the degree of ionization increases with increasing pH (see *Figure 5.7*).

At low pH, polymers adsorb in a dense layer of large mass with low adlayer thickness (d). In contrast, when fully ionized, anionic polyelectrolytes adopt an open coil configuration in solution because of intersegment repulsion. These highly charged species would adsorb in an open layer of low Γ_{ads} and high δ. The plane of charge ($\sigma 0$) and the steric interaction length (∂) are of critical importance. The double layer, vdW, and steric forces all originate at the polyelectrolyte-solution interface recently conducted, the first assuming that double layer, and vdW forces originate at the solid-polyelectrolyte interface and steric forces originate at the polyelectrolyte-solution interface. As an example, zirconia surfaces with adsorbed poly (acrylic acid), where the plane of charge is often located at some intermediate distance between the solid-polyelectrolyte and polyelectrolyte-solution interfaces. At low pH, the steric interaction length and calculated plane of charge (estimated from the normalized force versus separation distance curves) were coincident and occurred 1 nm away from the bare particle surfaces. As pH increased, there's a dramatic increase in the steric interaction length, with almost a tenfold increase ($d \sim 10\,nm$) at pH 9.

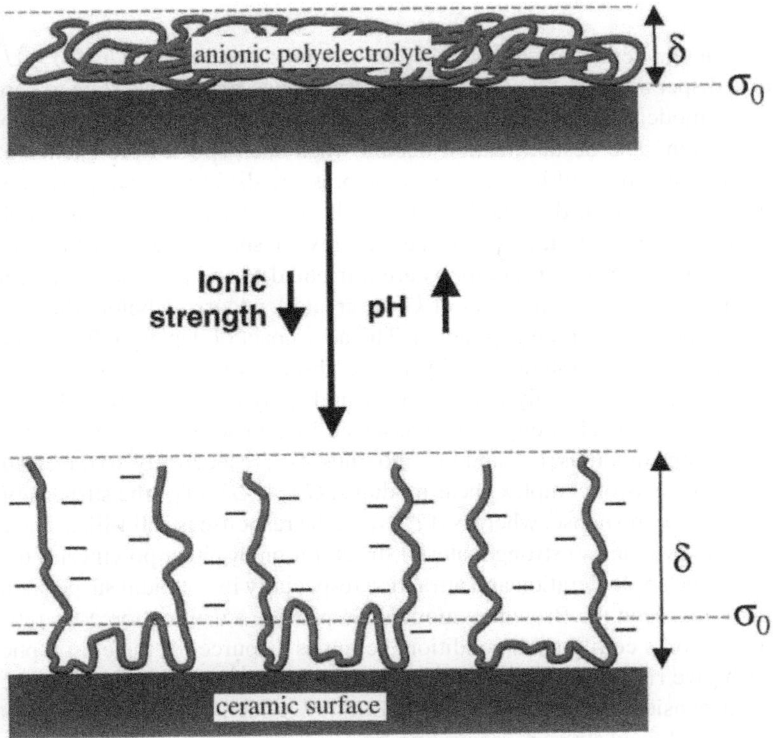

FIGURE 5.7 The schematic surface adsorption layer formation on ceramic particles for an anionic polyelectrolyte polymers.

The depletion forces which occur between large colloidal particles suspended in a solution of nonadsorbing, smaller species (e.g., polymers, polyelectrolytes, or fine colloidal particles) have severe effects on suspension stability. The depletants species may promote flocculation or stabilization of primary colloidal particles. The depletion effect denotes the existence of a negative depletant concentration gradient near primary particle surfaces. The concentration of rigid depletant species decreases at bare particle surfaces and increases to its bulk media value at some distance away from these surfaces. This distance, known as the depletion layer thickness, is of the order of the depletant diameter ($2a_{dep}$). The characteristic length scale of such interactions, which can be several nanometers or greater, depends on the effective depletant diameter. For uncharged depletant species, V_{dep} is given by:

$$V_{dep}(\lambda) = 0 \quad for\ h > 2a_{dep} \tag{5.6}$$

$$V_{dep}(\lambda) = \frac{a\phi_{dep}^2 kT}{10a_{dep}}\left(12 - 45\lambda + 60\lambda^2 - 30\lambda^3 + 3\lambda^5\right) \quad for\ 4a_{dep} > h \geq 2a_{dep}$$

$$V_{dep}(\lambda) = -\frac{3\phi_{dep} kT}{2a_{dep}} + \frac{a\phi_{dep}^2 kT}{10a_{dep}}\left(12 - 45\lambda - 60\lambda^2\right) \quad for\ h < a_{dep}$$

Where ϕ_{dep} is the depletant volume fraction in solution and $\lambda = 15\left(h - 2a_{dep}\right)/2a_{dep}$. The stable dispersions are known to undergo transitions from stable state to depletion flocculation mode, and finally the depletion restabilization, with increasing depletant volume fraction. The destabilization occurs when such species are excluded from the interparticle gap, resulting in an osmotic pressure difference that promotes flocculation. The effects on the stability of weakly flocculated, concentrated colloidal suspensions may include improving the stability of such systems with increasing depletant additions. Such observations are attributed to the presence of a repulsive barrier (estimated to be of the order of kT or greater), occurring before the exclusion of depletant species from the gap region. The net impact of depletion forces depends strongly on the initial system stability in the absence of such species. In ceramic suspensions, especially highly loaded ones can display a viscoelastic behavior. The quantification of viscoelasticity can be assessed by an oscillatory rheometer, which measures storage modulus, G', and loss modulus, G'', respectively, to the elastic and viscous components of complex shear modulus, $G*$. If $G'' > G'$, the suspension provides a liquid-like response, whereas if $G' > G''$, the response is solid-like. The elastic response is an evidence of strong material structuration by either polymeric entanglement or particle agglomeration and attraction (especially in colloidal suspensions). At low shear rates, even the Brownian motion can provide a driving force for returning the structure to its equilibrium condition, acting as a source for the elastic phenomena, which give rise to storage moduli in the oscillatory flow. In this situation, most ceramic suspensions follow the Herschel-Bulkley model, which relates stress τ to shear rate $\dot{\gamma}$ and yield stress τ_y, according to:

$$\tau = \tau_y + k\dot{\gamma}^n \tag{5.7}$$

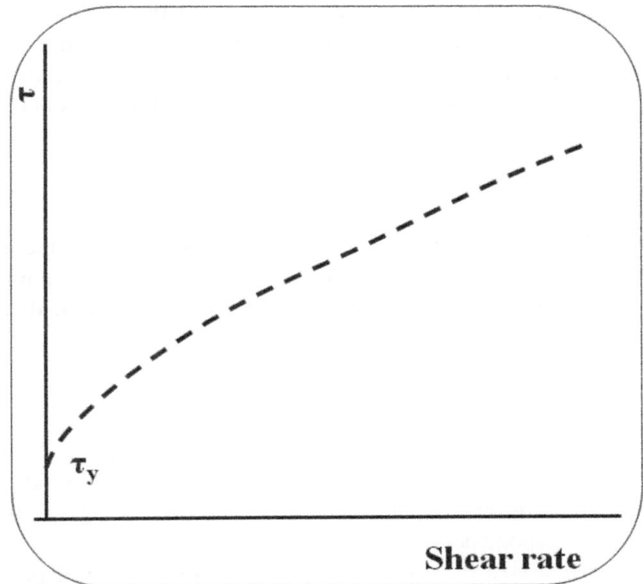

FIGURE 5.8 The schematic representation of a typical ceramic suspension.

The adjustable parameter K is called flow consistency index, and exponent n is the flow behavior index, representing a deviation from a Newtonian behavior (n =1). If n is lower than 1, it denotes a shear-thinning behavior; however, if it is higher than 1, it represents a shear-thickening behavior. (See *Figure 5.8*.)

It can be concluded that the ceramic suspension stability is closely related to rheology. The two main mechanisms that can negatively affect the stability are sedimentation due to gravity and flocculation due to interparticle attraction. Therefore, the terminal settling velocity of the particles is a result of the equilibrium of viscous and gravitational forces. According to Stokes's sedimentation law, which includes only the effects of gravitational and viscous forces, the particle terminal velocity can be evaluated for ceramic suspensions. The settling speed equation is valid for very diluted suspensions ($\phi < 1\,vol\%$). By including the solid fraction effect and validation of the equation for particles larger than 100 μm:

$$V_{settling} = V_{stokes}(1-\phi)^{4.65} = \frac{d^2\left(\rho_{solid} - \rho_{fluid}\right)g}{18\eta_{fluid}}(1-\phi)^{4.65} \tag{5.8}$$

The factors can reduce sedimentation in ceramic suspension: reducing particle size d, reducing density difference ($\rho_{solid} - \rho_{fluid}$), increasing the fluid viscosity η_{fluid}, and increasing solid fraction ϕ. By increasing solid fraction ϕ, at low concentrations, the average interparticle distance is large, resulting in a minor viscosity increase due to a small hydrodynamic disturbance. As the distance between particles decreases, the interparticle interaction becomes prominent. This fraction can cause severe changing in the critical shear stress and yield stress of the ceramic suspension as solid loading is inversely related to the suspension viscosity, finding an optimum compromise by considering buoyancy, hydrodynamic flow, Brownian motion, van der Waals interaction, and dispersion mechanisms. The equation can be written as:

$$\eta_r = \frac{\eta_s}{\eta_0} = (1 - \phi / \phi_m)^{-B\phi_m} \tag{5.9}$$

Where η_r is the relative viscosity, η_s is the suspension viscosity, η_0 is the medium viscosity, ϕ is the percentual volumetric solid loading, ϕ_m is the maximum volumetric solid loading (maximum packing), and B is Einstein's coefficient (also called intrinsic viscosity).

$$\eta_r = \frac{\eta_s}{\eta_0} = [a(\phi_m - \phi)]^{-n} \tag{5.10}$$

Where n (in general, n=2 for ceramic suspensions) thus adds another degree of freedom. The ϕ_m depends on packing fraction with various arrangements and can be estimated based on monodisperse sphere models (see *Table 5.3*).

By reducing particle size, as ϕ_m and B are dependent on particle shape and size distribution, the fine powders result in higher viscosity than coarser ones and are more prone to agglomeration due to their large specific areas. As the distance between particles decreases, the interparticle interaction becomes prominent. Notably, if the particles are smaller than 1 μm, the colloidal forces caused by Brownian motion and van der Waals dominate the suspension behavior. In dilute suspensions, the size distribution has no apparent effect as the average interparticle distance is significant. In high solid fraction suspensions, the viscosity increases toward its maximum packing and the particle size distribution effects are significant. The large particle size ratio enables smaller particles to flow in the interstices of the larger ones [12]. (See *Figure 5.9*.)

The suspension viscosity decreases significantly as the median particle size increases. For a given solids loading, suspensions containing larger particles will have a smaller amount of ceramic surface area in contact with the liquid. This leads to reduced influence from van der Waals and electrical double layer forces within

TABLE 5.3

The estimated ϕ_m values for different arrangement in suspensions.

Arrangement	Maximum packing fraction
Simple cubic	0.52
Minimum thermodynamically stable configuration	0.548
Hexagonally packed sheets with touching	0.605
Random close packing	0.637
Body-centered cubic packing	0.68
Face-centered cubic/hexagonal close-packed	0.74

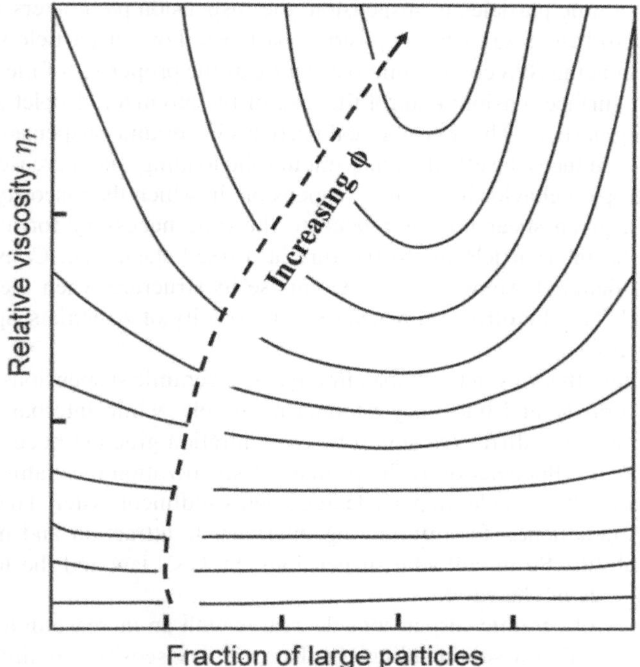

FIGURE 5.9 The schematic representation of viscosity dependency to solids content in different fractions of larger particles of a ceramic suspension.

the suspension. Reduced forces give the suspension a lower viscosity. The suspensions with a broader particle size distribution may show higher fluidity. The bimodal suspension of nanometric and micrometric particles directly changes the flowability of the ceramic suspension. A bimodal suspension with lower viscosity is preferable for SPS applications. The bimodal suspension creates less shear-thickening behavior than monomodal suspensions. Also, an optimal particle size distribution can be estimated, resulting in lower viscosity and maximum packing [4].

For multicomponent ceramic systems and beside particle size distribution, it may be desirable to work in a pH range where opposite charges are induced on different colloidal phases. This approach, termed heteroflocculation, prevents unwanted phase segregation from occurring during processing. For instance, in alumina-zirconia suspensions, when processed in a pH range below the respective IEPs of both phases (e.g., pH 2.5), the dispersed system underwent a dramatic phase segregation during centrifugal consolidation [13]. However, at an intermediate pH (~7) between their respective IEPs, aggregation led to an intimate mixture of these phases that was not disrupted during consolidation. The presence of nanoparticles not only increases the stability of a suspension made of coarse particles but also increases the viscosity of the fluid surrounding the coarser particles and reduces the settling velocity. It is worth mentioning that finer scale structures do not necessarily result from smaller median size ceramic particles in suspension; the suspension parameters and plasma conditions control the size of the depositing particles. Powder particle size is only indirectly influencing SPS coating microstructure as the properties of the suspension (viscosity and surface tension) control the size of the atomized droplet and thus of final depositing particle. The nonspherical particles in ceramic suspensions increase the aspect ratio, reduces the effective maximum solid loading, and increases viscosity.

The thixotropic behavior in ceramic suspension, in which the viscosity decreases over time at a given shear rate, is related to the time necessary for a full disaggregation of the interparticle microstructure at a fixed shear rate. Conversely, the structure also demands some time to recompose its structure when the shear rate is reduced [14]. The thixotropic flow curve and viscosity of zirconia suspension are shown *Figure 5.10*.

Based on these theories in the stable, fine-particle ceramic suspensions, ss particle size reduces, gravity and buoyancy become negligible, while interparticle forces, Brownian motion, and diffusion caused by concentration gradient become relevant. For small particles, flocculation is the primary destabilization mechanism, enhancing sedimentation because large particle aggregates sediment faster. Therefore, dispersants are fundamental for minimizing interparticle attraction and maintaining suspension stability. So in colloidal suspensions, Stokes's law and the Richardson-Zaki equation may not be valid.

The viscosity of ceramic suspensions decreases with an increase in temperature. For suspensions with low solid loading, the decrease in viscosity is mainly related to the decrease in the viscosity of the medium, which is proportional to the exponential function of the inverse of temperature, following an Arrhenius relation:

$$\eta = Aexp\left(\frac{E_a}{RT}\right) \tag{5.11}$$

$$\tau_r = \frac{n\dot{\gamma}a^3}{k_b T} \tag{5.12}$$

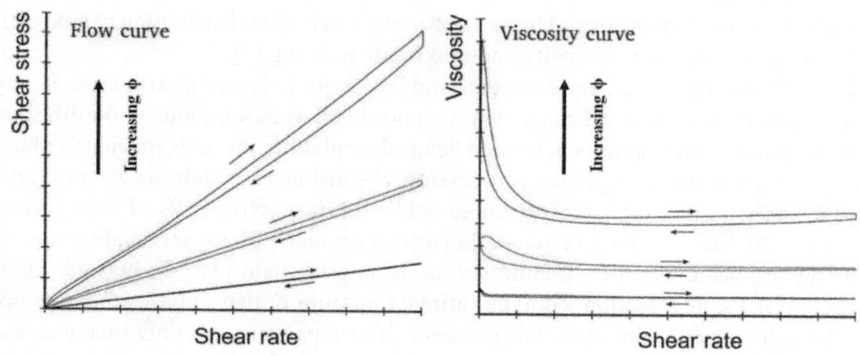

FIGURE 5.10 The schematic representation of shear rate dependency of a typical oxide ceramic suspension in different solids contents.

$$\frac{\eta - \eta_{(\dot{\gamma}\to\infty)}}{\eta_{(\dot{\gamma}\to 0)} - \eta_{(\dot{\gamma}\to\infty)}} = \frac{1}{\left(1 + \frac{\tau_r}{\tau_c}\right)} \tag{5.13}$$

Where E_a is the apparent activation energy, T is the absolute temperature, R is the universal gas constant, and A is a proportionality constant. The dimensionless shear stress τ_r is to compare suspensions with different temperature T and particle radius a. The suspension viscosity as a function of dimensionless shear stress can be plotted in a single curve with two Newtonian plateaus where b is a dimensionless, adjustable constant. Due to the high sensibility of the relative viscosity to the solid fraction for highly loaded suspensions, the thermal dilution effect is also significant. The thermal expansion of the liquid medium (monomers and diluents) is considerably greater than that of the ceramic particles, thus reducing the apparent volume fraction of the solid filler and decreasing viscosity.

Therefore, the increase in temperature decreases the viscosity of the suspension, changing its rheological behavior as the critical shear increases with temperature, shifting the onset of the shear-thickening behavior to higher shear rates. Increasing the suspension temperature seems to be an alternative for obtaining highly loaded suspensions with lower viscosity, although it would require printers with temperature control and resins with no volatile organic components.

The agglomerations also reduce the fluidity of the suspension and restrict the solid-phase content. Tight aggregates often exist in ceramic raw materials and are formed due to the chemical bonding of powder particles, while soft aggregates are formed by the van der Waals force between particles. The soft aggregates are usually decomposed in the preparation process of suspension because in most liquid mediums the attraction between the powder particles in water is far weaker than that in air. For example, the Hamaker constant of alumina is $3.67\times10^{-20} J$ in water, but $15.2\times10^{-20} J$ in air. The van der Waals attraction potential between the particles is directly proportional to the Hamaker constant. The soft aggregates in the suspension can be further eliminated by adding a dispersant and adjusting the pH values.

5.2 CERAMIC SUSPENSIONS AND SPS

The application of ceramic (oxide/non-oxide) powder (micrometer-/nanosized) aqueous and nonaqueous suspensions in SPS involves the transformation of heterogeneous monolithic bulk to flying droplets in plasma stream. Once droplets have reached their minimum size, d_m, the suspension media is vaporized. The particle vaporization is calculated with the help of the energy balance equations:

$$\frac{d(d_m)}{dt} = -\frac{Q}{\pi d^2 \rho_h L_v^l} \tag{5.14}$$

$$Q = \frac{\pi d_m^2}{4} h(T - T_e) \tag{5.15}$$

Where Q is the plasma heat transferred by conduction-convection. The plasma heat transfer coefficient h $(W / m^2 K)$ is calculated with the help of the Nusselt number, of course modified to account for temperature gradients within the boundary layer surrounding the droplet, the Knudsen effect, and the vapor thermal buffer. Plasma specific mass is kg / m^3 (see *Figure 5.11*).

The momentum of a droplet is related to its size by the third-power momentum dependence of droplet size:

$$Momentum = (\frac{4}{3}\pi(\left(\frac{D}{2}\right)^3))\rho_s v \tag{5.16}$$

Here, D is the droplet size after secondary atomization, ρ_s is the suspension density, and v is the mean droplet velocity.

In order to achieve necessary momentum, a critical solid volume fraction between the shear-thinning and shear-thickening behavior must be considered in an aqueous/nonaqueous oxide suspension. The solid volume fraction (SVF) should be optimal for the proper suspension. The solids volume loading in the suspension depended on high valence counterions and ion conductivity constants. While the micromechanisms of agglomerations' generation in ceramic suspensions activate, the plasma stream energy causes dispersion upon contact. The plasma plume temperature is reduced slightly with the injection of the liquid, and the thermal conductivity of the

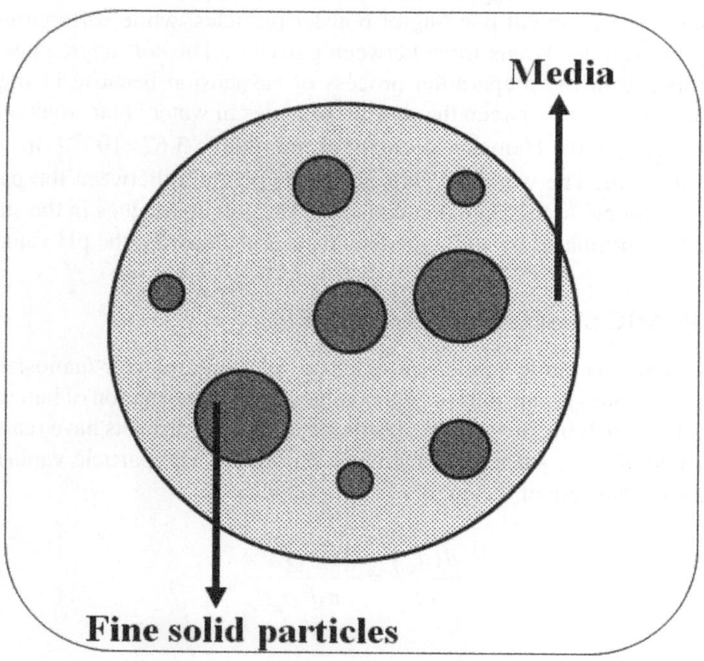

FIGURE 5.11 The schematic representation of a droplet in SPS process.

plasma is enhanced by ethanol or water media due to their dissociation, thus allowing better heat transfer to the particles in flight. It can be understood that water-based suspensions require approximately 3.2× greater energy to vaporize than organic media, such as ethanol. The larger water-based suspension droplets may require longer flight time in the plasma before the liquid is evaporated and the ceramic particle can be melted.

The sustaining of suspension requires control of chemical properties such as pH, which directly controls the ζ potential. The pH dependence of alumina and d zirconia micrometer suspensions are shown in *Figure 5.12*.

The isoelectric points (pH_{iep}) are about 5.8 and 4.9 for alumina and zirconia, respectively. At low pH, far from the pH_{iep} of either powder, only the alumina particles have a high positive zeta potential and are colloidally stable. In the vicinity of the pH_{iep}, the particles have a low zeta potential that may be either positive ($pH < pH_{iep}$) or negative ($pH > pH_{iep}$); suspensions prepared within this region are colloidally unstable and consist of large agglomerates. At high pH, far from the pH_{iep} values of both powders, the particles have a high negative zeta potential and are colloidally stable. The saturation zeta potential and the breadth of the *pH* range over which the potential is nearly maximized provide better colloidal stability at high *pH* for both powders. The viscosity of the slurries decreases gradually as the time of preparation (milling/stirring) increases. The absorption of the dispersant on the particles did not reach equilibrium, and the suspension was unstable until that time. *(See Figure 5.13.)*

Most oxide ceramic suspensions with high solids content (40–55 vol% solid loading) exhibit a shear-thinning behavior and relatively low viscosity (of less than 1 *Pa.s* at the shear rate of 10 s^{-1}). Higher concentration near 53–55 *vol%* suspensions possesses a thixotropy hysteresis with yield stresses. This phenomenon is associated with the shear-thinning behavior of the suspensions and indicates a flocculated state

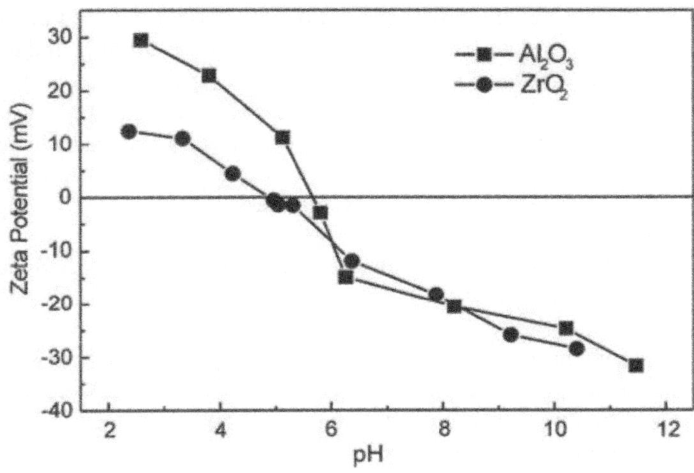

FIGURE 5.12 The representation of pH dependence of alumina and zirconia micrometer suspensions.

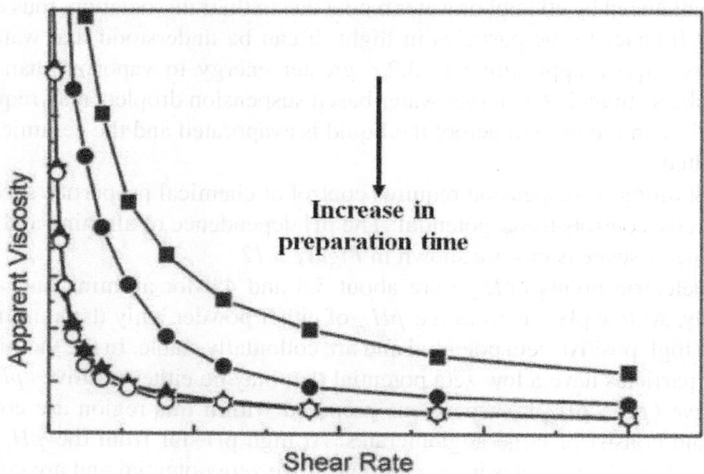

FIGURE 5.13 The representation of rheological properties dependency to preparation time.

of particles within the liquid. The decrease in thixotropy, viscosity, and degree of shear thinning with the decrease of the solid loading of a suspension implies that the degree of powder agglomerate decreases. The thixotropy disappears when the solid loading of the suspension is less than 50 vol%. The concentrated colloidally stable suspensions exhibit shear-thinning behavior in steady shear because of a perturbation of the suspension structure by shear. At low shear rates, the suspension structure is close to equilibrium because thermal motion dominates over the viscous forces. At higher shear rates, the viscous forces affect the suspension structure, and thus shear thinning occurs. At very high shear rates, the viscous forces dominate and the viscosity plateau measures the resistance to flow of a suspension with a completely hydrodynamically controlled structure. The degree of shear thinning and the viscosity at high shear rates increase with an increase in the volume fraction of solid. The properties of the suspension change drastically at a certain critical particle concentration, Φ_g, which corresponds to the formation of a space-filling particle network:

- At $\Phi < \Phi_g$, the suspensions have no yield stress.
- At $\Phi = \Phi_g$, the suspensions show yield stress.
- At $\Phi > \Phi_g$, the suspensions can sustain a stress before yielding and the elasticity may be significant.

The high solids content results in the average separation distance between particles in suspension becoming shorter and forms a space-filling particle network, making flow more difficult. When a stress larger than the yield stress is applied, the networked structure is broken into smaller units (flocs), which then move past each other. If the floc attrition is affected by the strength of the hydrodynamic and attractive forces, pseudoplastic behavior prevails and viscosity decreases with shear rates.

The strong shear forces at high shear cause the flow units to be smaller, and thereby the flow is facilitated. The destruction of flocs releases a constrained liquid, which results in a decrease of the effective volume fraction of the flocs. This phenomenon results in thixotropic behavior in the system.

It is obvious that the yield stress of the suspensions increases with increasing solid volume fraction; high concentration suspension shows obvious shear-thickening behavior. Since shear thinning is the result of breakage of particle aggregates in suspensions, a higher yield stress of the oxide suspension corresponds to a higher strength of the particle aggregates in it. When sheared, particle aggregates in suspension could not be broken entirely before shear thickening occurs, which indicates the formation of hydrodynamic aggregates. Therefore, the microstructures suspension is far from homogeneous when compared to that of suspensions with less solid volume fraction.

The agglomerations in suspension focuses on two aspects: (I) the effects of reunion on the rheological characteristics of suspension and (II) the dispersibility of the powder, which can be effectively improved by using dispersing agents and by other methods. The agglomeration-generation mechanism in an alumina ceramic suspension can be summarized as follows. When the solid-phase loadings are less than $\phi 0$, a loose agglomeration is produced. When the solid-phase content is higher than $\phi 0$, a tight agglomeration is formed. When the solid-phase content is equal to $\phi 0$, the particles in suspension take on a uniform and stable dispersion, but no reunions appear. If the solid-phase content is too low in a ceramic suspension used for colloidal molding, it causes not only a heavy shrinkage but also an uneven shrinkage of the ware during latter heat treatment.

According to rheological theory, macrorheological behaviors of a suspension depend on the microinteraction forces between particles. In an aqueous medium, the interaction forces between particles usually include the van der Waals attractive force and the electrostatic repulsive force, which depends on ionic valence number and concentration that affect the thickness of the electrical double layer and the critical coagulation concentration.

The *IEPs* of oxide increases with hydration time. When hydration time reaches certain periods, the *IEP* of oxide becomes a constant. This is because a metal monohydrate layer on oxide particle is formed in an aqueous medium. This hydration process is due to the thermodynamical instability of oxide in water. The *IEP* of oxide increases with the hydration time. When the hydration process reaches equilibrium, the *IEP* of the oxide gradually becomes stable.

Therefore, the ion concentration in the suspension is a controlling parameter. The ion concentration (n_t) in the suspension consists of ion concentrations (n_m) of suspension mediums and the soluble ion concentration (n_p) in the ceramic powders, as follows:

$$n_t = n_m + n_p = \left(n_{m1} + \cdots + n_{mi}\right) + \left(n_{p1} + \cdots + n_{pi}\right) \tag{5.17}$$

Where n_{mi} is the ion concentration of the *ith* suspension medium and n_{pi} is the soluble ion concentration of the *ith* ceramic powder.

The presence of *high-valence counterions* can also indicate suspension properties. The counterions are the reverse-charged ions to particle charge. The highly valence counterions mean counterions whose charge number is larger than or equal to 2. The critical coagulation concentration depends inversely on the sixth power of the charge numbers (ze) according to:

$$n_c = \frac{107\varepsilon^3 (kT)^5 \gamma^4}{A^2 (ze)^6} \tag{5.18}$$

Where ε is a dielectric constant, k is the Boltzmann constant, T is the absolute temperature, γ is an interfacial energy, A is a Hamaker constant, z is an electrovalent number, and e is an electrical charge. If there were soluble high-valence counterions in the ceramic powder, they would increase with an increase in the solids content of the suspension, and the critical coagulation concentration would drop dramatically. In other words, the van der Waals attractive potential would increase and the electrostatic repulsion between the particles would reduce, and to some extent, the particles would occur to agglomerate and the viscosity of the suspension would rise rapidly. Thus, a highly concentrated suspension cannot be prepared. If high-valence counterions were removed from the ceramic suspension, the critical coagulation concentration would not drop rapidly with increase in the volume loading. Thus, there is a possibility of preparing a concentrated suspension.

The high concentration of extra ions requires a special attention to *ion conductivity constants* and their effect on suspension. With increase in the volume loading, the ion concentration in the suspension would increase. The ion conductivities increase linearly with solids volume loading of seven kinds of alumina powders containing different ion concentrations. For a concentrated suspension, these lines have a greater deviation. This is because the free ions in the suspension are constrained by the overlapping of the electrical double layers around the particles. The inclination slopes of these lines were given according to:

$$K = \frac{\Delta C}{\Delta \phi_V} \tag{5.19}$$

Where K is a constant, whose unit is *ms.cm*$^{-1}$ *(see Figure 5.14)*.

The K parameter is defined as the ion conductivity constant. For a given ceramic powder, because the ion concentration of the suspension medium does not change with solids volume loading, the ion conductivity constant never varies even though the suspension medium used is different. For a fixed starting material with the same particle size, distribution, shape, specific surface area, etc., if there were different ion concentrations, the ion conductivity constants would be different. Thus, it can be used to characterize the soluble ion concentration in ceramic powders. It is a very important parameter related to the successful preparation of a concentrated suspension.

There exists a relationship between ion conductivity constants and solid volume loading. After removing high-valence counterions, with the increase in volume loading, the ion concentration in the suspension would not immediately cause the

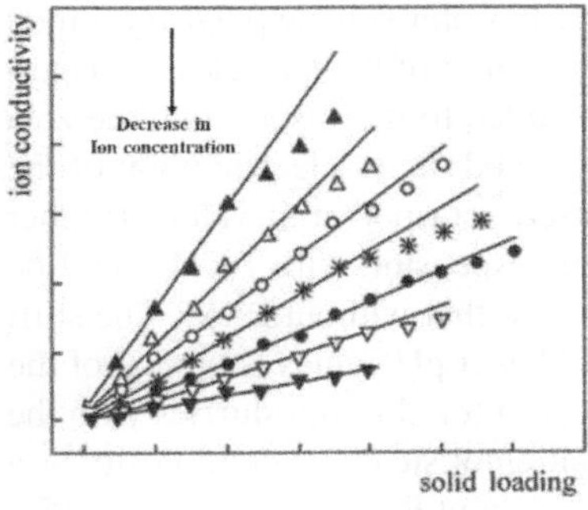

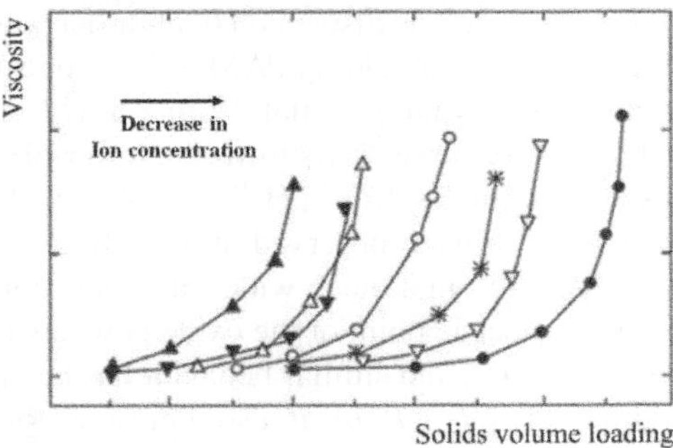

FIGURE 5.14 The representation of ion conductivity dependency to solid loading (left) and viscosity (right) for different ion concentrations.

reduction of critical coagulation concentration. If the ion concentration in the suspension is very *low*, the electrical double layer would be very thick, resulting in the increase of the viscosity of the suspension due to overlapping between the electrical double layers. If the ion concentration is very *high*, the electrical double layer would be very thin, resulting in particle agglomeration due to the increase of attractive potential and reduction of electrostatic repulsion. The two previous situations cannot lead to the successful preparation of a highly concentrated suspension with low viscosity. Only with a suitable ion concentration leading to a thin electrical double layer that would maintain enough repulsion to disperse particles can a highly concentrated suspension with low viscosity be achieved. As seen in *Figure 5.14 (right)*, at an acidic pH, oxide powders suspensions with different ion conductivity constants used to prepare aqueous suspensions with varied solids volume loading, the following statements can be made:

(1) *IEPs* of oxide are a function of hydration time of the particle surface. In an acidic medium, *IEPs* of oxide changes gradually from near neutral to basic and then stabilizes after enough hydration time.
(2) Ion conductivities increase linearly with solid volume loading of suspensions. Its slopes are defined as ion conductivity constant, which is an important parameter to characterize the quality of ceramic powder directly related to the preparation of concentrated suspension.
(3) At acidic conditions, the particle surface of alumina is charged positively; a highly concentrated suspension cannot be prepared due to the presence of high-valence counterion.
(4) When the ion conductivity constant (K) reaches a certain value, a concentrated suspension of high solids content with a moderate can be successfully prepared.
(5) A novel way to prepare a highly concentrated suspension of ceramics is to control the ion conductivity constants suitably to reduce the thickness of the electrical double layer and to maintain repulsion between particles, after removing high-valence counterions.

For instance, the dispersion of alumina powder with water-soluble polyacrylamide (PAM) shows that adding PAM affects the surface charge of the particles in dispersion and their zeta potential. With 2.8 *wt.%* PAM added to the suspension, the zeta potential–pH curve shifts toward a lower-pH region, and the isoelectric point of the suspension moves from pH 7.4 to pH 3.6. Increased zeta potential values (higher than 30 *mV*) were observed at $pH > 10.5$, and the suspension with PAM could be well dispersed in a much wider pH range compared to that without PAM. The shift of the isoelectric point of the oxide powders toward lower pH values is typical of the PAM function, and similar behavior has been observed for alumina slurries with the addition of $NH_4 - PMA$. It also represents the variation of steady-state viscosity as a function of shear rate for the suspensions with and without PAM. *(See Figure 5.15.)*

It can be concluded that the suspension without the addition of PAM displays relatively strong shear-thinning behavior. The degree of shear thinning decreases significantly after the addition of PAM. Using PAM, highly concentrated suspensions can be prepared, but their fluidity is likely to be compromised. The concentrated

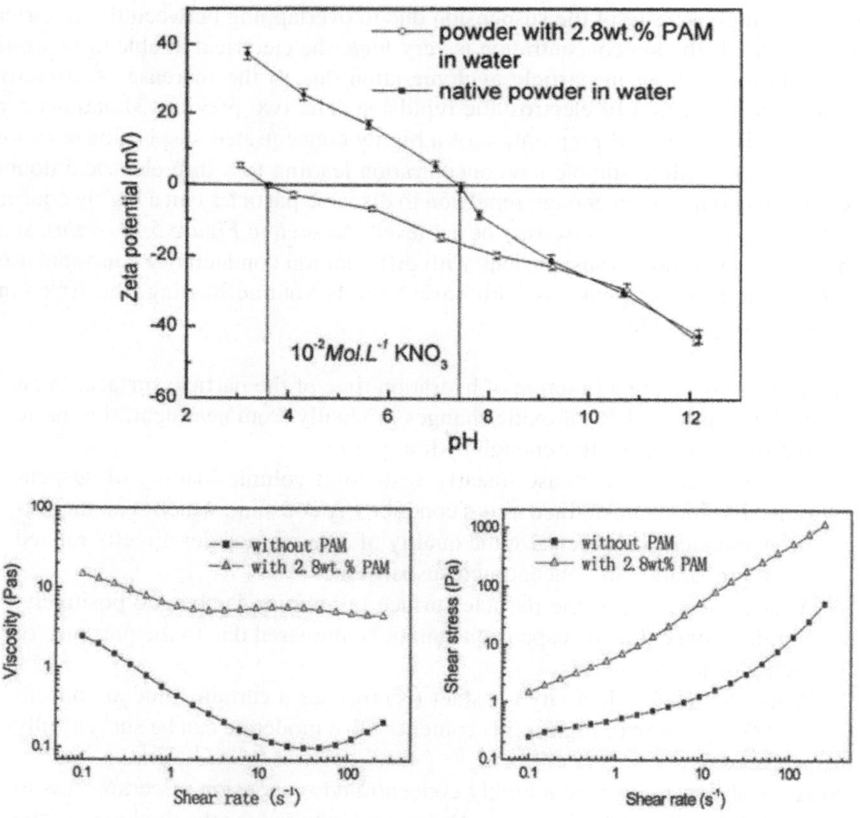

FIGURE 5.15 The representation of zeta potential dependency to pH for PAM-added suspension (up) with the effects of adding PAM (down) for viscosity and shear stress.

stable suspensions usually exhibit shear-thinning behavior due to perturbation of the suspension structure by shearing, and the suspension with the addition of PAM is the shear thickening (dilatancy). At high shear rates (depending on solid loading), the viscosity increases as the shear rate increases. The shear thickening is a consequence of an order-to-disorder transition of the particle microstructure. Also, the corresponding variation of shear stress as a function of shear rate for the suspensions with and without PAM is detectable. The suspension with the addition of PAM exhibits a higher yield stress τ_y compared to the suspension without PAM, which is in a good conformity with the Herschel-Bulkley model. The suspension yield stress (τ_y) scales with the solids concentration (ϕ):

$$\tau_y \sim \left(1-1.5\alpha\zeta^2\right)\left(\frac{A}{24S_0^{\frac{3}{2}}}\right)\left(\frac{1}{R^{d-\frac{3}{2}}}\right)\phi^m \tag{5.20}$$

Where α is a constant related to the Debye thickness (k^{-1}) and the surface separation (S_0) between the particles, ζ the zeta potential, A the Hamaker constant, R the particles radius, d the Euclidean dimension, and $m=\left(d+X\right)/\left(dD_f\right)$, with D_f and X the fractal dimension of the clusters and the backbone of the clusters, respectively. It can also be concluded that the shear stress (τ_y) increases with the decrease in the zeta potential given the solids concentration.

The polyethylene glycol (PEG) in alumina suspension can also control the rheological behavior. The high-purity α alumina powder with micron particle size with improved dispersion of the powder and fluidity of the suspension by an organic base, ammonium citrate, as a dispersant is a common suspension formula. The zeta potential with or without PEG in alumina suspension shows that the zeta potential for alumina powder only ranged from 38 mV at pH 3 to ~ 40 mV at pH 12, with an IEP of pH 7.3, while that in the case of the suspension containing PEG ranged from 24 mV at pH 3 to ~38 mV at pH 12 with an *IEP* of pH 7.1. The *PEG* will not dissociate and has hardly any effect on the alumina surface charges. Shear thinning occurs at low shear rates, and shear thickening occurs at very high shear rates. The suspension without *PEG* will show lower viscosity, and the critical point from shear thinning to shear thickening occurs at higher shear rate, while the suspension with *PEG* exhibits higher viscosity and the critical point from shear thinning to shear thickening occurs at lower shear rate. *(See Figure 5.16.)*

As a result, the concentrated, colloidally stable suspensions display shear thinning because of a perturbation of the suspension structure by shear. At low shear rates, the suspension structure is close to equilibrium because thermal motion dominates over the viscous forces. At higher shear rates, viscous forces affect the suspension structure, and shear thinning occurs. At very high shear rates, viscosity increases as the shear rate increases.

The basic behavior of most commercially used oxide suspension for SPS applications are as follows:

The atomization of a water-based YSZ suspension tends to produce a significantly larger average and a tighter distribution of droplet. This effect is linked to the surface

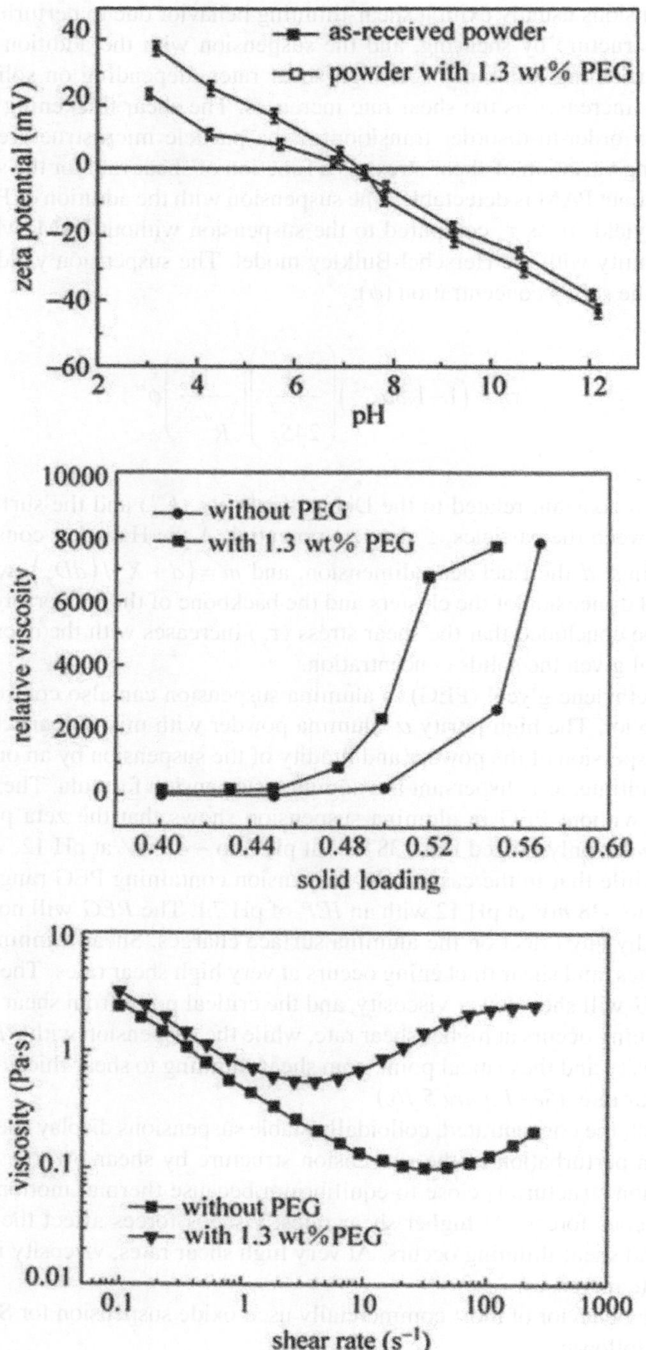

FIGURE 5.16 The representation of zeta potential, viscosity dependency to pH, and solids content/shear rate for with or without PEG additives in suspension.

tension of the suspension media. Water-based suspensions have a greater surface tension than ethanol and thus resist atomization in the plasma stream more strongly [15].

The water-based suspensions injected into plasma tend to produce faster particle speeds with some reduction in particle temperature. The increased speed is attributed to the higher momentum of a larger droplet being more able to penetrate into the central, fastest part of the plume and therefore attain greater acceleration from the plasma.

The aqueous *non-oxide* ceramics suspensions usually include alkali cleaning and calcinations of the aqueous ceramic such as silicon nitride suspension. It is believed that non-oxide ceramics, such as SiC-, Si_3N_4-, and Si_3N_4-bonded SiC, are important and promising advanced ceramics [16]. The application of non-oxide ceramics has still been limited mainly owing to their low reliability and high cost. A typical suspension in these systems involves deionized water, moderate solids loading, and dispersant (i.e., tetramethylammonium hydroxide).

For instance, the aqueous Si_3N_4 involves hydrolysis of Si_3N_4:

$$Si_3N_4 + 6H_2O \Leftrightarrow 3SiO_2 + 4NH_3$$

And hydrolysis of byproducts, such as:

$$Si + 6H_2O \Leftrightarrow H_4SiO_4 + 2H_2$$

The isoelectric point (IEP) of alkali-cleaned powder shows a slight shift to the right, and the zeta potentials of the two powders are close to each other when the pH value is highly basic. The suspension shows pseudoplastic fluids behavior. *(See Figure 5.17.)*

The mutual reaction with Na^+ and consequent ion exchange causes the powders to be deionized by washing with deionized water, which directly controls ζ as an important factor influenced by the update. Electric double layer is the surface charge of the particles. The reason behind changes in surface charge in the $Si_3N_4 - H_2O$ system is the presence of $SiNH^{3+}$, $SiOH^{2+}$, and SiO^- in acidic situation and $Si(OH)_{n+1}^{n-1}$ and $Si(OH)_nO^{(4-n)-}$ in alkaline situation. The location and density of these surface groups change along with the variation of the pH value, which determines the charge density and charge character of the surface of Si_3N_4. Another important factor influencing the electric double layer is the species and concentration of ions in the suspensions, which affects the height of potential barrier and distance of interaction between the particles. The abovementioned two factors completely determine the interaction between the particles. The advantage of ion exchange is that the ion-exchange resin can exchange the high-valence counterions, such as high-valence cations for Na^+. This can greatly decrease the harmfulness of the counterions. The conductivity of the supernatants of the suspensions (μScm^{-1}) will indicate the progress of washing and ion exchange. The deionization decreased the concentration of the ions in the suspension and thus changed the density of the ions at the particle surface through adsorb-resolve equilibrium. As a result, the zeta potential of the suspensions drops after deionization. After dispersant is added to the suspension, the viscosity of the suspension of deionized powder becomes less than one-tenth of that of the

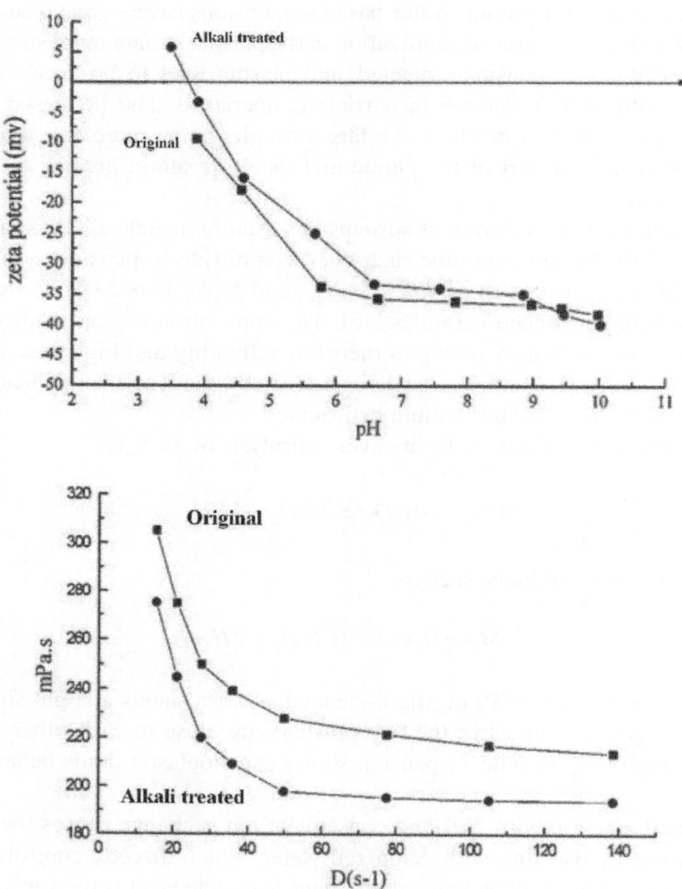

FIGURE 5.17 The representation of zeta potential to pH (left) and viscosity-shear rate (right) for original and treated non-oxide powder in suspension.

unprocessed powder with the same solid volume concentration. This may be owing to the fact that more dispersant molecules are adsorbed and repulsion between the particles is improved when more room is obtained in the Stern layer after deionization. It can be seen that because of the rise in the zeta potential and the consequent increase in the repulsion between the particles, the apparent viscosity of the suspension drops significantly (over 90%). *(See Figure 5.18.)*

It can also be seen that the pH_{iep} of powder suspensions dropped after deionization. This is owing to the decrease in the density of the counterions cations at the surface of the particles. In the suspensions with relatively low solid volume concentration, the distance between the particles is large when compared with the particle size, and enlargement of the radius of the slide plane may not lead to evident increase in the resistance of the particles to rearrange and slide against each other. Moreover,

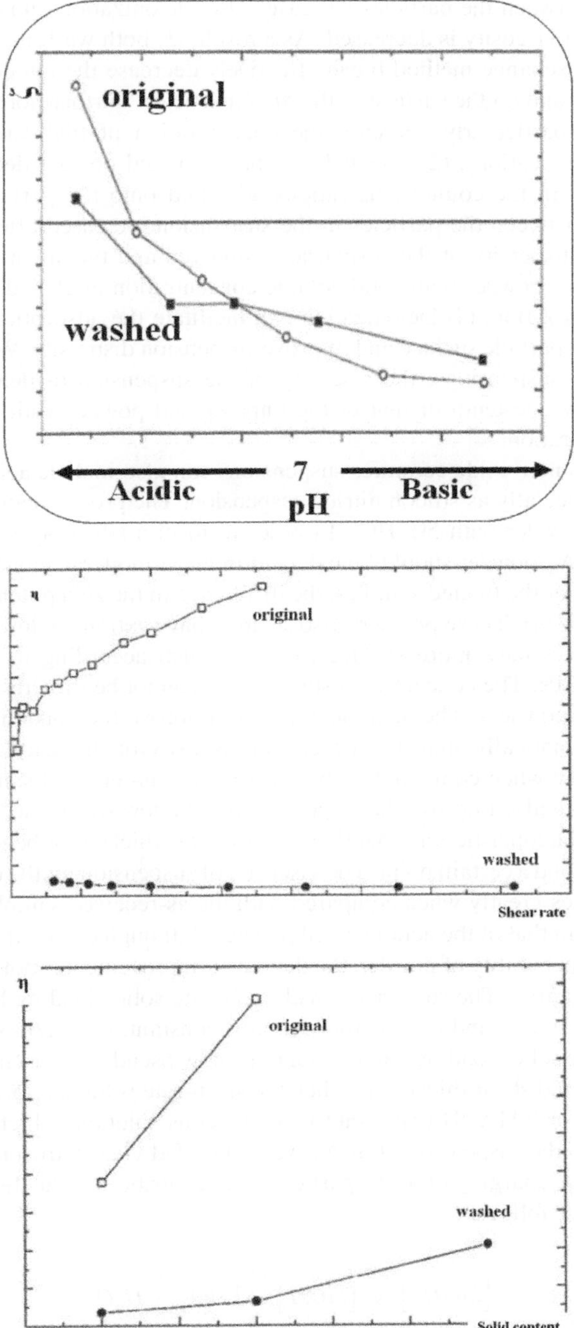

FIGURE 5.18 The representation of zeta potential dependency to pH (left) and changes in viscosity with shear rate (right) and solids content (below) before and after deionization in Si_3N_4 suspension.

the repulsion between the particles increases after deionization; thus, the dispersity is improved and viscosity is decreased. As a result, (1) both washing with deionized water and ion-exchange method b can effectively decrease the ion concentration in the suspensions and on the surface of the Si_3N_4 particles. Furthermore, ion-exchange method b can particularly decrease the concentration of high-valence counterions–high-valence cations. (2) The pH_{iep} of the deionized powder decreased, owing to the decrease in the counterions-cations adsorbed onto the particle surface. (3) The repulsion between the particles in the suspension increased after deionization. Therefore, the dispersity of the suspension improved and the apparent viscosity of the suspension of powder with solid volume concentration of 20% decreased below 10% after deionization. (4) Deionization can facilitate the adsorption of dispersant molecule on the particle surface and improve suspension dispersity. When dispersant is added to the suspensions, the viscosity of the suspension of deionized powder decreases below one-tenth of that of the unprocessed powder with the same solid volume concentration.

The aqueous non-oxide ceramics suspensions can also include acid cleaning and calcinations especially as silicon nitride suspension. The process can involve a mixture of Si_3N_4 powder with 5% HCl. In order to form a silica layer on the particle surface, the Si_3N_4 powder should be calcined in air at moderate conditions, such as 600°C for 6 h. For the treated samples, the difference in the zeta potentials is not distinctive at a high pH. These powders also seem to have extremely low *IEPs* (<pH 2). Calcined powders have more surface oxygen, which according to Hackley et al. indicates low IEPs. The concentrated suspensions cannot be directly prepared from the as-received powders. The acid-cleaning can improve the aqueous dispersibility of powders dramatically, and the suspension viscosity of the acid-leached powder decreases greatly when compared with its respective as-received sample. Although acid-leaching can also improve the dispersibility. At a low shear rate, the suspension behaves as a pseudoplastic fluid, while it shows shear-thickening behavior when the shear rate is beyond a certain point. The viscosity of suspension for the direct-calcined powder decreases greatly when compared with the as-received sample, but it is still much higher than that of the acid-cleaned powder. Although calcination can improve the aqueous dispersibility of powder, the degree of improvement is not as pronounced as with acid-cleaning. The suspension with moderate solids loading behaves almost as a Newtonian liquid, and its viscosity remains constant. Furthermore, the suspension with higher solids loading can be found to show pseudoplastic characteristics at low shear rates and shear thickening when the shear rate is higher. *(See Figure 5.19.)*

The TMAH or TMAOH dispersant is an aqueous solution, which is an organic alkali, and thus, the suspension of Si_3N_4 was stabilized via electrostatic repulsion at basic pH. Surface charging of Si_3N_4 particles may contribute to the dissolution of the surface groups as follows:

$$[SiOH_2]^+ \xleftarrow{H^+} [SiOH] \xrightarrow{OH^-} [SiO]^- + H_2O$$

$$[Si_2NH_2]^+ \xleftarrow{H^+} [Si_2NH] \xrightarrow{OH^-} [Si_2N]^- + H_2O$$

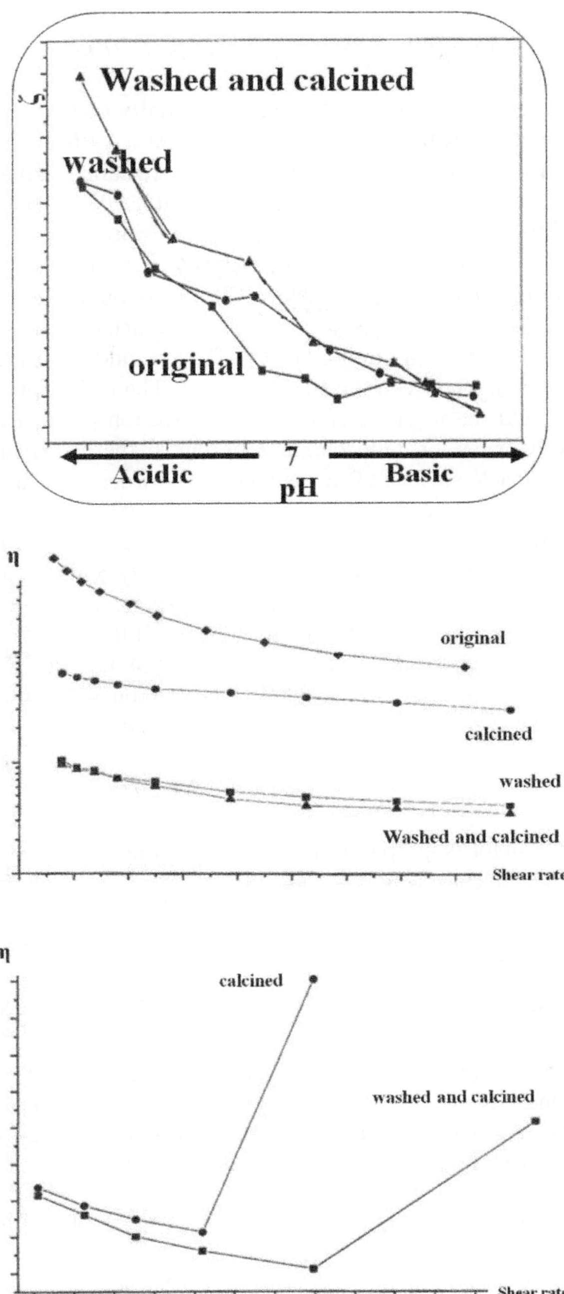

FIGURE 5.19 The representation of zeta potential dependency to pH (left) and changes in viscosity with shear rate (right) and the effect of calcination (below) before and after deionization in Si_3N_4 suspension.

$$[SiNH_3]^+ \overset{H^+}{\leftarrow} [SiNH_2] \overset{OH^-}{\rightarrow} [SiNH]^- + H_2O$$

As seen, these ions are pH-dependent and essentially equivalent to the specific adsorption of protons or hydroxyls. The silanol group, $Si-OH$, is favorable for surface charging. If the silanols are more, then the electrostatic repulsion between the particles becomes greater, with other conditions remaining constant. From the viewpoint of preparing highly concentrated slurries, the substitution of $Si-OH$ for amine structures is beneficial to improve the dispersibility of Si_3N_4 powder. It is difficult to prepare highly concentrated aqueous slurries if the amount of amine structures or other hydrophobic groups is dominant on the particle surface. The stable esters that are hydrophobic with the configuration $Si-O-C-R$ would be formed on the Si_3N_4 surface by the ion of alcohol with the silanol groups. The CH_3 groups of the alcohol are believed to shield the negative charge of the electron pairs at the oxygen ion, which are the active sites for deflocculant coupling. If the hydrophobic groups are dominant on the particle surface of the as-received powder, it is easy to understand the poor aqueous dispersibility.

The nonaqueous non-oxide ceramics suspensions are common in particle. The suspension usually involves deionized water, analytically pure ethanol, and isopropanol and tetramethylammonium hydroxide as a dispersant at highly basic pH. In contact with DM water, the $Si-OH$ surface group, and the siloxane surface group, $Si-O-Si$ will be formed. The surface oxygen content of the as-received Si_3N_4 powders is neglectable, and there are excess of $Si-O-Si$ groups on the particle surface, which could be hydrolyzed gradually in the aqueous solution as follows:

$$Si-O-Si+2H_2O \rightarrow 2Si-OH$$

The effect of ethanol isopropanol includes the amine structures (Si_2-NH and $Si-NH_2$) on the particle surface:

$$Si-O-H+HO-CH_2CH_3 \rightarrow Si-O-CH_2CH_3+H_2O$$

$$Si-OH+HO-CH(CH_3)_2 \leftarrow Si-O-CH(CH_3)_2+H_2O$$

The emergence of the $C-H$ bonding and the disappearance of the isolated $Si-OH$ groups and $-Si-O-Si$ groups do not match with ethanol. The isopropanol results in unstable and easily formed $Si-OH$ groups in aqueous solution, at the amorphous oxygen-rich layer on the Si_3N_4 powder:

$$2Si-O+Si-H+HO-CH-(CH_3)_2 \rightarrow Si-O-Si+Si-O-CH-(CH_3)_2+H_2O$$

There is no ion between the amine structures of dispersant and the liquid media, ethanol, and isopropanol. The apparent viscosity of in $DM-H_2O$ will reduce, while the apparent viscosity of ethanol and M11-isopropanol increases because hydrophobic $Si-O-C-R$ groups are formed on the particle surface of the powders. This hydrophobic surface group deteriorates the aqueous dispersibility. *(See Figure 5.20.)*

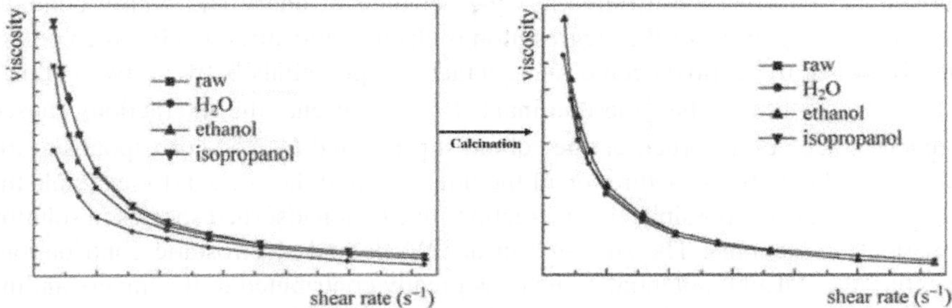

FIGURE 5.20 The representation of viscosity dependency to shear rate before and after calcination in Si_3N_4 suspension.

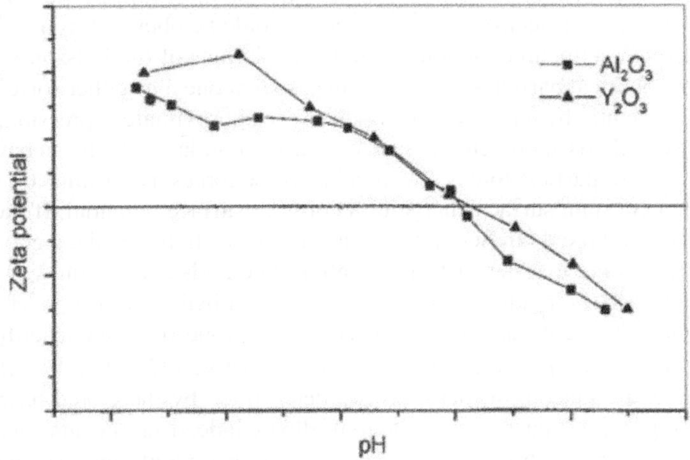

FIGURE 5.21 The representation of zeta potential dependency to pH in binary oxide systems.

The binary oxide systems, such as aqueous medium $Y_2O_3 - Al_2O_3$, show their own specific behavior. The zeta potentials trends show that unlike any curve of the pH-dependent zeta potential for Al_2O_3, Y_2O_3, they show different dependency. *(See Figure 5.21.)*

As most SPS suspension include at least two ceramics, the heteroaggregation in ceramic suspensions is a place of concern. While two types of colloids are dispersed in a suspension, if these colloids acquire opposite charges, this may induce aggregation driven by electrostatic interactions, with subsequent flocculation and precipitation. As mentioned before, the heteroaggregation process is quite complex and depends on a series of parameters, such as solute volume fraction, suspension composition, pH, size ratio between the two types of colloids, etc. This behavior is the main difference between suspensions of unary and binary colloids. The colloid-colloid interactions in suspension differentiate the behavior of unary and binary colloids. The main driving force for this aggregation is electrostatic attraction between unlike colloids. When the colloids come into contact, the potentials between two colloids U_{1-2}^{steric} and $U_{1-2}^{\text{structural}}$ may become dominant. U_{1-2}^{steric} represents the interactions caused by macromolecules adsorbed on the colloid surface, and $U_{1-2}^{\text{structural}}$ corresponds to the interactions due to the structuration of the liquid around the colloid (for example the hydration force) and, possibly, the interactions due to nonadsorbed species in solution (depletion interactions). The sum of van der Waals and electrostatic contributions constitute the (DLVO) potential, which has greatly contributed to the understanding of the behavior of colloidal suspensions. The van der Waals term is generally attractive for the ceramic suspensions considered in this paper. The electrostatic interaction depends on the surface potential induced on the interacting colloidal particles by the formation of charged groups on their surface when they are inserted in the medium. At the same time, the medium causes also a screening of the Coulomb interaction, which is thus exponentially damped and becomes of short-range type. The magnitude of these screened electrostatic interactions should be obtained from the solution of the Poisson-Boltzmann equation. Analytical solutions of the Poisson-Boltzmann equation for spherical particles is not available, so that one may either solve the equation self-consistently by numerical methods or use approximate expressions.

When two colloids approach, the surface charges regulate, which can have several effects on the interparticle forces. To calculate these forces, two limits are generally considered: a constant surface charge or a constant surface potential. If the charges on colloids are not fixed but fluctuate in equilibrium with the media, a good approximation is to assume that the surface potential of colloids is constant. For example, oxide particles (MO) in aqueous suspensions present hydroxyl groups on their surface ($M-OH$). These groups have some acid-base properties, and according to the pH, they can become positive ($M-OH^2+$) or negative ($M-O^-$). The charges of oxide particles in aqueous suspensions are determined by these acid-base equilibria. Assuming a constant surface potential allows indeed taking into account the variation in the surface charges which can occur when colloids approach each other. The approximation of the true surface potential by identifying the surface potential with the zeta potential (ζ) can be carried out by electrostatic interaction potentials. The electrostatic interaction potentials describe the electrostatic interactions derived from the superposition of the diffuse double layers which are around the colloids. These potentials are deduced from the Gouy-Chapman theory, which only deals with the diffuse layer (the Stern layer is not considered), implying that the surface tension has to be interpreted as the values of the diffuse layer potential. The potential that can be accessed experimentally is the zeta potential, which is measured in the diffuse layer on the slip plane. This is not exactly the diffuse layer potential; however, it has been shown that it is a good approximation to it, especially when the ionic strength is small and when the potentials are of small magnitude. In a unary colloidal system, all colloids acquire charges of the same sign so that electrostatic interactions are always repulsive. Their magnitude can be used to counterbalance the effects of van der Waals attraction in such a way that the suspension can become stable. At increasing electrostatic interactions (i.e., at increasing surface potential), the suspension may change its character from strongly flocculating to weakly flocculating and to stable.

In a suspension containing binary colloids (A and B), the situation may be quite different, since electrostatic interactions can be of different signs. The behavior of the suspension thus depends on a complex interplay between the attractive van der Waals interactions, the repulsive A-A and B-B electrostatic interactions, and the A-B interactions, which may be either attractive or repulsive, depending, for example, on pH. As an example, in alumina and silica binary system, for pH larger than neutral, both colloids are negatively charged. All electrostatic interactions are repulsive and contribute to stabilizing the suspension. In the pH-neutral range, both zeta potentials are significantly large and of opposite sign, positive for alumina and negative for silica. In this case, the homogeneous alumina-alumina and silica-silica interactions are repulsive, with negligible secondary minimum, while heterogeneous alumina-silica interactions are attractive because electrostatic and van der Waals contributions are both of the same sign [17]. In this case, the forces between colloids share some resemblance with those between different ions in ionic crystals, like NaCl or CsCl, with the main qualitative difference being that electrostatic interactions in colloids are screened. *(See Figure 5.22.)*

For smaller pH values, electrostatic potentials are still of different signs, but the potential on silica becomes negligible. In the absence of organic additives, the DLVO potential is generally sufficient to describe the colloidal suspension behavior. However, because of the van der Waals interactions, the DLVO potential$\rightarrow -\infty$ for surface-to-surface separation. At a very short distance, the interaction should intuitively be repulsive, because colloids cannot penetrate each other. In an aqueous

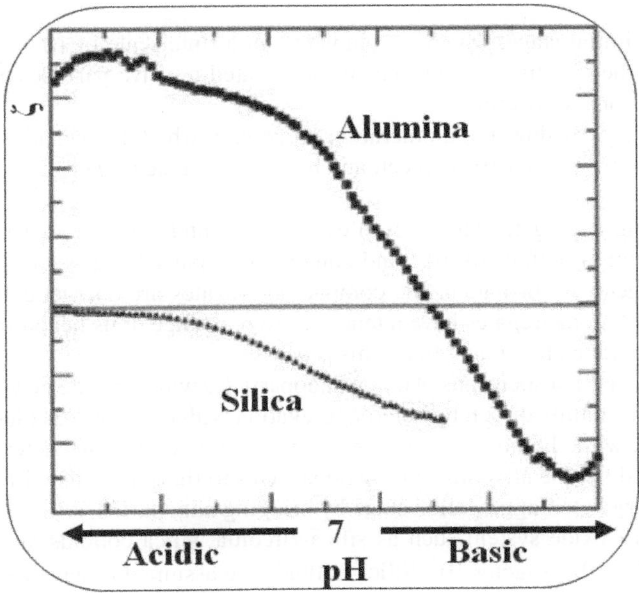

FIGURE 5.22 The zeta potential of alumina and silica in aqueous suspension plotted as a function of pH.

medium, a repulsive hydration force ($U_{1-2}^{\text{structural}}$) can be considered, but in practice, it is quite difficult to estimate it quantitatively from basic principles. This repulsive force phenomenological repulsion is proportional to r^{-m},where m is a positive exponent which is chosen to avoid significant superposition of colloids. The resulting potential well depth at contact, the minimum potential, can be estimated by measuring the equilibrium adsorption of small silica colloids on the surface of large alumina colloids, a quantity which depends on the ratio $u_{min} / (k_B T)$. The best agreement between the results given by the potential and the measured adsorbed quantities was found for $14 k_B T \leq u_{min} \leq 16 k_B T$ at room temperature. Superposition of colloids can be avoided also by imposing different types of constraint. The aggregation processes of the colloid-colloid interactions (heteroaggregation) are dependent on the total volume fraction ϕ_S, the composition R of the suspension, and the size difference between the two types of colloids.

The heterocolloid oxides with moderate- and small-size mismatch can act as either binary and semiunary suspension. In the case of the almost equally sized colloids (radii), th e κ_a will be $\kappa_a >> 1$, i.e., the Debye screening length was much smaller than the particle size. In heterocolloids oxides with large-size mismatch, the small mass ratios of smaller colloids are sufficient to induce the aggregation of larger, oppositely charged colloids. This increase is attributed to the strengthening of the particle-particle interactions as the particle size grew larger. These colloids acquire opposite charges at pH range between 6.5 and 8. The large, positively charged colloids present quite irregular shapes, while the small negative colloids will be spherical to a very good degree of approximation. The Brownian simulations reveal that the aggregation process consists of two steps:

(a) A fast initial stage, which is completed on a time scale of $10^{-2} - 10^{-1} s$, in which the negative colloids adsorb on isolated positive particles and reach the stationary coverage.
(b) A subsequent slower agglomeration process, in which the negative-covered positive particles diffuse, meet, and begin to form aggregates.

Concerning aggregate shapes, both experiments and simulations show that it is possible to obtain both chain-like and compact interparticle aggregates in this system with huge-size mismatch. The compact aggregates are energetically favorable so that chain-like aggregates have a tendency to rearrange, thus becoming compact. However, kinetic factors may hinder this process.

Similarly, in the suspensions of oxide nanoparticles with smectite clays, nanoparticle behavior is mainly driven by heteroaggregation with clay colloids (due to electrostatic forces), while homoaggregation remains negligible. Primary heteroaggregates can be formed via the attachment of nanoparticles to the clay. This will be followed by a secondary heteroaggregation stage by bridging nanoparticles.

In a binary oxide system such as silica-zirconia, silica colloids are larger than zirconia colloids by a factor 10^2. Silica colloids are assumed to have zero potential, while zirconia colloids are strongly charged and induce local opposite charge when they adsorb on silica. The resulting attraction is extremely short-ranged compared to the size of the silica colloids, since its range is estimated to be about 1 / 600 of the silica colloid size. These facts render coadsorption of zirconia on silica extremely unlikely so that the rings cannot form. In alumina-silica, the attraction range is about 1 / 40 of the alumina size.

5.3 SUSPENSIONS AND MICROSTRUCTURE FORMATION DURING THERMAL BARRIER COATINGS VIA SPS

It is worth mentioning that the microstructure formation in SPS coating can be as much suspension-dependent as plasma mechanohydrodynamic. Besides preparation/stabilization of suspension powder, the injection mode of liquid phase, interaction between liquid and plasma jet, microstructure of as-sprayed coatings, and the corresponding deposition mechanism [18]. Similar to many ceramic microstructures, the porosity formation is a major concern. The discrimination of these pores in terms of their size and shape distribution, anisotropy, specific surface area, etc., is critical for the understanding of processing, microstructure, and properties relationships. The control of porosity size/content and oriented structure by using submicron and nanometric size of powder particles is directly related to suspension. The main sources of porosity are the interoriented spacing and region in the vicinity of the interoriented spacing [19].

As expected, the microstructures of SPS coatings are heterogeneous, such as highly dense, highly porous, and columnar-like or vertically cracked. Because of the improved performance of column-structured coating, it is desirable to produce this type of deposits [20]. It is within reason to propose that the nature of coating buildup is governed by the droplet trajectories and the resulting angle at which the molten droplets impact these asperities [21]. The molten droplets are less likely to be affected by the plasma drag, resulting in close-to-normal deposition (i.e., no normal velocity component [$v_{\perp}=0$]). On the contrary, the lower momentum molten droplets of suspensions are most affected by plasma drag and result in shallow deposition (i.e., with relatively lower parallel velocity component [$v^{//}$]). Here, the velocity components are mentioned with respect to the axial centerline of the plasma plume. The droplet with highest momentum impacts the bond coat very nearly along the axial centerline, and a droplet with the least momentum impact farthest from the axial centerline. Along with coating thickness, the microstructure is also affected by change with distance from the axial centerline [22]. Thickness also decreases significantly with increase in off-center distance, with thickness decreasing sevenfold from close to the axial centerline to small distance from the axial centerline [23]. It is interesting to note that the microstructure is found to be orientated and textured with cracks close to the axial centerline, where the deposition is from the core of the spray plume but gradually changes to a completely orientated-type microstructure as the distance away from the axial centerline increases. Therefore, the microstructure illustrates that coating deposition from the spray plume is the result of molten droplets being deposited under different conditions in the core of the spray plume (close to axial centerline) compared with the outer edges of the spray plume (farther away from the axial centerline). Furthermore, while the microstructure remains columnar

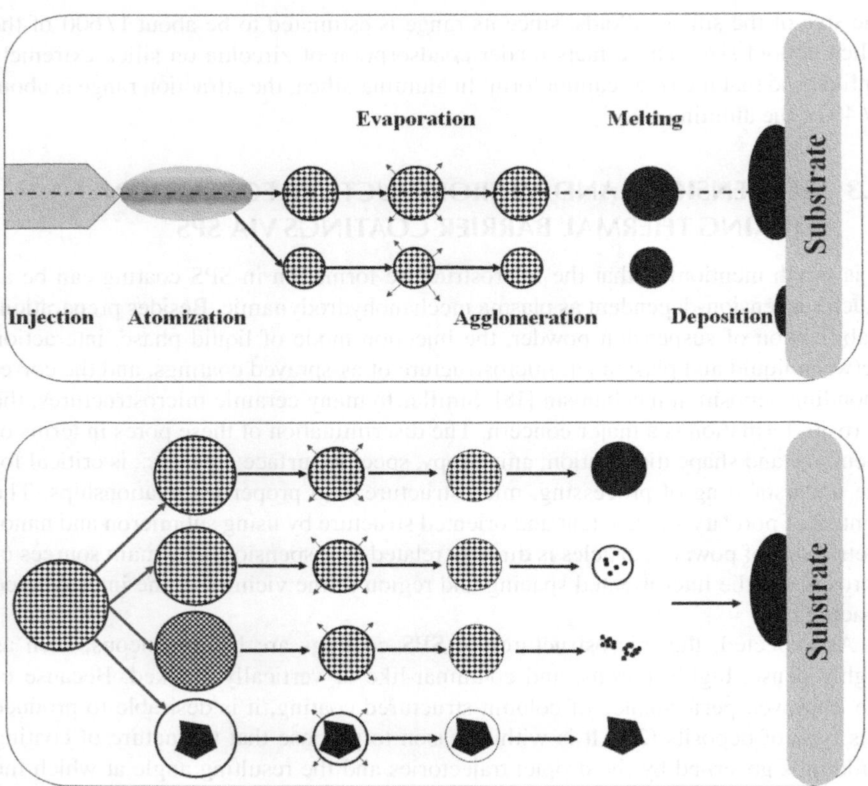

FIGURE 5.23 The schematic behavior of droplet trajectories in plasma stream upon impact and deposition of substrate (up) and the possible stages of droplet transformation before deposition (down).

after a certain distance away from the axial centerline, the angle between the bond coat–top coat interface and the interorientated space/crack remains the same (see *Figure 5.23*). The plasma decreases from nearly 90° at the axial centerline to about 40° at the very far end. The change in angle corresponds to the change in column built-up direction and reflects the variation in the angle of impact of the molten droplet on the surface asperity [24]. As discussed previously, such a significant deviation in droplet trajectories is attributable to changes in droplet momentum and thus highlights the influence of droplet momentum on resulting coating microstructure, as schematically illustrated in *Figure 5.24*.

The drag force (F_D) pushes fine solid particles parallel to the substrate surface, whereas adhesion force (F_A) causes adherence of particles to the substrate. In case of relatively big particles, there is inequality $F_A > F_D$ because of its moving in the center region of plasma jet, which results to good heat treatment and being fully molten. On the other hand, the small particles are moving on the periphery of plasma jet and, in this case, are inverse inequality $F_A < F_D$, which results in its moving parallel to the

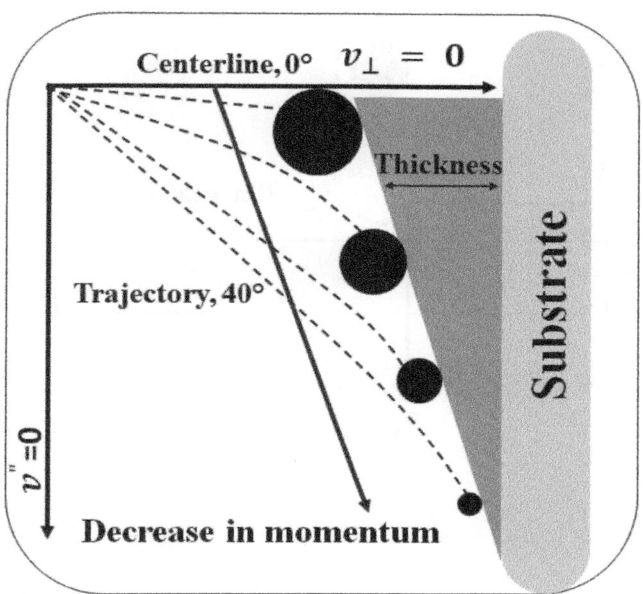

FIGURE 5.24 The schematic behavior of droplet trajectories in plasma stream. The momentum dependency of coating thickness and the angle of trajectory are indicated.

substrate surface. Then small, fully or partially molten particles adhere to the peaks of substrate roughness *(see Figure 5.25)*. In the first step, due to the single surface irregularities, single splats start to create a root of columns. Then, in the second step, particle trajectories depend on the relations between F_D and F_A, and it is similar to the case of roughened substrate. Third step is connected with growth of columns, which is supported by shadowing effect [25].

The progress in coating process depends on the deposition efficiency, corresponding to the ratio between the mass of resulting YSZ coating and the mass of YSZ injected into the plasma jet (where x represents deposition efficiency and where n is the different measurements) [26]:

$$\Delta x = \left[\frac{\Sigma(x-\overline{x})^2}{(n-1)}\right]^{1/2} \tag{5.21}$$

The suspension effect on structure formation indicates the droplet morphology and possible shadowing effect, which directly affects the oriented and textured coating microstructure. As an example, the large droplets without the parallel component of velocity are barely affected by the plasma gas. After the first split of droplets in plasma stream, the evaporation of media stage results in the formation of attracted solid particles, sintering of some fine solid, formation of flying aggregates in the stream, final evaporation, and melting before deposition at impact [27] (see *Figure 5.26*).

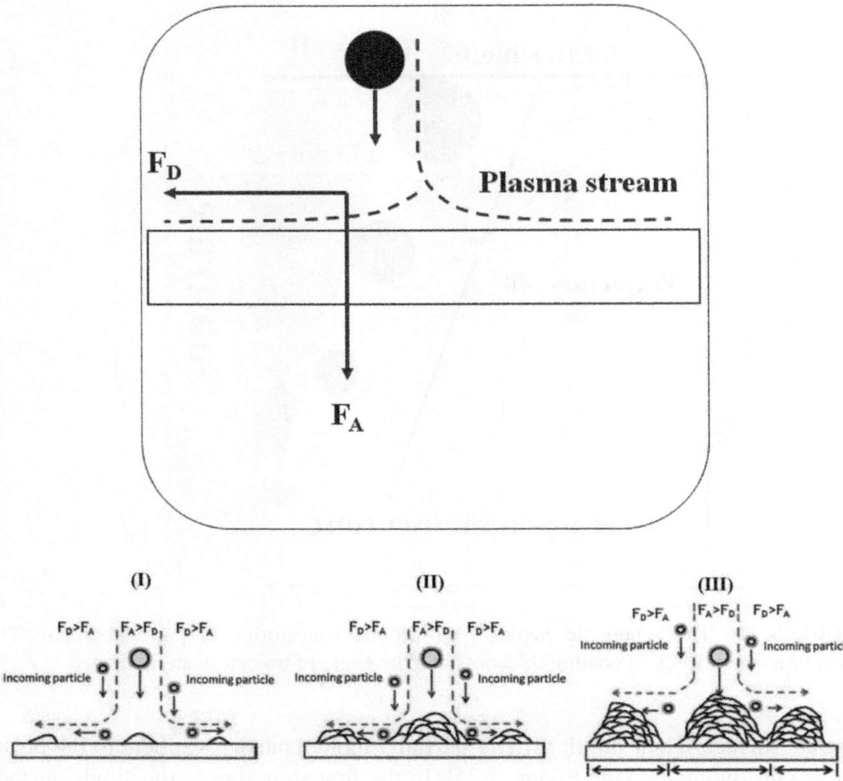

FIGURE 5.25 The schematic behavior of droplet trajectories in plasma stream (up). The momentum dependency of coating thickness and angle of trajectory are indicated. The structure formation based on $F_A - F_D$ relations.

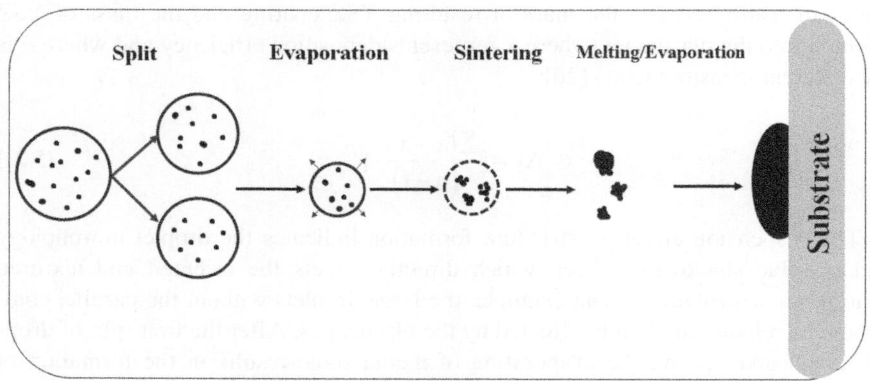

FIGURE 5.26 The schematic behavior of droplet trajectories in plasma stream. The momentum dependency of coating thickness and the angle of trajectory are indicated.

The suspension concentration indicates other droplet properties, such as possible shell formation or homogeneous melting. The splits in highly concentrated initial droplets lead to particle droplet breakage, which is followed by formation of smaller and highly particle-concentrated droplets that turn to solid (aggregated) particles before melting and deposition. In low-concentration suspensions, the initial droplet breaks up smaller ones with particles gathering at outer surfaces that turn to particle shells before impact. (See *Figure 5.27.*)

The droplet impact is simultaneously affected by size, concentration, and angle of contact (as a function of droplet speed and level of deviation from plasma stream) [6], which in fact dictates the level of coverage by SPS process. During collection of splats and almost-spherical distributed particles, those that have traveled in the hot zones of the jet are well melted and form splats, while those that have traveled in the jet fringes form tiny spheres sticking all around the central zone. (See *Figure 5.28.*)

The deposited material during normal impact plasma stream is presumed as the highest suspension droplet efficiency during SPS process, which can produced basic coatings. The first type includes ceramic suspension droplets moving with the plasma through the initial turn at substrate impingement. As a result, the substrate parallel velocity component of these droplets dominates the substrate normal component. However, as the plasma flows across the substrate, the inertia difference between the plasma and small ceramic suspension droplets causes some of the entrained droplets to be unable to follow the more sudden directional changes produced as the plasma moves around surface asperities. Initially, deposits formed on substrate asperities grow both laterally and vertically. During this time, the lateral growth of taller deposits may engulf shorter deposits. Among the remaining deposits of similar heights, the increase in drag forces generated by the decrease in spacing separating these structures reduces the plasma flowing between them. This process eventually causes the lateral growth to stop, leaving an interdeposit gap. As spraying proceeds, the ongoing vertical growth of the deposits and interdeposit gaps produces columnar structures that span the coating cross section normal to the substrate. Spraying will

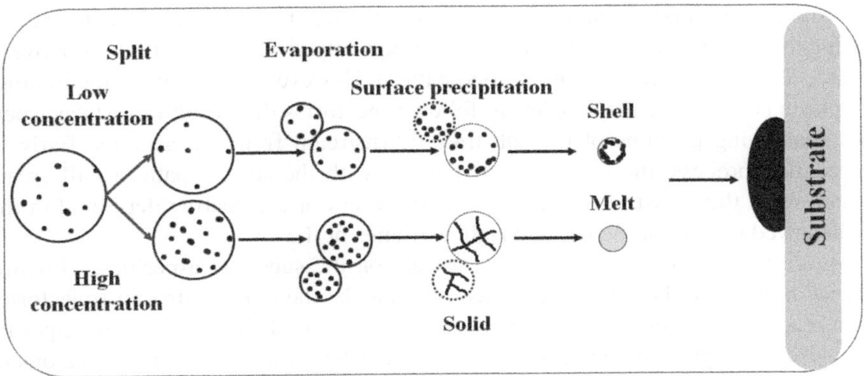

FIGURE 5.27 The schematic behavior of droplet trajectories in plasma stream for high- and low-concentration suspensions.

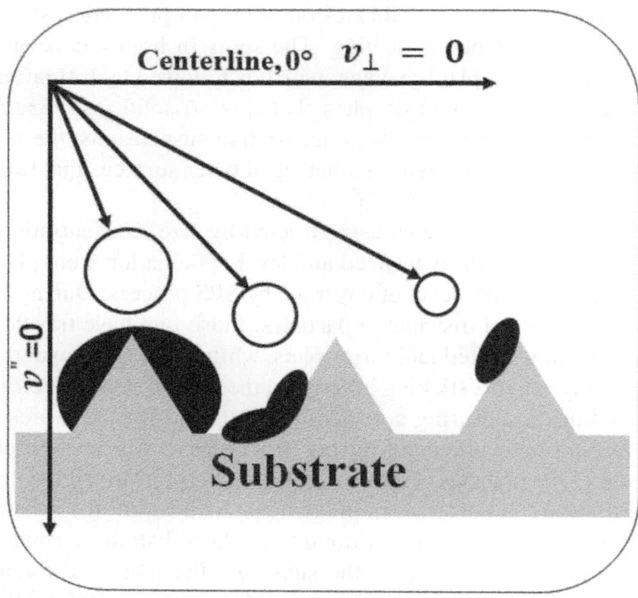

FIGURE 5.28 The schematic behavior of droplet trajectories in plasma stream at contact.

naturally yield localized nonuniform distributions of droplet impacts, which can lead to asperities in the coating surface [28]. These coating asperities can become sites of preferential site. Consequently, the coating is expected to have a surface featuring multiple deposits separated by valleys; these deposits are the cluster formations comprising the cauliflower-like surface structures mentioned previously [29]. The second pattern of coating requires conditions in which the droplet retains a velocity directed primarily perpendicular to the substrate, yet the substrate parallel component still affects deposition. In this case, the droplets are more massive than those depositing according to the first mechanism, and they separate from the plasma before completing the substrate impingement turn. Consequently, deposition in regions between surface asperities increases with second pattern. However, the effects of the impinging plasma drag cause the droplets to follow trajectories that result in surface asperities shadowing portions of the substrate downstream from the asperity. Early in the coating process, the shadowing occurring with the second pattern will create variations in the growth rate, causing the thickness of the coating deposited in the unshadowed regions at the left and right extremes of the

Figure to be growing twice as fast as that on the substrate asperities. This difference is because the left substrate asperity blocks the oblique droplet trajectories from reaching two regions of the substrate to the right of the peak on this asperity. Likewise, the right substrate asperity blocks droplets from reaching two regions of the substrate to the left of the peak on this asperity. As the height differential between the coating in the unshadowed regions and the material deposited on top of the substrate asperities decreases, the deposition on the side of the former will increase.

This increase causes the unshadowed sections of the coating to grow toward that, covering the substrate asperities. Therefore, given a long-enough spray time, the coating building from the initially unshadowed sections will overgrow that above the substrate asperities. Due to this overgrowth behavior, a coating microstructure formed by the second pattern should exhibit a convergence of the porosity bands separating columnar structures. As with the first pattern, the droplet trajectories in the second pattern mechanism promote cluster formations on coating surface asperities. However, these structures exhibit less growth with the second pattern than with the fist pattern because droplets with velocities mainly normal to the surface are less likely to deposit preferentially on surface asperities. Therefore, cluster formations will be less distinct on the surface of a microstructure produced by the second pattern than by the first pattern. The third pattern is proposed as describing conditions in which plasma drag forces do not affect the deposition characteristics, meaning, the droplets yielding three third-pattern SD are more massive than those causing the second pattern. As a result, the third pattern is characterized by deposition on the entire substrate surface during each plasma gun pass, i.e., droplets do not preferentially impact on surface asperities, and no shadowing occurs from these asperities. The droplet trajectories required for this complete coverage would be approximately normal to the substrate surface. Thus, this suggests that the third pattern would occur when plasma is sprayed with suspension droplets having micrometer diameters. The complete substrate coverage possible with the third pattern prevents the formation of columnar structures separated by porosity bands. In addition, the momentum of the droplets in this mechanism should preferentially drive liquid into the valleys in the coating/substrate surface. Thus, coating asperities generated by nonuniform third pattern will not grow into cluster formations; instead, the coating surface will tend to become more planar than the grit-blasted substrate surface. (See *Figure 5.29.*)

During SPS coating, the suspension is firstly fragmented into liquid drops, followed by the liquid evaporation. Then two different processes occur: (I) The single particles and agglomerates injected into the plasma jet are released and melted, then impact onto the substrate. Single particle yields ultrafine lamellae or nonflattened spherical particles. Due to low inertia, most individual particles reach the substrate with a lower velocity and cannot be flattened. The agglomerates may form either small lamellae or unmelted sintered aggregates. (II) The large-size aggregates may form in the plasma jet. Some of them may break up, resulting in the formation of single particle or small agglomerate, which is similar to case I, while others may produce large lamellae or sintered aggregates [30].

The formation of columnar structures can be connected to the interaction of atomized droplets with the plasma jet. If the sizes of droplets in the plasma are in the range from 1 to 5 μm or smaller in diameter, the impact trajectory of in-flight particles will be severely affected by the plasma gas stream. Melting and solidification of ceramic powders results in dense splats and, eventually, a dense coating. The low-concentration powder suspension creates droplet breakup, liquid evaporation, and suspended surface precipitation and shell formation and, as a result, is porous and contains semireacted particles [31]. The high-concentration solutions facilitate droplet breakup, media evaporation, then solute volume precipitation and pyrolysis. The small particles in the plasma jet will participate in pyrolysis completion and

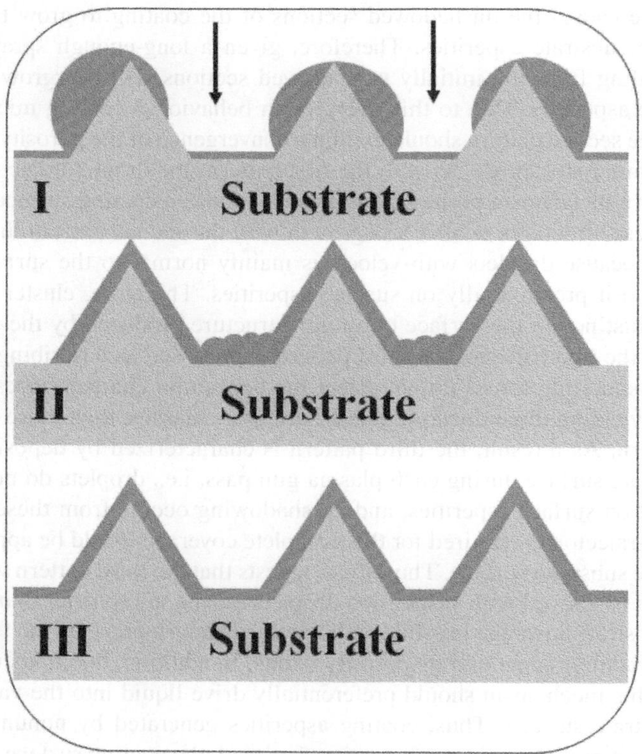

FIGURE 5.29 The schematic of deposited material during normal impact plasma stream.

melting, forming small splats on the substrate. The large particles in the plasma jet also form large splats with microcrack or large nonflattened clusters. Some clusters may contain unreacted powder, which may decompose during the process of coating formation. The unreacted powder deposited on the substrate begins to react when the temperature exceeds the powder reaction temperature. When the total tensile stress that results from the decomposition of powder surpasses the tensile strength of the coating, cracks may form on the surface of the coating. These cracks further extend into the coating along with the powder decomposition. Due to the agglomeration and aggregation of the initial nanosized grains, the behavior upon liquid vaporization may differ. It seems that the bigger agglomerates or aggregates explode upon liquid evaporation, resulting in molten particles and smaller particles evaporation, which mainly result from agglomerated particles [11]. The droplet's initial fragmentation and pyrolysis results in the formation of fully solid, hollow particles (HP), spherical particles (SP), and remnant unpyrolyzed mass (UM). With the stack of lamellae, unpyrolyzed mass is incorporated in the coating. Due to the smaller splat size and pyrolysis stress, cracks are easily initiated at locations where multiple regions of unpyrolyzed mass are distributed in close vicinity. As for the coatings prepared with a relatively low precursor feed rate, the reduced pyrolysis stress leads to a low crack density. With increasing the liquid feed rate, many microcracks and pores are observed in the coating, and thus the pyrolysis stresses are accommodated with them. In addition, the formation of skewed cracks is also induced by the unpyrolyzed mass. It can be concluded that the net microstructural evolution in SPS [21] involves combinations of suspension conditions, droplet size, angle, and also the material supply by continuous coating passes.

If higher surface tension favors the reduction of fragmentation in the jet fringes, for example by using water instead of ethanol, the specific heat and latent heat of vaporization are larger with the former. Under the same conditions, water droplet will vaporize two times larger than ethanol. Water as media, which increases the surface tension of drops, reduces after droplet vaporization and transformation into plasma, the energy available to melt suspension particles. In such conditions, particles contained in ethanol are melted, while those in water are very poorly melted. Once the media is completely evaporated, the treatment of particles contained within the droplet will depend on their size distribution and morphology.

The nature of coating buildup in SPS is governed by the angle at which the molten droplets impact the asperities existing on the already built-up coating surface (which can be bond coat surface asperities during the deposition of the first pass of the ceramic). If the droplets trajectory is not affected by plasma drag, as in the case of droplets having high momentum, the droplet impacts nearly at normal incidence, which results in uniform splashing over as well as in between the surface asperities. Such a normal deposition after few passes (depending upon the surface roughness) fills up the gaps between the adjacent asperities, resulting in more planar deposition and progressively reducing the irregular surface profile originally present on the as-sprayed bond coat surface. Further increase in number of passes thereafter serves to build up the coating on this planar profile and does not result in columnar microstructure. On the contrary, if the droplet trajectories are significantly affected by the plasma drag, as in the case of droplets having low droplet momentum, then the droplets from the very first pass have a shallow impact on the surface asperities. As the number of passes increase, continued shallow deposition can result in so-called shadowing effect and give rise to tiny columnar-type features building up on the surface asperities. Further increase in number of passes results in these features growing into a columnar-type microstructure. The formed structure will have low columns density, high intercolumnar porosity, and medium intercolumnar gaps, which is not ideal for low thermal conductivity and the high cyclic lifetime that is expected for TBC purposes (see *Figure 5.30*).

In these structures, the ores may be of nanometer, submicrometer, or even micrometer size, having various shapes and forms of connected or nonconnected networks; morphology may vary from dense and homogeneous one, through vertically cracked up to fully columnar; deposits may be formed by fully molten splats, sintered particles only, or by both as so-called two-zone structures; roughness of coating's surface may vary in a very broad range (as reported in, roughness values may be between 1.6 and 14.1 μm); the thickness may be as low as a few micrometers and reach hundreds of micrometers; the lamellar microstructures, which oriented parallel to the surface of substrate, are obtained by the stacking of splats; the columnar structures or vertically cracked (VC) cross the thickness of the coatings from highly dense, highly porous or

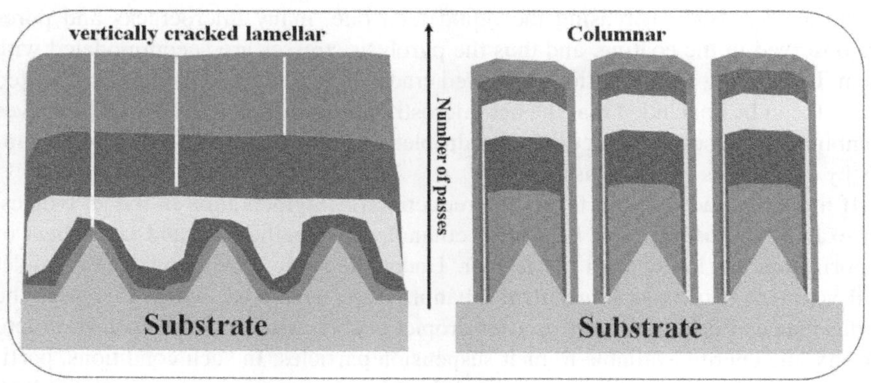

FIGURE 5.30 The schematic evolution of two main SPS structures.

segmented by vertical cracks; the substrate roughness can greatly affect the coating microstructures as conventional plasma sprayed TBCs are reported to be very rough (about 11–12μ*m Ra*), and the roughness is crucial in a thermally sprayed coating as it provides mechanical anchoring; coatings from low concentration suspensions are porous and consist of aggregated solid particles; the high concentration suspensions create denser coatings and primarily consist of fine splats; the medium with a low surface tension and a low boiling point results in a relatively high dense coating; small droplets with strong component of velocity create the impact trajectory that is strongly dependent on the plasma jet and results in the formation of vertical cracks; higher droplet momentum can result in normal deposition, giving rise to lamellar and/or vertically cracked–type coatings, whereas lower momentum can result in shadowing effect, giving rise to columnar structures; low droplet size and low droplet speed can produce columnar microstructures; high droplet size and high droplet speed can produce vertically cracked–type microstructure; vertical cracks, spacing between columns (intercolumnar spacing [IC]), interpass porosity (IP) bands, and branching cracks are the basis for these structures; and at a microscale, coatings show features such as fine pores and cracks (interconnected or unconnected).

Based on the microstructure at a macroscale, these coating structures can be mainly categorized as follows:

1. Vertically cracked structure
2. Highly porous structure
3. Columnar structure

The cross section of an SPS-TBC within a column and near intercolumnar spacing shows various microstructural features such as columns, intercolumnar spacing, spherical particles, fine cracks, nanopores, submicron pores (SP in green), and micron pores (MP in white) in a closed packed setting.

The effect of the suspension powder concentration and rheology can be summarized as controlling the size of the coating forming particles; reducing the suspension solids loading reduces the suspension viscosity; allowing easier atomization and increasing the production of the smallest suspension droplets during fragmentation at reduced viscosity; increasing the mass of powder contained within each suspension droplet produced during the fragmentation process by increasing the powder solids load within a suspension.

The relationships between various important process parameters and droplet momentum with columnar microstructure type shows the importance of suspension quality. The typical columnar structures include fine columnar, coarse columnar, and feathery armed columnar structure. (See *Figure 5.31.*)

These three exemplary morphologies, named fine, coarse, and feathery, are shown, which vary with respect to the columnar microstructure. The cross-sectional include numerous single crystalline columns and nanosized feather arms along the complete column periphery. The top views show different columnar tips, which also vary with respect to size and shape. The overall porosity of the top coat can be attributed to the gaps between columns, feathery arms, and gaps in the feather tips.

The relation between total porosity content and feathery columnar structure is directly associated to intercolumnar gaps. As an example, in YSZ (8 wt. % yttria partially stabilized zirconia, $\phi: 25\%$, $D_{50}: 300$ nm in ethanol) suspension, these gaps increase porosity from 11% to 16%. (See *Figure 5.32.*)

The pyramidal tip angle of a typical is around 30–45 degrees and arms are located at 40–60 degrees. The total column diameter (D_{TC}) and the reduced column diameter (D_c), the columnar diameter excluding the feather arm length, are the expressive parameters of a typical feathery arm. The "reduced column diameter" is commonly termed simply as "column diameter." Similarly, the reduced column radius, which is half of the reduced column diameter, is termed as "column radius" along columnar gaps. The columnar gaps may show almost-regular width along the column length with a low tortuosity over the column height, or they might be highly irregular, with high tortuosity contour along the column length. The densely packed columnar/

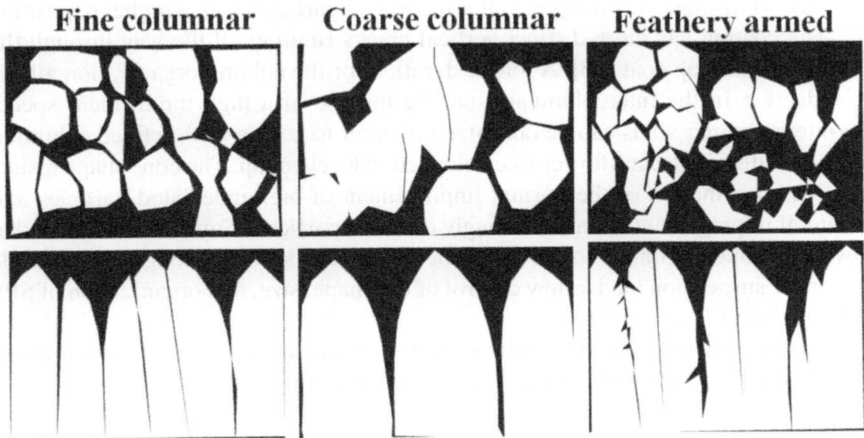

FIGURE 5.31 The schematic top (up) and cross-section (down) views of main SPS structures.

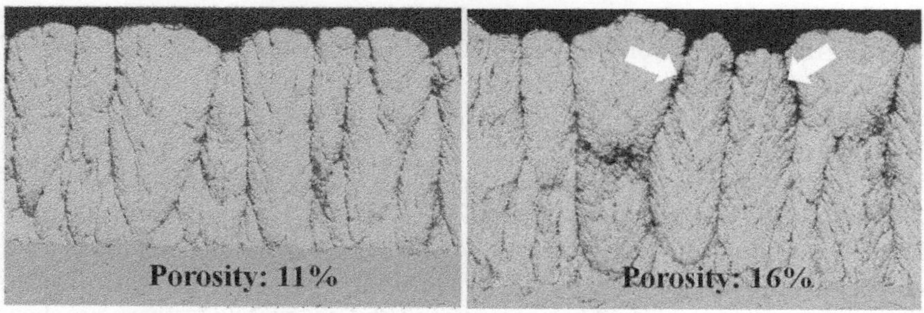

FIGURE 5.32 The effect of intercolumnar gaps on porosity content in feathery columnar structure.

compact columnar is the other morphology in this system. The governing parameters are total length of the column and the column length (L_C). A number of possible morphological variations can be obtained by gaps between two columns, the columnar gap (G_C). The thickness of each feather arm is named as the feather arm width or feather thickness (T_F), the horizontal length of the feather arm along the column width is the feather arm length (L_F), and the gap between individual feather arms is considered as void between feather arms (V_F) and will be termed shortly the feather gap [32]. Further, the inclination of feather arms with horizontal axis is considered as the feather inclination angle (θ_F); the vertical angle of the pyramidal tip is the feather tip angle (θ_T) [24] (see *Figure 5.33*).

The columnar microstructure by the SPS process is responsible for TBC lifetime [5]. The low size of particles (submicron- and nanosized) for the SPS process allows them to be deviated by the plasma flow in the vicinity of the substrate. It leads to a combined normal and lateral growth of columns around the substrate asperities, which is assumed to be the explanation of these typical cone-shaped columns. The columnar structure with a denser coating is attributed to the use of lower feed rate and standoff distance, resulting in better melting of particles with a higher deposition rate. The coating has more distinct vertical cracks running all the way through the thickness of the top coat [26]. A longer duration for the column organization allows the reduction in the intercolumnar voids by the use of a high torch linear speed. The intercolumnar voids can be optimized in order to keep a well-defined columnar structure with a high cauliflower-like compaction level on top. The cone shape results from the combination of the normal impingement of big, undeviated particles and the lateral impingement of small, strongly deviated particles. Some studies show that relevant parameters linked to the SPS process, such as substrate roughness, particle size, or the suspension load, allow control of the shape, size, and organization of SPS columns. (See *Figure 5.34*.)

The column density can be stated as unit for coated surface per unit length, which is defined as the number of columns present in the ceramic top:

$$Columnar\,Density = \frac{Number\,of\,column\,boundries\,intersecting\,a\,line}{True\,length\,of\,line} \tag{5.22}$$

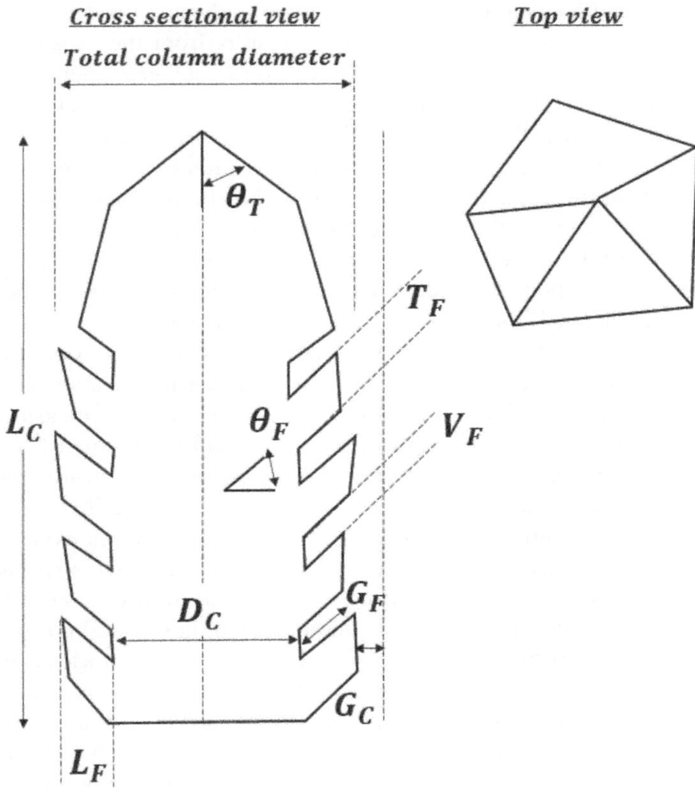

FIGURE 5.33 The effect-detailed dimensional parameters of a feathery-armed columnar structure.

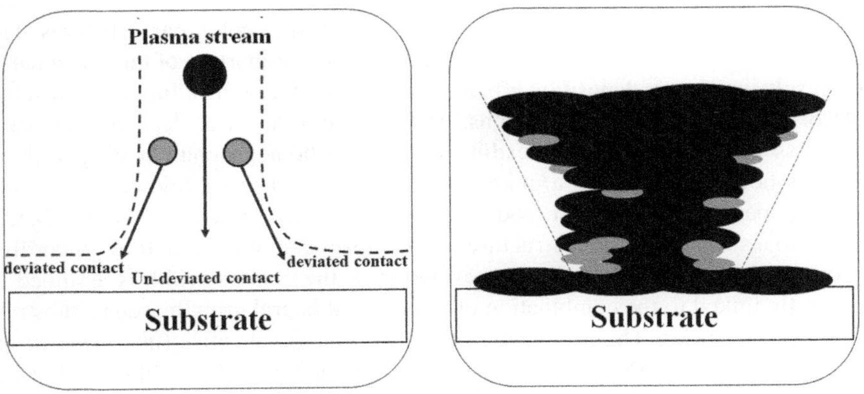

FIGURE 5.34 The feathery-armed columnar structure formation via different particles in plasma stream.

The well-populated columnar structure with high density is referred to as compact columnar structure. The compact shape results from the greater thickness deposited by each pass of the plasma torch due to the higher suspension feed rate and higher particle concentration in suspension, which can be used toward a rapid overlapping of columns. The control of columnar microstructures is dependent on the size of the droplet generated during the atomization of the suspension as this decides the size of the impacting particle that will form the coating [4]. The influences from plasma drag at the substrate surface, though not sufficient to generate a fully columnar coating, leading to more uniform columnar structures. During this segment, the suspension undergoes stronger fragmentation by the plasma, producing smaller particles. Therefore, the columnar structure seems to generate in three domains: in the first domain, coating presents an "initiation area," where columns are not fairly defined but in an early formation stage comparable to nucleation sites. The column mean diameter is difficult to be measured at this stage; in the second domain, coating exhibits cone-shaped columnar features separated one from another by intercolumnar voids (V); and in the third domain, the heads of columns are clearly visible. They exhibit a typical cauliflower (C) shape. This structure can be provided by the difference of trajectories experienced by particles in the vicinity of the substrate, where the plasma flow is strongly deviated, which leads to columns presenting a typical cauliflower shape. This specially arranged structure results from a combination of lateral and normal coating growth velocities. They are provided by the difference of trajectories experienced by particles in the vicinity of the substrate. The observed typical cone shape of columns leads inevitably into growth perturbations during coating development. The coalescence of two adjacent narrow columns can results in big cluster formation. Shadowing effect, by big columns or clusters, provides new intercolumnar voids clearly visible on the top surface. It leads to new column growth sites on the top of former partially shadowed ones and are named as secondary columns, forming secondary clusters of cauliflower zone. An accentuated shadowing effect between two big columns or clusters can also results in a reversed cone shape (see *Figure 5.35*) [21].

The *erosion effect* is the other determinative process in columnar structures. The lower enthalpy plasma flow, and, as a result, the higher proportion of unmolten particles, results in a lower deposition efficiency and can act as a grinding media responsible of microstructural modifications. The overspray material deposited between each pass can be cleaned up, leading to a rather-homogeneous coating without interpass boundaries despite a quite high value of $v_{\perp}$. The cauliflowers at the top surface appear to be smoother, and intercolumnar voids are significantly reduced, leading to a sustainable microstructure of a compact columnar structure. According to column development theories, the evolution of the column width is assumed to be directly linked to the combination of normal and lateral growth around substrate or coating asperities. The normal ($v_{\perp}$) and lateral ($v_{//}$) velocities (in *μm per pass*) are sufficient to describe analytically the cone-shaped growth of columns. For the coating thickness (τ), the increase of column means diameter (D) during coating buildup, where t is the spraying time, D_a the shape parameter according to substrate asperities, and A_{exp} the curve slope regarding column expansion. The biggest molten

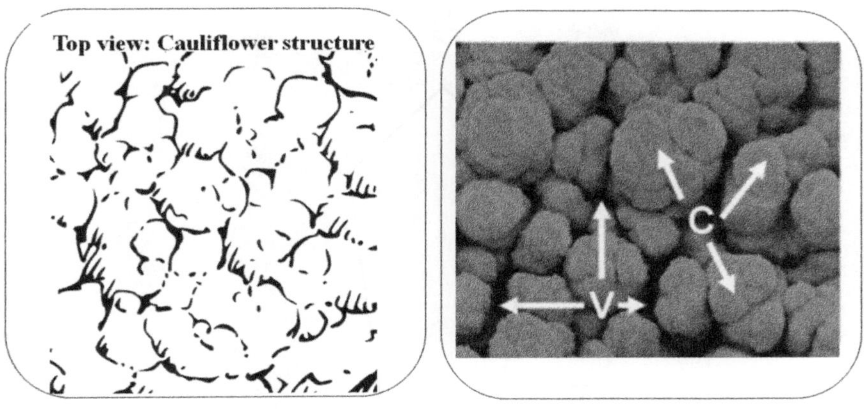

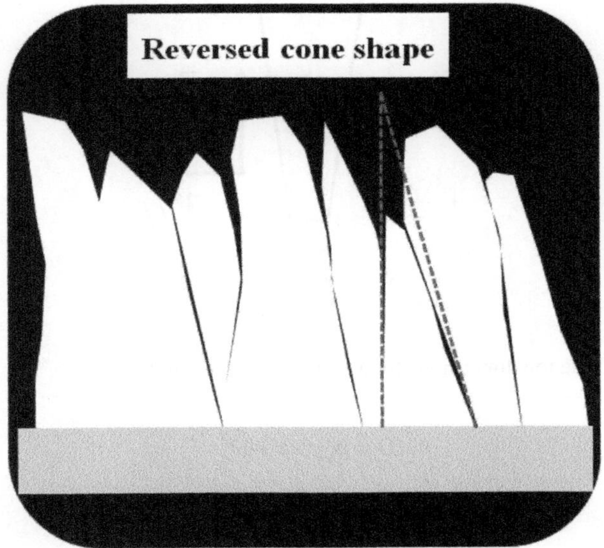

FIGURE 5.35 The top view of compact columnar structure with a cauliflower shape (up) and a schematic or reverse cone in columnar structure.

particles, undeviated at the vicinity of the substrate, mainly contribute to $v_\perp$, whereas the smallest ones are responsible for $v_{//}$ [24] (see *Figure 5.36*).

$$v_\perp = \frac{d\tau}{dt} \tag{5.23}$$

$$v_{//} = \frac{dD}{dt}$$

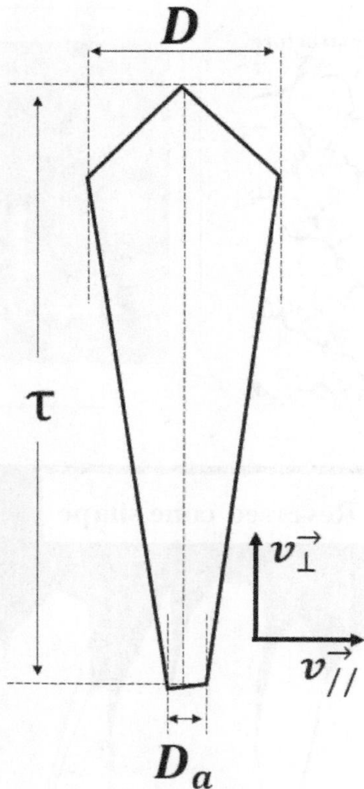

FIGURE 5.36 The top dimensions for a typical columnar unit.

$$D = A_{exp} \times \tau + D_a$$

$$\frac{dD}{dt} = A_{exp} \frac{d\tau}{dt}$$

$$v_{//} = A_{exp} \times v_{\perp}$$

By multiplying to pass during coating time:

$$v'_{//} = A_{exp} \times v'_{\perp} \text{ and } v^{'}_{//} = v^{'}_{//} \times \left[\frac{\Delta A_{exp}}{A_{exp}} + \frac{\Delta v^{'}_{\perp}}{v^{'}_{\perp}} \right] \frac{dD}{dt}$$

As a consequence of the increase of the suspension load, the intercolumnar voids tend to disappear; the increase of $v^{'}_{\perp}$ is responsible for a rapid development of the coating, which prevent intercolumnar voids. As the $v^{'}_{\perp}$ increases, the $v^{'}_{//}$ is increased, too, leading to rapid recovering of columns and the compact aspects of them. The increase in ceramic particles in suspension creates modified microstructures, schematically shown in *Figure 5.37.*

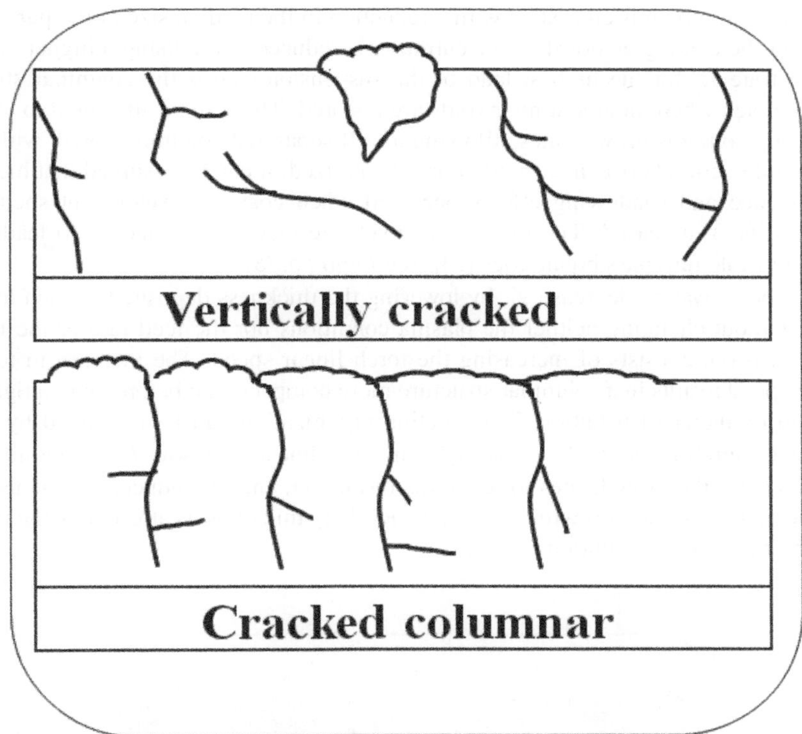

FIGURE 5.37 The schematic presentation of vertically and cracked columnar structures.

The *cracked columnar* structure is the resultant of larger particles which will experience less influence for the plasma drag within the boundary layer at the surface of the sample due to their higher momentum, thus leading to less columnar coatings. It is considered a transitional structure between fully columnar and segmented/vertically cracked microstructures [20]. The formation mechanism of vertical cracks is based on powder reactions, shrinkage, and tensile stress and, as a result, the tensile stress induced by the decomposition of solution powder during deposits preparation.

In binary oxide suspension such as YSZ (8 wt. % yttria partially stabilized zirconia, $\phi : 25\%$, $D_{50} : 45$ nm in ethanol) suspension, the droplet experiences some influence from plasma drag at the substrate surface, though not sufficient to generate a fully columnar coating with 5–10% porosity. Increasing the powder concentration from 25 wt. % to 33 wt. % increases the deposition efficiency. The increase in thickness per pass and deposition efficiency (55% thicker, with 30% increase in solid load) would be of interest to increase the processing rate of SPS on components. However, it is clear that increased solids load would result in different coating microstructures by using YSZ (8 wt. % yttria partially stabilized zirconia, $\phi : 33\%$, $D_{50} : 200$ nm in ethanol) suspension. Using YSZ (8 wt. % yttria partially stabilized zirconia, $\phi : 17\%$, $D_{50} : 200$ nm in ethanol) suspension, the microstructure of the coatings shifts from denser and vertically cracked to progressively columnar in nature. The change in microstructure type is consistent with a reduction in the median size of the particles forming the coating as powder concentration is reduced. Even using a higher value of substrate R_a for the highest load of the suspension causes the columnar structure to reach where intercolumnar voids are restored. This can be attributed to more separated columns growth sites, allowing a well-separated columns growth without strong interaction between each other as schematized. It can be assumed that higher coalescence and shadowing effects occurred when coating development sped up ($v_{\perp}$) and ($v_{//}$) increased. The increase of load rate of the suspension also leads to clearly visible interpass boundaries [33] (see *Figure 5.38*).

The other way to decrease $v'_{\perp}$ by lowering the thickness deposited at each torch pass, without changing neither the plasma conditions nor the feed rate or the load of suspension, consists of increasing the torch linear speed. The increase in torch linear speed results in a columnar structure more compact than before, with a tighter column diameter distribution. This coating organization can be explained by the combined modifications of $v'_{\perp}$ and substrate roughness. The low *Ra* value allows to keep close the growth sites for columns formation, and the reduced $v'_{\perp}$ induces a less perturbed column growth with a quite good organization, as attested by the tight column diameter distribution.

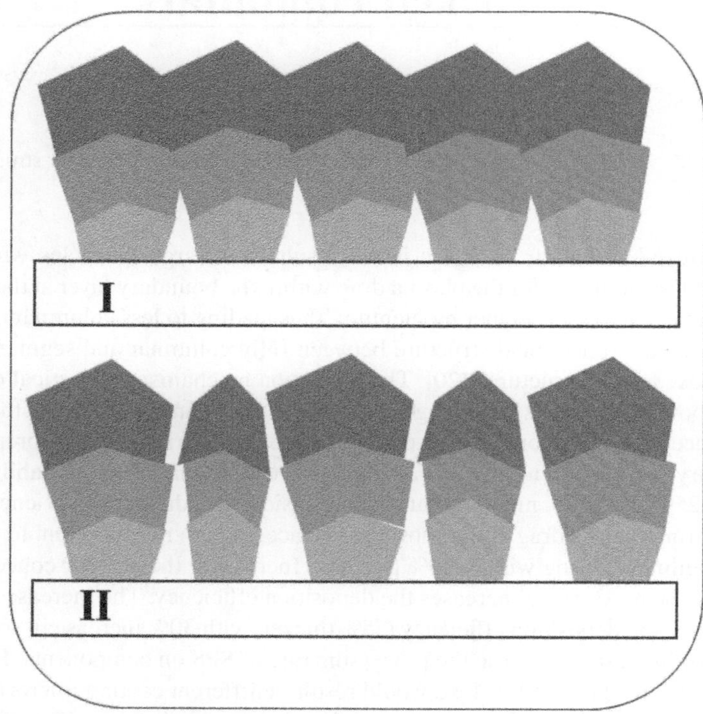

FIGURE 5.38 The difference in column arrangement from compact (I) to intercolumnar gapped (II) structure.

The *interpass porosity boundary (IPB)* layers are formed by particles treated in the periphery of the plasma core, deposited at the leading and trailing edges of the plasma plume as it traverses the substrate. The particles which are treated in the cooler plasma periphery region will have lower velocity and temperature than material treated in the plasma core. As a result, this material deposits in the semimolten or resolidified state and lower velocity, leading to layers of increased porosity. This structure is produced by the deposition of material treated in the plasma plume periphery rather than in the plume core. Suspension droplets entrained in the plume periphery are more likely to be slower and less molten than ones entrained in the plasma core because interactions with the surrounding atmosphere cause the temperature and velocity of the plasma to decrease with radial distance from the plume centerline. These particles are less effective at penetrating into the plasma core. The possible cause of the lower penetration into the plasma core may be related to the small median particle size. The structure causes the decrease of thermal conductivity and is described as disadvantageous for thermal lifetime due to promoting lateral cracks close to the interface between the ceramic and the bond coat by tensile stress development [34]. The interpass boundary defects can be explained by overspray deposition between each layer. The particles circulating in periphery of the plasma jet are incorporated as "dust" (unmolten and/or resolidified for example) in the coating between two passes of the torch. This phenomenon is greatly impacted by the $v'_{\perp}$. For instance, in YSZ (8 wt. % yttria partially stabilized zirconia, $\phi: 25\%$, $D_{50}: 400$ nm in ethanol) suspension, while the percentage of both large- and small-scale pores is great, a greater quantity of material can be treated in the plasma periphery rather than the plasma core, creating a 16–18 wt.% porosity.

In *vertically cracked* structure, due to overall larger and faster ceramic suspension droplets such as water-based suspensions, a densely packed columnar structure with some horizontal cracks starting from the column gaps can be formed [35]. These horizontal cracks, called also branching cracks, are typical for dense TBCs and are a consequence of the high-energy release during spraying and solidification of the molten oxides, such as zirconia particles. Also, switching media from an ethanol to a water-based produces YSZ particles that are larger and less susceptible to plasma drag influence during deposition. These suspensions have larger average and a tighter distribution of droplet sizes than an alcohol-based YSZ suspension with the same powder concentration having a threefold higher viscosity. As an example, ↑ the droplets produced when spraying the (YSZ:8 wt % yttria partially stabilized zirconia, $\phi: 25\%$, $D_{50}: 400$ in water) suspension with 5–10% pore percentage shows horizontal cracks and a more planar top surface with more tortuous and/or higher aspect ratio (i.e., more crack-like).

As increasing the impacting droplet sizes (due to higher solids load) also reduces the coating total porosity, decreasing the solid load might cause formation of *highly porous* structures. The structure involves large-scale porosity features, such as large pores or column gaps. In a *bilayer* top coat, the columns in the second layer continued to grow on top of the first layer columns. A denser first layer was produced than the second layer, as intended during spraying, creating a dense, porous stacking in systems such as (YSZ:8 wt. % yttria partially stabilized zirconia, $\phi: 25\%$, $D_{50}: 500$ in ethanol) suspension with 4–11% pore percentage. (See *Figure 5.39.*)

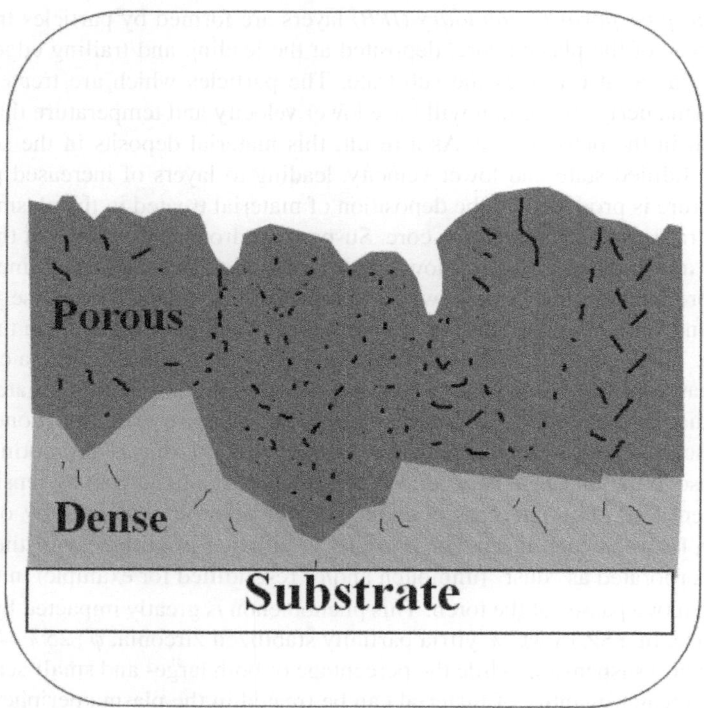

FIGURE 5.39 The bilayer formation of dense and porous layers.

The size and distribution of pores can be structure specific, with evident physico-mechanical properties. The VC coating has a relatively smoother surface with more cracks with uniformly distributed porosity that lowers the thermal conductivity of coating and better strain tolerance due to the existence of through-thickness vertical cracks. The higher porosity content decreases thermal conductivity of the top coat, and the lifetime will be decreased with the increase in total porosity. The pore can have porosity ranging from micrometric to nanometric scale. The reduction in fine porosity is due to densification of the pores present within the columns. The typical defects can include (I) vertical cracks, (II) fine cracks, (III) branching cracks, (IV) independent closed pores, (V) inert-pass porosity, and (VI) connected closed pores (see *Figure 5.40*) [36].

Porosity is one of the key characteristics of the ceramic TBC. In SPS coatings, several of the abovementioned different microstructural features, both at large scale and small scale, contribute to the total porosity content of the coating. These pores are nothing but empty voids mostly filled with air at ambient conditions. Porosity can be open when the pores are interconnected (accessible by a fluid if injected externally) or closed when they are isolated from each other (nonaccessible by a fluid from outside the coating). The definition of *open porosity* and *closed porosity* may vary depending on the usage of the porosity term in various applications, but

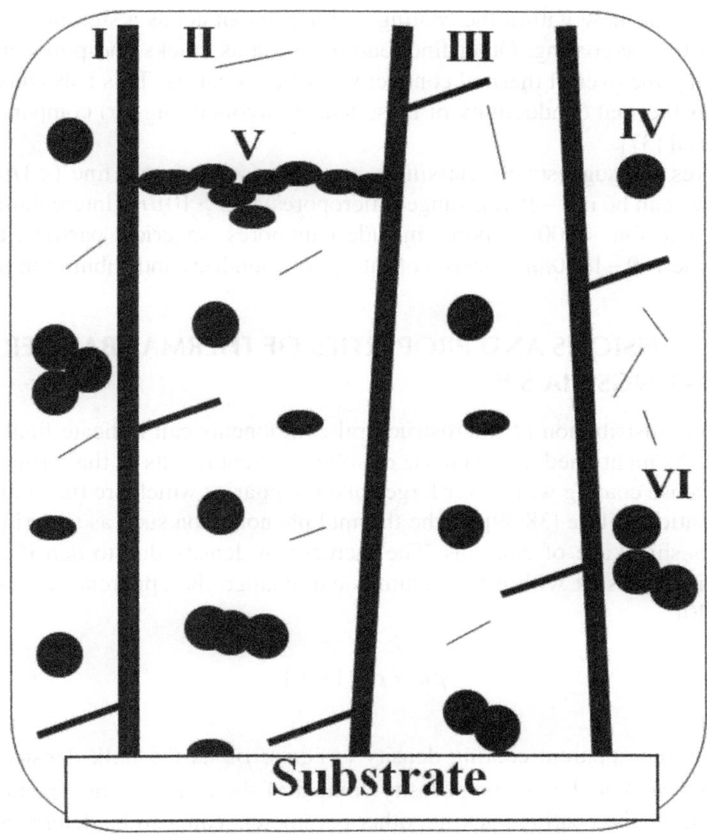

FIGURE 5.40 The porosity and crack in a typical SPS-TBC microstructure.

in case of TBCs, it is dealt with a fluid accessibility in pores. The open porosity involves vertical cracks, intercolumnar spacing, interpass porosity bands which are typically connected with the vertical cracks or intercolumnar spacing. Also, in some cases, there can be branching cracks, cracks which originate at the vertical cracks or intercolumnar spacing and grow parallel to the substrate. The closed porosities are smaller- and larger-scaled globular pores which are independently present in the coating, independent clustered pores which are connected with each other but not with any of the features shown in the open porosity so that an external fluid cannot reach them. Also, in some cases, fine cracks which are present in the coating but not connected with any of the features shown in the open porosity. They can significantly affect both the thermal insulation nature as well as the lifetime of the coating. Vertical cracks or intercolumnar spacing are through the thickness of the TC. These cracks, if present in large number (higher vertical crack density in the coating), can increase the overall thermal conductivity of the coatings. However, branching cracks and interpass porosity bands are perpendicular (or close to perpendicular) to the direction of heat flow within the coating and hence can act as a significant thermal barrier within the coating. Other fine features, such as cracks and pores, also help in decreasing the overall thermal conductivity of the coating. This is because of the much lower thermal conductivity of these features (containing air) compared to the bulk material [37].

The pores are suggestively classified as coarse ($> 1\mu m$) and fine ($< 1\mu m$). The coarse pores can be in $1-10\mu m$ range (micropores) and $> 10\mu m$ (intercolumnar and verticals). The fine $< 100nm$ pores include nanopores, spherical partilce, and fine cracks, while $100-1000nm$ consists of interpass boundries and submicron pores.

5.4 SUSPENSIONS AND PROPERTIES OF THERMAL BARRIER COATINGS VIA SPS

The size and distribution of microstructural components can indicate final coating properties. As mentioned, the increase of solids content results in the formation of a relatively dense coating with fewer large splat boundaries which are the main reason of delamination failure [38,39], or the thermal phenomenon such as sintering of top coat causes shrinkage of columns. The increase in density due to densification of fine porosities present within the columns can change the apparent density of the coatings [40].

$$\rho a = \rho b(1-P) \tag{5.24}$$

Here, ρa is the apparent coating density (g/cm), ρb is the bulk density of coat material (g/cm^3), and P is the porosity content of the coating (area fraction). Due to densification and coating packing, other parameters can also be determined. The intercolumnar spacing (vertical cracks) indicates how a columnar structure with well-separated columns shows higher lifetime as compared to a compact columnar structure. After long exposure at high temperatures, the intercolumnar gaps are widened. The increase in intercolumnar gaps due to reduction in fine porosity after densification within the columns results in shrinkage of columns and opening up.

The *sintering* of TBS is very much similar to the compact powders; ceramic TCs in the TBCs also contain porosity in the form of pores, delaminations, and cracks, which at higher temperature can be altered due to sintering, affecting coatings' thermal and mechanical properties. The major driving forces for sintering in general are the curvature of the pore (surface energy), an externally applied pressure, and chemical reaction. The first two are the major driving forces. As a result, following situations are probable. Firstly, the densification which mainly decreases the porosity due to the closure of pores or healing of cracks (also referred as). The mechanism of densification is reported to be due to the combination of surface and grain boundary diffusion. The mechanical properties such as hardness, E-modulus, and fracture toughness are also found to be increased due to the reduction in the porosity because of densification. The reduction in porosity due to densification can increase the thermal conductivity significantly. Secondly, the pore coarsening causes coalescence of pores and/or widening/opening of column gaps. Due to several powder compacts during solid state, sintering is caused by localized transport of atoms/molecules due to diffusion and/or bulk particle rearrangement. The coalescence of initially (in as-produced state) isolated pores or by nucleation and growth during the recrystallization of the ceramic coating is also perceivable [41]. The widening of column gaps TBCs ought to be occurring due to the thermal expansion mismatch between the TC and the substrate (especially when higher thermal expansion coefficient for substrate than the ceramic TC was noticed). Thirdly, the grain growth happens with the increase in the grain size of a solid, which occurs in both dense and porous polycrystalline solids at higher temperatures. Due to the conservation of matter, an increase in the average grain size is accompanied by the disappearance of smaller grains. The driving force for grain growth is the decrease in free energy that accompanies reduction in the total grain boundary surface area. The excessive sintering of the ceramic top coat during the thermal cyclic testing may cause a reduction in strain tolerance of the coatings; an increase in the stress level within the TBC system; the closure of pores and cracks, which in turn results in reduction in total porosity of the coating; an increase in thermal conductivity; weak intercolumnar bonding due to shrinkage of the columns; and even providing easier paths for oxygen ingress as well as crack propagation in the top coat between the columns [42]. During thermal cycling test, the microstructure evolves and is probably submitted to a sintering effect, leading to a more rigid coating and a loss of strain tolerance, which can also affect TBCs performance significantly. Many vertical cracks and uniformly distributed pores are the sources of superior strain tolerance and higher thermal cycling life. The $> 10\mu m$ and $1-10\mu m$ pores are the most susceptible to the pore coarsening, especially $> 10\mu m$ porosity, which increases mainly due to the widening of intercolumnar spacing during heat treatment, whereas the very fine pores, i.e., $< 100nm$ porosity, are more prone to undergo densification. The $100-1000nm$ pores are most stable, showing very little change after heat treatment, whereas the $1-10\mu m$ pores show a mixed behavior, i.e., both increase as well as decrease but mostly increase [43].

The SPS-TBCs are regarded as better thermal shock resistance and lower thermal conductivity. As the low-viscosity suspensions are mainly prepared by dissolving submicron or nano solid particles in a solvent, which is water or organic liquid such as ethanol, creating highly packed coatings is achievable. The suitable dispersants which adsorb on particles surface are usually added to the suspensions to maintain their stability and prevent the solid particles from agglomerating and precipitating. For example, polyacrylic acid and phosphate ester are used on water-based suspensions and ethanol-based suspensions.

As ceramic suspensions indicate the particle size and the surface chemistry of powders, solid loadings, and percentage of dispersant, low-concentration ceramic suspensions create porous and consist of aggregated solid particles coating, while high-concentration ceramic suspensions form denser and primarily consist of fine splats coatings. As a result, the top coat consists of a very fine, feathery columnar microstructure. The thickness of the columns decreases from top to bottom, and the size of the porous channel reduces accordingly. At the bottom, very dense and fine thin columns are generated. Very fine nanopores can be identified at this zone [44].

Thus, the microstructure-dependent physicomechanical properties of SPS-applied ceramic TBCs can be summarized as: SPS-columnar microstructure creates lower thermal conductivity, lower fracture toughness, and limited lifetime expectancy [30]; SPS vertically cracked and columnar creates significant improvements in thermal cyclic fatigue (TCF) lifetime [45]; SPS-TBS creates the increase in the thermal insulation performances of the ceramic top layer of TBCs, which was identified as a main issue in the increase in the operating temperatures. Thermal properties, such as thermal conductivity of coatings, largely depend on the microstructure, namely, vertical cracks and phonons transfer through structure. As thermal conductivity depends on thermal diffusivity, specific heat and apparent density of the coating are determinative. There are several possible contributing modes of heat transfer, such as

- conduction through the solid YSZ,
- conduction through the gases in pores,
- radiative heat transfer, and
- some contribution from convection if the segmented cracks are present.

The thermal conductivity generally decreases with increasing total open porosity and inversely with coating density. For instance, the thermal conductivity of SPS-YSZ is about 0.3–1.6 $W/m{\bullet}K$, which is between 30 and 40% less than APS (atmospheric plasma spraying) and EB-PVD (electron beam physical vapor deposition) formed coatings. The thermal conductivity at ambient conditions is due to the phonon conduction. In reality, since the coating is not fully dense and consists of many microstructural features such as pores and cracks, the overall thermal conductivity is lower. This is because the microstructural features interrupt the phonon conduction by scattering at these microstructural features' boundaries. Hence, more scattering interfaces are preferred in the form of microstructural features in coating to get a lower thermal conductivity. The introduction of interpass boundaries in TBCs can significantly reduce the thermal conductivity (about 0.6 $W/m{\bullet}K$), which is due to the existence of lots of micropores in SPS coatings. The heat transport in this system is dominated by phonons below 1,450 K, and the micropores enhance the phonon scattering and thereby effectively reduce the thermal conductivity of coatings, which surpasses the effect of vertical cracks on the thermal conductivity. The large mass of vertical cracks provides direct channels for heat flow and result in the increase of thermal conductivity. Despite the fact that thermal transformation limits YSZ-TBS application, a 6–8 wt.% YSZ at 1,200°C experiences a 0.7 $W/m{\bullet}K$ conduction [22]. The higher content of porosity in the case of SPS coatings also increases the thermal resistance through the thickness and decreases thermal conductivity. Due to the presence of fine ($< 1\mu m$) submicron- or nanosized microstructural features in these coatings, phonon scattering enhances, which then results in lowering the overall thermal conductivity. Since the pores present in such coatings are in submicron or nano range, conductivity in the gas within the pores can also be significantly lower. The thermal conductivity can be calculated as:

$$\lambda = \alpha \rho C_P \tag{5.25}$$

Where λ is the thermal conductivity, α the thermal diffusivity, ρ the coating density, and $C_P$the heat capacity. As a result, the thermal diffusivity is:

$$\varphi(t) = -\lambda S \frac{dT(t)}{dt} \tag{5.26}$$

Where φ, T, and S represent the heat flux, the temperature, and the illuminated surface. For both SPS coatings, the thermal conductivities present an almost-flat profile over the tested temperature range. The diffusivity is mainly attributed to the difference of porosity volume and distribution and is temperature-sensitive. (See *Figure 5.41*.)

As depicted in the following equation, the higher the void content and the more elongated the void (longest axis oriented perpendicular to the heat flux), the lower the coating apparent thermal conductivity:

$$\frac{k_{coating}}{k_{bulk}} = \frac{1}{1+(2/\pi)\times VC\times(b/a)} \tag{5.27}$$

Where $k_{coating}$ is apparent thermal conductivity of the coating, k_{bulk} the thermal conductivity of bulk material $W/m{\bullet}K$, VC the void content %, and b/a the void aspect ratio. The intercolumnar voids observed in columnar structures of SPS due to its elongated shape can increase radiation contribution to thermal conductivity. (See *Figure 5.42*.)

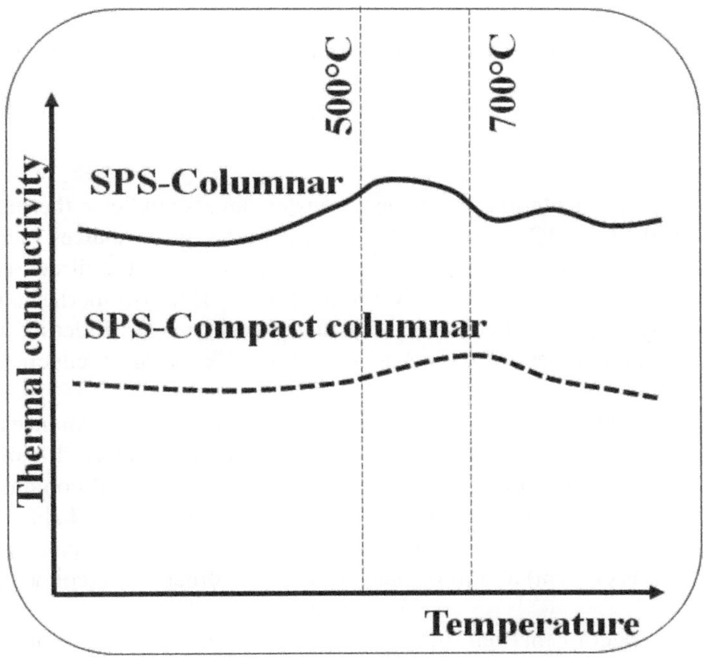

FIGURE 5.41 The structure dependency of a typical YSZ-SPS TBC system.

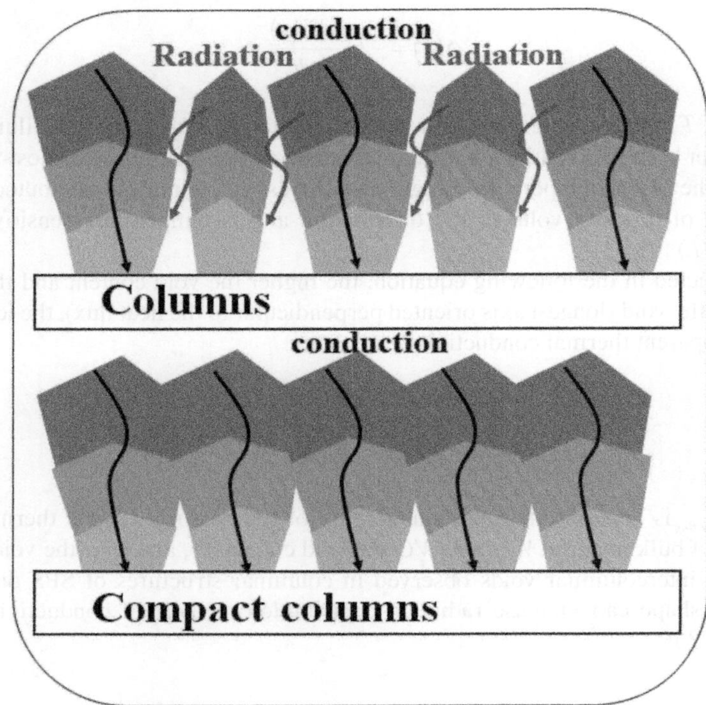

FIGURE 5.42 The heat transfer mechanisms dependency to structure of a typical SPS-TBC system.

The pore/crack orientation toward heat transfer can also indicate thermal behavior of a typical SPS-TBC. The porosity that most effectively reduces thermal diffusivity is that with the most area oriented perpendicular to the direction of heat transfer. Therefore, higher porosity level in the form of IPBs within the coating and which run roughly perpendicular to the primary heat transfer direction will have increased the effectiveness of the porosity within this coating at reducing thermal conductivity.

Thermomechanical behavior and thermal insulation performances of the ceramic TBC coatings are mostly related to the void architecture. Its quantification is hence fundamental and usually includes (i) the void total content, (ii) the void size distribution by size class, (iii) the discrimination by void shape (globular, crack, delamination), and (iv) the void connectivity to the surface on the one hand (open voids content) and to the substrate on the other hand (connected voids content).

The other aspect of thermal diffusivity relies on chemical composition dependency. Decrease in thermal diffusivity of coatings with substitution of lower valance cations as $M_{1M_2}^{'}$ such as incorporation of yttrium atoms into the zirconia lattice

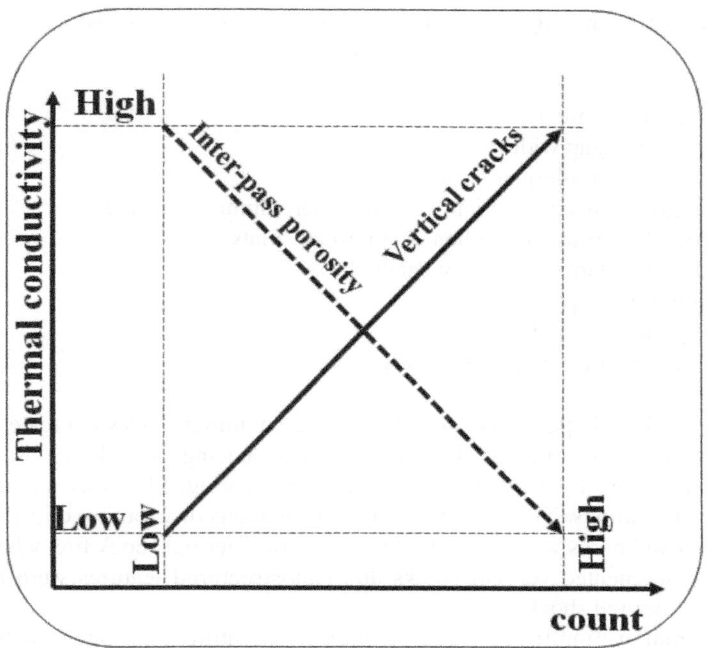

FIGURE 5.43 The thermal conductivity relation with microstructural features levels.

requires the generation of oxygen vacancies. These point defects produce scattering of the lattice waves (phonons) that transport the thermal energy through the coating. The medium dependency of thermal diffusivity indicates aqueous/nonaqueous selection on the suspension media. There are larger, faster, and followed paths in water-based suspensions, which are more orthogonal to the substrate than those produced from ethanol-based suspensions [46]. As a result, faster impact speed of larger droplets causes better cohesion within the coating, generating more fine-scale pores and less large-scale pores. The combination of these properties causes better heat conduction through the thickness of the ceramic layer. The individual significance of some of the microstructural features in a typical SPS microstructure on thermal conductivity is shown in *Figure 5.43*. The vertical cracks, intercolumnar spacing, if very large or wide, may ease the passage of hot gases to flow through them, which can increase the thermal conductivity of the coating. However, inter-pass porosity bands are almost perpendicular to the direction of heat flow, so they can help in decreasing the thermal conductivity. Other fine-scale pores also help in reducing the thermal conductivity due to the enhanced phonon-scattering interfaces in the microstructure.

The thermal shock-resistance properties of SPS-TBCs have an inverse relation with volumetric expansion coefficient and direct dependency with thermal conduction. The high strain tolerance that results from the vertical cracks of the columnar structures improves the thermal cycling life. The high thermal shock performance is affected by optimum $>10\,\mu m$ and $100-1000\,\mu m$ porosity sizes and less dependent on other porosities. Overall, it can be improved by:

- Optimum column density
- Low column gap width
- Optimum intercolumnar spacing
- Negligible pores in the vicinity of the intercolumnar spacing
- Negligible partially unmolten or improper splats
- Negligible interpass porosity bands
- Optimum intersplat porosity
- Negligible spherical particles
- Negligible intrasplat porosity [47]

A structure of largely parallel columns with intercolumnar spaces to create $>10\,\mu m$ porosity, with the vicinity of the intercolumnar spacing devoid of porosity and intra-columnar porosity in the $100-1000\,\mu m$ range, is designed for thermal shock performance applications. The presence of inhomogeneous distribution of pre-existing microcracks and pores were found to deteriorate the thermal shock life, whereas the presence of segmented vertical cracks and nanostructured features were found to improve the thermal shock life.

The thermal cycling lifetime and the high-temperature atmospheric behavior of SPS-TBCs are followed by general and zone-restricted oxidations. The longer thermal exposure at high temperatures results in crack propagation, which is further assisted by the growth of thermal grown oxide (TGO) that grows on the top of the bond coat [48]. This layer is positioned between ceramic top coat (thermal insulator) and metallic bond layer (bonder and oxide protector), as shown in *Figure 5.44*. The growth of the TGO occurs between the bond coat and top coat and is mechanically constrained by both. This is because the formation of the oxide is accompanied by a volumetric expansion strain (~25% for alumina) which has to be accommodated at this location. A small fraction of this strain (~1%) is directed in the plane of the oxide layer, but the remainder occurs perpendicular to the bond-coat surface. If this surface were flat and no edge constraints existed, then the top coat would simply be displaced upward, but in practice, of course, the surface is nonplanar. As a consequence, there will be a variation in the upward (x_2) displacement along the bond coat surface simply from geometric factors even though the intrinsic oxidation rate is everywhere constant. Local, out-of-plane displacement rates due to oxidation around a bond coat protuberance. This displacement, g_n, normal to the bond coat surface, is constant everywhere, but because of the curvature of the protuberance, the out-of-plane vertical component, g, in the $x_2$direction will vary with location [49]. It will, in fact, be a minimum at the flanks of the protuberance. If the top coat is to retain continuity and adherence to the TGO and bond coat, then it will experience imposed out-of-plane tensile continuity strains. These strains, and associated stresses, will be a maximum, where the vertical displacement is at a minimum, i.e., at the flanks of the protuberance [48]. It is established that the oxidation weight gain of double- and triple-layered SPS-TBCs is apparently lower than that of single-layered TBCs [50].

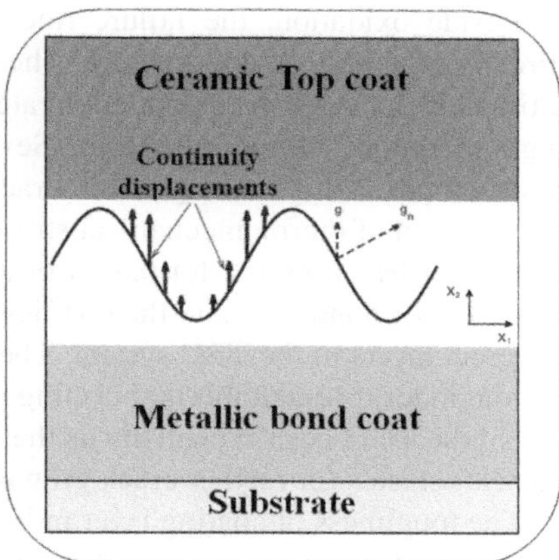

FIGURE 5.44 The schematic of layer arrangement in typical SPS-TBC (left) and growth of TGO at bond coat interface (right).

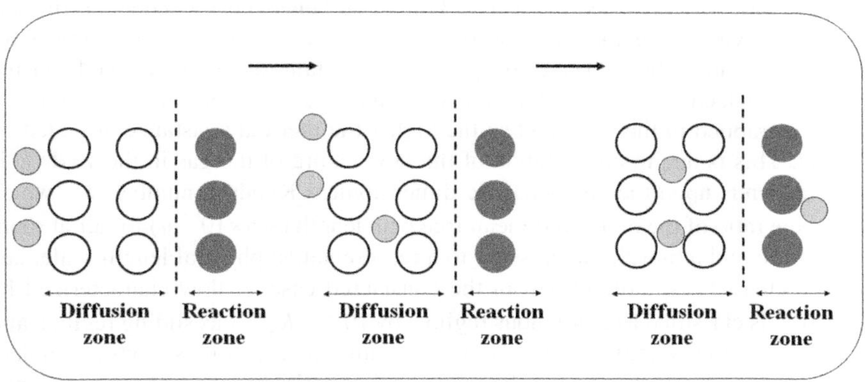

FIGURE 5.45 The schematic of layer oxidation in typical SPS-TBC, showing the atomic diffusion-reaction along interface.

Generally, the oxidation and elemental diffusion are a continuous and related process. After TBCs operate at the elevated temperature, under the driving of the difference in chemical potential between the external reservoir, e.g., O_2 and coating, the guest atoms, e.g., O, leave the external reservoir and insert into coating at the boundary. Subsequently, the guest atoms continue to diffuse forward in the coating until they reach the reaction region. When the guest atoms encounter the outward alloy elements, e.g., Al and Ni, new oxides form and significant stress occurs due to the surrounding constraint. The induced stress not only causes the failures of TBCs but also affects the diffusion and oxidation processes. Thus, oxidation and diffusion, as a whole, are responsible for long-term failure mechanisms of TBCs in heavy-duty gas turbines [51] (see *Figure 5.45*).

Beside oxidation, the failure (thermal cycling failure [TCF]) of SPS-TBC is directly related to microstructure. The thermocyclic fatigue performance of a TBC ultimately depends on the oxidation rate of the bond coat, with the ceramic layer having a secondary effect on lifetime. Several failure mechanisms of TBCs have been reported previously, among which cracking at the top coat–bond coat interface due to induction of thermomechanical stresses is the most common one [52]. The failure is mainly driven by the formation of a thermally grown oxide (TGO) layer due to oxidation of bond coat and the coefficient of thermal expansion (CTE) mismatch of different layers in the TBC system. The results of modeling the same TGO interface region under repeated thermal cycling have shown crack initiation at surface asperities in the TGO layer repeatedly as the layer itself grows. These can act as the initial nucleation sites for further crack growth across the interface.

The toughness of coating layer indicates TCF behavior, such as more strain tolerance by PSZ than fully stabilized zirconia via less crack susceptibility. These stresses are induced during thermal cyclic testing because of mismatch in layers. The vertically cracked microstructure is clearly less strain-tolerant than the columnar microstructures with similar chemical composition. The cracking will only propagate into the ceramic top coat in systems where fracture toughness of the ceramic coating is relatively low. The presence of such IPBs may be beneficial for reduction in thermal conductivity of the coating; however, they also act as a low-energy pathway for crack propagation. The void size also plays an important role in the induced thermal resistance. Indeed, the smaller the void size, the lower the thermal conductivity of the gas entrapped in the void, and so the higher the thermal resistance provided by the void. This is due to an evolution of the flow nature of the gas in the void. This nature depends upon the characteristic dimensionless Knudsen number, K_n, which depicts the ratio of the molecular mean free path length ($\sim 6\times10^{-8}\,m$) for air at room temperature and atmospheric pressure to a representative physical length scale, i.e., the void characteristic dimension in the considered case. A flow characterized by $K_n \ll 10^{-2}$ is classified as continuous regime, by $10^{-2} < K_n < 1$ as sliding regime, and by $K_n \gg 1$ as rarefied regime. At a given temperature (i.e., an almost-constant molecular mean free path), decreasing the void characteristic dimension leads to change in the regime from continuous to rarefied, resulting ultimately in an increase in the void thermal resistance:

$$k_{void} = \frac{k_{gas}}{1+\left(2.5\times10^{-5}/d_{void}\times p\right)} \tag{5.28}$$

Where k_{void} in the equivalent thermal conductivity of void ($W/m{\bullet}K$), k_{gas} the thermal conductivity of gas entrapped in void ($W/m{\bullet}K$), d_{void} the void diameter, and p the pressure of gas in void (Pa). The heat flux is transmitted by a flux of phonons and corresponds in a first approximation to the sum of the momentums of phonons. The so-called Umklapp process is depicted as an anharmonic (oscillation which does not follow a simple harmonic motion) phonon-phonon (or electron-phonon) scattering process [53]. From Umklapp process results a phonon with a norm of its momentum vector larger than the characteristic dimension of the Brillouin zone of the considered material at the considered temperature. As a result, the momentum of the phonon is decreased due to a decrease in its mean free path, and since the phonon thermal conductivity is a function of this mean free path, its thermal conductivity is decreased also. Ultimately, the apparent conductivity of a material into which Umklapp process takes place is decreased. The phonon scattering takes place predominantly at crystal boundaries, crystal defects, etc. Reducing the characteristic dimension of the crystals leads to an increase in the boundary density, promoting the Umklapp scattering.

For instance, an oxide coating (YSZ:8 wt. % yttria partially stabilized zirconia, $\phi: 25\%$, $D_{50}: 300$ in ethanol suspension with 5–10% pore percentage) undergoes considerable sintering, and the failure of the coatings occurs in the TGO layer. Similar mode of failure along with the widening or opening of the column gaps (indicated by arrows) due to sintering can be observed in all similar coatings. The difference in expansion and contraction of top coat and bond coat due to their CTE mismatch exerts shear forces on the TGO layer, which results in the failure of coatings due to cracking within the TGO layer [54]. The study of the effect of TGO growth around the convex region on the stress evolution in TBCs shows that the fast growth of nonprotective oxide can enhance the development of stress. The fast growth of local mixed oxide (MO) induces the cracking and interfacial delamination in the TC. The structure may form (MO-spinel) phase, and the suppression of local MO growth can prolong the service life of TBCs. The local growth of the oxide layer induces extra tension (see *Figure 5.46*) which can cause interracial tensile stress and debonding (I) due to thermal expansion mismatch between ceramic and MO, jacking up ceramic layer and inducing tensile tress (II), and interfacial delamination (III) with the points of stress concentration shown as red arrows [30]. Perovskites, zirconates, and aluminates [22,51] have less lifetime performance than YSZ coatings mainly due to lower mechanical properties, such as the lower toughness of zirconates, for example.

The generated cracks during TCF can even enhance thermal compliance and allow the coating to resist to several additional cycles of TCF. The schematic microstructural behavior of columnar and compact columnar during TCF are compared in *Figure 5.47*. In the columnar structures, columns and a quite-homogenous initiation zone at the base of columns will be segmented during the test ($t_{cracks,col}$), and as a result, the coating is suddenly debonded from the bond coat and fails ($t_{ail,col}f$). In the compact columnar structures, initial microstructure is not really able to accommodate thermal mismatch ($t_{0,col/comp}$) between substrate and top coat which leads to large cracks probably appearing between compact columns ($t_{cracks,col/comp} < t_{cracks,col}$) and allowing thermal compliance, creating cracked morphology and sintered, leading in the end to a similar failure whereby the top coat is suddenly debonded from the bond coat ($t_{fail,col/comp} < t_{fail,col}$).

The media dependency of TCF is focused on the media physical properties. For ethanol-based suspensions, the tendency for columnar microstructure formation increases with reducing suspension viscosity due to stronger atomization of the suspension and resultant smaller particles in the plasma plume. The organic media in suspension causes an increase in viscosity and, as a result, increases the coating thermal conductivity and tends to reduce the thermocyclic fatigue lifetimes [55]. For water-based suspensions, the media produces a shift in the coating microstructure morphology from columnar to vertically cracked due to higher suspension surface

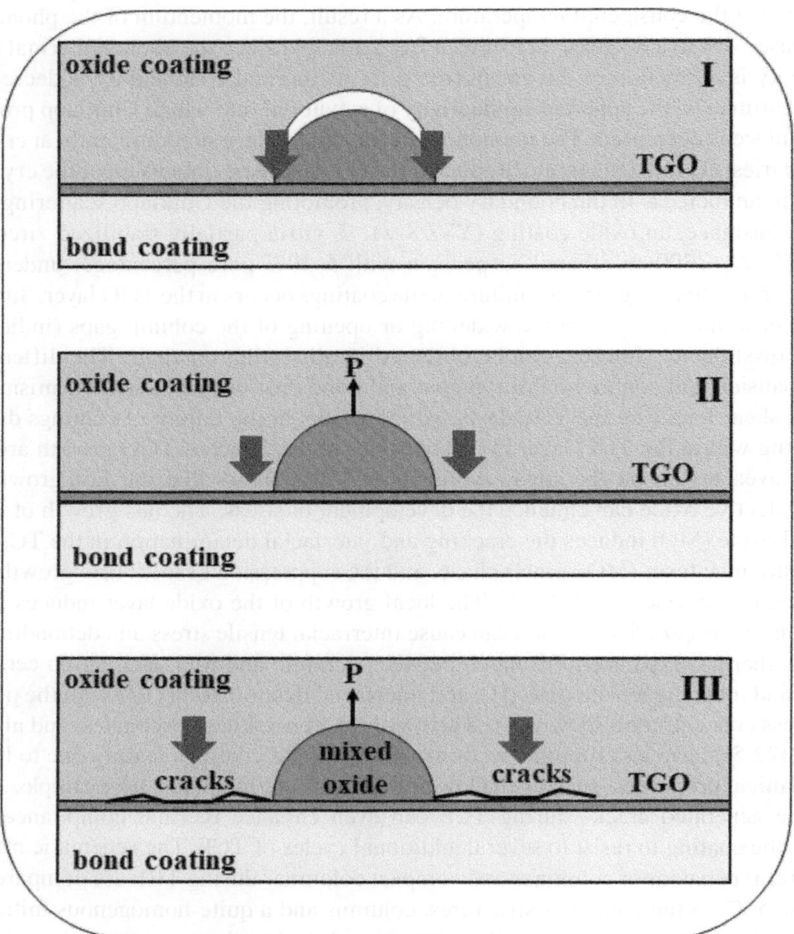

FIGURE 5.46 The schematic of layer oxidation in typical SPS-TBC showing local growth of oxide layers and stress accumulation.

tension, as moving from ethanol to water as a solvent has the tendency to dramatically increase atomized droplet size.

Usually, at the ceramic interface with the bond coat, there is a fairly uniform thermally grown oxide (TGO) layer that consists primarily of alumina with minimal amounts of mixed oxide phases present. The cracking appears to be initiated within the TGO layer and propagated across the coating interface predominantly within the TGO layer itself. The cracking is driven by stresses built up due to oxide growth and thermal mismatch within the interface region during cyclic exposure for leaving an uneven layer of ceramic attached to the TGO beneath. In this case, there is a lot of very angular cracking within the ceramic SPS coating at the base of the columns, for

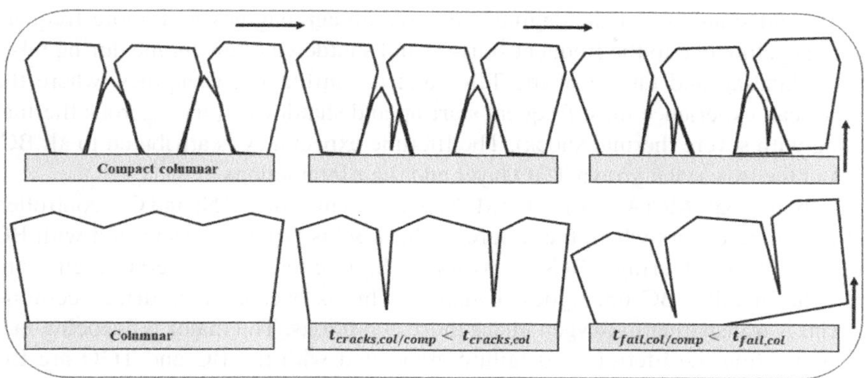

FIGURE 5.47 The schematic of coating debonding after TCF for columnar and compact columnar structures.

instance in YSZ:8 wt. % yttria partially stabilized zirconia, $\phi : 25\%$, D_{50} : 200 in ethanol suspension with 12–13% pore percentage system after 350 thermo-cyclic fatigue cycles at 1,100°C for sixty minutes. Also, the failure might have resulted from cracking through the first interpass porosity band. This cracking leaves an almost-uniform layer of oxide top coat attached to the TGO that shows evidence that the cracking might have initiated within the TGO or at the TGO–bond coat interface (8 YSZ:8 wt. % yttria partially stabilized zirconia, $\phi : 25\%$, D_{50} : 50 in water suspension with 7–10% pore percentage system after 150 thermocyclic fatigue cycles at 1,100°C for sixty minutes).

Therefore, the three layers, i.e., TC, BC, and TGO, are crucial and should be considered while designing the SPS TBC for superior TCF performance. Factors such as surface roughness and chemistry of the BC, the TGO composition, and the nature of TC microstructure, TC phases, and TC sintering behavior were identified as to be the most crucial. In order to design a high TCF performance columnar SPS-TBS, the following criteria should be addressed: a bond coat with minimum roughness and β instead of γ structure; alumina-based TGO layer with minimum spinel content; top coat with dense columnar structure with minimum spherical particles and medium porosity (100–1,000 nm) content; the nontransformable phase like tetragonal phase in zirconia; and minimal sintering and avoiding pore coarsening/grain growth/densification in the top coat [52].

The extent of the thermomechanical load (Isothermal Cyclic) can depend on the gas turbine application (for example, military aeroengine or land-based power generation IGTs or commercial civilian aeroengines) and the type of component in the gas turbine (for example, combustor or turbine blades and vanes or afterburner). The TBCs applied on the gas turbine components for any of the abovementioned applications can experience high stress loads which are relaxed by forming new cracks or propagating the pre-existing ones, ultimately leading to delamination according to application, such as the land-based power generation IGTs, which run over a long period of time (long-running cycles) under isothermal conditions without frequent start-up and shutdown of the engines, the civilian aeroengines with more frequent thermal cycling due to the frequent start-up and shutdown of an engine during take-off and landing and failure of the TBC; and the military aeroengines, where the engines can experience most frequent start-up and shutdown (very rigorous thermal cycling with severe thermal shock). The lifetime expectancy is attributed to all BC, TC, and the in-service grown TGO layer and their interactions.

The bond coat (MCrAlY alloy for M: Ni, Co, or a mixture of Ni and Co) controlled cyclic lifetime occurs where the failure in the TBC is typically associated with BC oxidation (TGO). During TBCs' exposure to high-temperature operating environment, the metallic BC undergoes oxidation. This is because the surface ceramic material is transparent to oxygen under those conditions. The major influencing factors controlling the lifetime and failure associated with the BC and TGO are BC chemistry, BC roughness, and TGO composition. The primary and stable oxide is alumina, because it is the least formation energy among other possible oxides of constituent metals such as Ni, Co, and Cr present in the BC, about -1675.7 kJ / mol. The dense alumina acts as a protective oxide layer for further oxygen diffusion and prevents formation of other detrimental mixed oxides.

This BC layer consists of β- (Ni [Co]Al) (aluminum-rich phase and main aluminum reservoir) precipitates in a continuous γ- (Co/Ni/Cr) (gamma phase) matrix. During high temperatures, the aluminum starts to diffuse toward both the BC-TC interface as well as the BC-substrate interface, consuming the aluminum reservoir (beta phase). The beta phase–depleted region close to both the interfaces becomes rather clear, and this region is referred to as "depletion zone." These depletion zones are formed in the BC due to the incomplete diffusion of the aluminum from the BC [56]. As the aluminum level within the BC is reduced due to its continuous depletion, thermodynamically stable α-alumina slowly grows first as a TGO layer and continues to grow to a certain critical thickness. The α-alumina layer grown on a highly wavy BC profile cracks on cooling as a result of high local curvatures and thermal expansion mismatch between BC and TGO (α-alumina). These cracks can now provide an easy ingress of oxygen to the underlying BC. This may then further continue the α-alumina formation if there is still some aluminum left; otherwise, they start forming mixed oxides of other underlying BC elements, which in this case are Ni, Co, Cr, etc. It has been shown that when aluminum is depleted to below 2.5 wt. % Al in the BC, no more formation of the protective α-alumina occurs. The formation of other mixed oxides such as Ni/Co/Cr spinels in the TGO is considered as detrimental for the TBC life due to their faster growth rate, which leads to the rapid local volume change in TGO that generates excessive stresses. It has been identified as impossible to completely eliminate the formation of such detrimental spinels. The spinel phases have lower fracture toughness than the alumina, which then eases the cracking and crack propagation through TGO. Also, the roughness, however, creates a wavy profile with hills and valleys on the BC. The TGO is typically formed following nearly the BC profile. In case of conventional TBCs, this wavy profile creates a different stress distribution along the profile in the TGO. Due to the wavy profile of the TGO geometry, out-of-plane stresses are produced, resulting in tensile stresses at the hills and compressive stresses at the valleys. These tensile stresses at the hills act as a driving force for the crack propagation through the BC-TGO interface. These cracks are formed because of the volume change near the TC-BC interface due to the formation of an extra layer (TGO). As the TGO layer is formed due to BC oxidation in this case, such a crack formation cannot be easily avoided. The major cause of failure in TBCs associated to BC oxidation is related to the stress state at the BC/TGO/TC interfaces. These stresses are introduced in the TBC due to mismatch in CTE (mismatch stress) between different layers (BCTGO-TC) and/or due to the growth of a new layer in the form of TGO in between BC and TC (growth stress). These stresses are relaxed by forming cracks close to the BC-TGO-TC interface, ultimately delaminating the ceramic TC and resulting in failure of the TBC. In order to minimize the failure associated with BC and TGO and increase the lifetime of the TBC, the BC with sufficiently high aluminum reservoir is necessary, and smoother bond coats can also minimize the cracking issue at the BCTGO interface and hence increase the lifetime of the TBC.

The significant top coats (oxide ceramic) in controlling lifetime of the TBC can be as crucial as BC and TGO layers. Most common failure mode observed in the conventional SPS-TBCs is cracking in the ceramic top coat close to the TGO-TC interface. Such crack formations in the TC are also associated with the TGO formation. The crack initiates at the region where the curvature of the BC roughness profile changes from concave to convex, as reported in case of conventional TBCs. The interfacial fracture toughness of the TC/spinels interface is lower than that of the TC/α- alumina interface, which can increase the possibility of fracture between the top coat and TGO. Such failure can be minimized by increasing the fracture toughness of the top coat at the TGO-TC interface. This was achieved by several researchers by adding a thin, dense YSZ layer close to the TGO-TC interface. The intrinsic failure of TBCs, and the growth of TGO layer, is the key issue of coating durability [57]. However, the thickness and morphology of TGO vary with the increase in service time at high temperature. Therefore, it is necessary to study the evolution of TGO layer during thermal exposure. Influenced by the diversity of the high-temperature oxidation mechanism, the oxidation diversity of TBCs is extremely complex, including the composition of materials, the preparation process, and the thickness of coatings. Therefore, different thicknesses and shapes of the TGO layer are formed, which is of great significance to stress distribution around the TGO layer and to the life prediction of TBCs [58]. During the thermal shock cycle, growth of TGO is observed, and the multilayer accumulation of TGO is observed in the protrusion of the surface roughness of the TBCs. The accumulation of TGO results in the cracking of the top layer. The residual stresses during heating-cooling stages can cause crack formation and propagation from these defects (see *Figure 5.48*).

During the thermal shock cycle, growth of TGO can be observed, and the multilayer accumulation of TGO is observed in the protrusion of the surface roughness of the TBCs. The accumulation of TGO results in the cracking of the top layer. The growth rate of TGO near the peak region is higher than that near the valley region. The TGO grows unevenly at the peaks and valleys of the rough top coat/TGO interface after isothermal exposure. The phenomenon that the growth rate of TGO near the peak region is higher than that near the valley region. The TGO grows exponentially near the peaks and valleys, and TGO is thick at the peaks, where there are large tensile stresses, uniformity, and unevenness. There is a significant difference

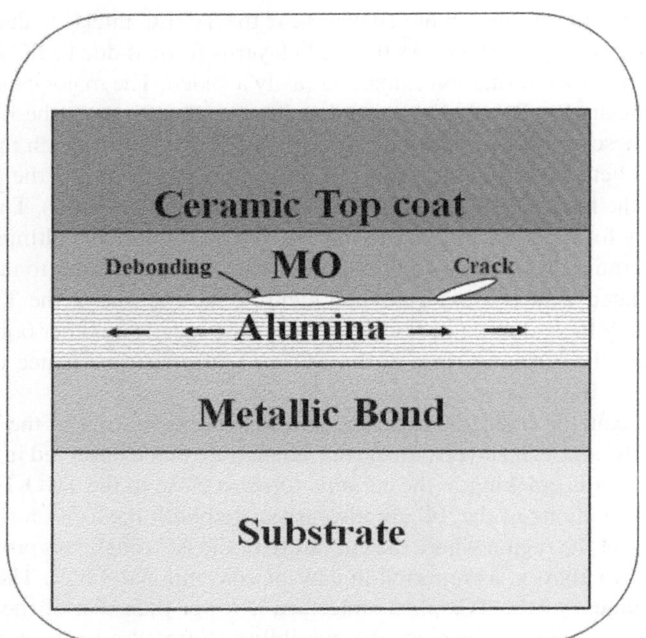

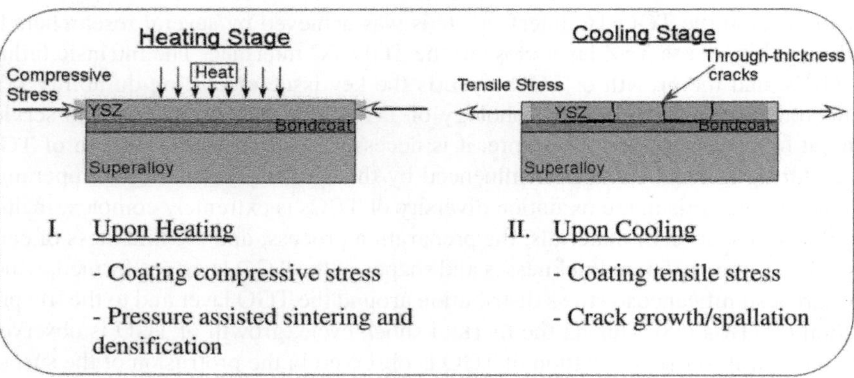

FIGURE 5.48 The schematic of coating failure by TGO growth (up) and crack propagation from these defects due to thermal energy exchange (down).

in the distribution of stress between uneven and uniform TGO. For uneven TGO, the maximum out-of-plane stress of peak and valley increases 200%, and the maximum tensile stress along the interface of the TGO/surface layer is smaller than that of uniform TGO. Moreover, compared with TBCs which were serviced under a uniform temperature field, in the case where TBC is subjected to a thermal gradient, the gradient will affect the growth rate and stress distribution of TGO [39]. Different from the dense microstructure of α-alumina, MO exhibits a porous microstructure and is brittle; accordingly, cracks and interfacial delamination are prone to occur there. Thus, the growth of uniform MO and the induced catastrophic stress are responsible for failures of TBCs in heavy-duty gas turbines, e.g., the cracks in α-alumina and TC layers, and the delamination at the α-alumina/MO interface [51].

The resulting lifetimes for the two optimized SPS columnar structures (up to 1,100°C): up to two thousand cycles to failure, which is representative of the lifetime. The most design parameters include the bond coat type: $\beta-(Ni,Pt)Al$ and with a $Pt-\gamma/\gamma'$, the bond coat roughness, which was identified as a key parameter in order to modulate the size of SPS columns and their organization (a loss of performance happens when the substrate mean roughness Ra is increased). In fact, an increase in mean roughness has a strong impact on column morphology. The columns will become larger, and their size is much dispersed. A broader distribution of column diameter is formed at the highest value of substrate roughness. The bigger the initial column diameter, the more perturbed coating buildup becomes due to shadowing and coalescence effects. The cracks may appear after cycling. The vertical cracks and columnar structure are able to bring thermal compliance to the top coat, allowing a higher thermal cycling resistance. Therefore, a homogenous structure presents a lower thermal compliance compared to vertically cracked structures. For instance, between two YSZ-SPS columnar structures on a $b-(Ni,Pt)Al$ bond coat, the coating performed on the rougher substrate ($Ra = 1.5\,\mu m$) presents larger columns. The loss of organization between columns, as well as the increase in column sizes, could induce an inhomogeneity of thermal compliance. This leads to an increase in internal stress at some locations which could speed up debonding of the YSZ-SPS top coat.

The failure associated with severe thermal gradients through the ceramic TC in TBCs can also be analyzed by studying TBCs' thermal shock behavior. Under thermal shock conditions, TBC experiences a severe thermal gradient across the ceramic TC layer compared to the isothermal cyclic conditions. Higher thermal gradient across the ceramic TC layer can affect the BC temperature, which in turn can affect the TGO formation and thermal mismatch stresses in the ceramic TC near the BC-TC interface, affecting the lifetime. The thermal gradient is of course dependent on coatings' thermal conductivity, which in turn depends on TC microstructure and porosity. At higher thermal conductivity of the ceramic TC, less thermal gradient across is conceivable, meaning the BC can experience higher temperature, which can result in lower lifetime, and the higher strain tolerance, which is a major characteristic of the columnar structure, can accommodate the strain introduced in the TC during the frequent thermal cycling in both isothermal as well as thermal gradient conditions. The increase of strain tolerance in TBCs is known to be beneficial for improving their thermal cyclic lifetime. The fracture toughness of the ceramic TC is also considered to be one of the crucial factors in determining the lifetime of the TBC. Higher fracture toughness is needed not only close to the TC-TGO interface but also throughout the ceramic TC, as cracks can initiate anywhere in the coating especially under sever thermal gradient conditions. The influence of fracture toughness on TBCs' lifetime results in an improvement in the fracture toughness of the ceramic TC and is beneficial for the lifetime.

The chemical aspects can also influence the behavior. The rare-earth zirconates, complex perovskites, and lanthanum aluminates can present a lower bulk thermal conductivity compared to YSZ. The rare-earth zirconates ($RE_2Zr_2O_7$) are mainly used as a top coat due to their lower thermal expansion coefficient [59]. The $La_2Zr_2O_7$ (LZ) shows low thermal conductivity, high melting point, high sintering resistance [60], and high phase stability, which influences the surface tension and viscosities, which plays a critical role in the formation of splats [61]. The gadolinium zirconate (GZ), which also shows high melting point, high temperature phase stability, low thermal conductivity, and good calcia-magnesia-alumina-silicate (CMAS) resistance, is a good example in this system. The application of multilayered TBCs leads to a lower thermal conductivity and a higher thermal life. The rare-earth doped TSZ (dissolving different amounts of rare-earth nitrates into YSZ suspensions) forms cauliflower-like structure with interpass boundaries. The Nd_2O_3 / Yb_2O_3 TSZ systems shows how the increase of Nd_2O_3 / Yb_2O_3 concentration can cause the porosity of coatings to increase and the thermal conductivity to be decreased, similar to 3Y-zirconia/graphene oxide(GO) [15].

The microstructural aspects on TBC can also severely affect the mechanical properties, as the presence of various microstructural features such as pores and cracks are necessary for providing the required thermal insulation in a TBC. These features can also affect the coating's mechanical properties, such hardness, E-modulus, and fracture toughness. The E-modulus and toughness are considered as the two most crucial mechanical properties for TCs. The E-modulus is a measure of coatings' compliance or strain tolerance, whereas toughness is the measure of coatings' ability to resist crack formation and propagation. The lower the E-modulus, the higher the strain tolerance, and the higher the toughness, the higher the crack initiation and propagation resistance. Hence, from a design perspective, a coating with low E-modulus and high fracture toughness would be preferred. High porosity is reported to be beneficial for reducing the E-modulus; however, the toughness of the coating was found to deteriorate if coating possesses high porosity. Also, different microstructural features in the ceramic TC microstructure can affect these properties in a different way. In fact, E-modulus is found to be more sensitive to the presence of various microstructural features due to its anisotropic nature, since different features in the ceramic TC microstructures can be present along different directions. The understating of these effects leads to the motivation for developing the dense vertically cracked (DVC) TCs to lower the in-plane E-modulus, which can improve the thermal cyclic performance by improving the coatings' compliance as it is in the case of columnar TCs. This has further motivated the thermal spray community to develop DVC or columnar coatings by SPS process. The higher porosity and unmolten zones in SPS TCs are found to lower the E-modulus, whereas the presence of dense zones is found to increase the E-modulus. Similar to the E-modulus, the toughness of the TC can also be reduced due to the presence of delaminations or horizontal cracks in the microstructures as the crack propagation along these features can be easier. For SPS-TBC sprayed microstructures, the presence of branching cracks, interpass porosity bands, or high porosity, etc., can also substantially reduce the toughness due to easier crack propagation. The toughness is identified to be more important mechanical property than E-modulus for lifetime and has been considered as the crucial thermal cycling, life-determining factor in case of SPS TCs. The Young modulus (E-modulus) (three- or four-point bending and indentation method) and fracture toughness (double torsion method, double cantilever bending, indentation method) and hardness (indentation method) can be obtained via the instrumented hardness (HIT) (convertible into Vickers Hardness HV) and instrumented E-modulus (EIT) (obtained from the unloading slope):

$$H_{IT} = \frac{P_{max}}{A_P} = \frac{P_{max}}{constant \times h_c^2} \tag{5.29}$$

$$E_{IT} = \frac{1-v^2}{\left\{\left(\frac{2\sqrt{A_P}\left(\frac{dh}{dp}\right)_{P_{max}}}{\sqrt{\pi}}\right) - \left(\frac{1-v_i^2}{E_i}\right)\right\}}$$

Where P_{max} (N) is the maximum applied indentation load; A_P (mm^2) is the projected (cross-sectional) area of contact between the indenter and the test specimen determined from the force-displacement curve and a knowledge of the area function of the indenter; h_c (mm) is the contact depth; *Constant* is constant for a given indenter; v is the Poisson ratio for coating material; v_i and E_i are the Poisson ratio and E-modulus of the indenter, here for indenter, respectively. The cracks $(0.25 \le l / a \le 2.5)$, without significantly damaging the coating, can be generated for fracture toughness measurements using similar Vickers indentation technique. Cross-section polished TBCs should be indented at a maximum load:

$$K_{IC} = 0.018\left(\frac{E}{H_{IT}}\right)^{\frac{2}{5}} H_{IT} a^{\frac{1}{2}} \left(\frac{a}{l}\right)^{\frac{1}{2}} \tag{5.30}$$

Here, K_{IC} is the mode, I is the indentation fracture toughness $(MPa.m^{1/2})$, E is the elastic modulus (MPa) of the coating determined from $EIT\left(E = EIT * (1-v2)\right)$, a is the indentation half-diagonal length (m), and l is the crack length (m).

For instance, a YSZ:8 wt. % yttria partially stabilized zirconia, $\phi : 25\%$, $D_{50} : 500$ in ethanol suspension with 10–11% pore percentage system shows that the fracture toughness yields the lowest fracture toughness value of 1.58 ±0.03 $MPa.m^{1/2}$. There are several factors influencing the erosion performance of TBCs, such as microstructure, porosity content, and fracture toughness. Higher fracture toughness allows more kinetic energy of the erodent to be absorbed by the TBC before the onset of delamination cracks. The aforementioned erosion results are in agreement with the literature, with the lower fracture toughness being plausibly responsible for its inferior erosion resistance. On the other hand, the superior erosion resistance can be attributed to its lower porosity content, higher hardness, and higher fracture toughness. The open column-like microstructure with higher column density and column gaps along with inferior fracture toughness is more prone to erosion by tunneling mechanism, which results in maximum material loss, compact column-like microstructures along with relatively higher fracture toughness, the and TBCs possess relatively superior erosion resistance [62].

The combination of chemical and thermal behavior is recorded for hot corrosion/oxidation properties [63]. For a YSZ:8 wt. % yttria partially stabilized zirconia, $\phi : 25\%$, $D_{50} : 500$ in ethanol suspension with 10–11% pore percentage system, the molten salt infiltration/hot corrosion evaluation can be carried out with a mixture of vanadium pentoxide (V_2O_5) and sodium sulfate (Na_2SO_4) in the ratio 55:45 wt% at 900°C for 8h. TBC with higher column density, lower column width, and higher coarse porosity, due to presence of a greater number of column gaps, aids molten salt infiltration throughout the top coat with relative ease. This can be attributed to the fact that tensile stresses in the TBC during heating cycle accelerate the molten salt infiltration with relative ease through the relatively higher number of column gaps. The TBCs with lower column density, higher column width, and lower number of column gaps are relatively more resistant to infiltration through column gap. Additionally, within columns, lower porosity and minimal interconnected pores are desirable to mitigate molten salt infiltration, especially the porosity content within columns [63] (see *Figure 5.49*).

According to the necessary parameters to be evaluated, the characterization and process parameter-properties relations in SPS-applied TBCs include methods to indicate the amount and effects of porosities on propenoates. (See *Table 5.4.*)

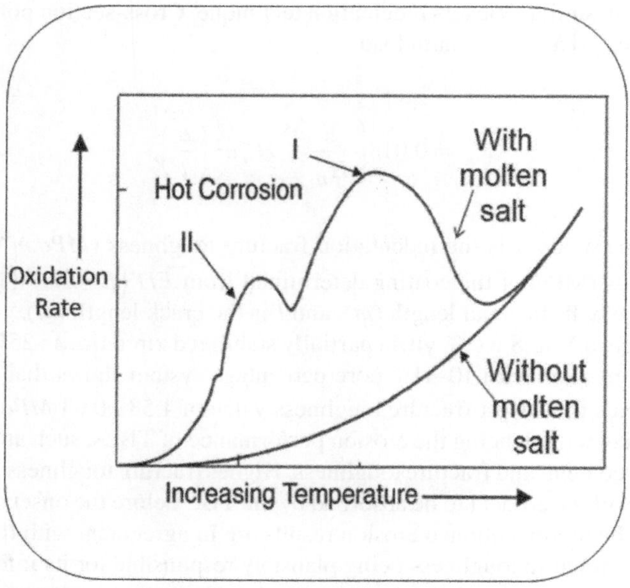

FIGURE 5.49 The hot corrosion induced oxidation with and without molten salet media.

TABLE 5.4
The physical characterization of SPS-TBC.

Method	Defects
Archimedean porosimetry (AP)	Closed porosity
Mercury intrusion porosimetry (MIP)	Open porosity
Pycnometry (P)	Open porosity
Stereological (S)	Voids-Cracks-Delamination
Electrochemical impedance spectroscopy (EIS)	Voids near substrate

These physical methods are based on the intrusion of a liquid into the void network. It can be water or other suitable liquid (Archimedean porosimetry) or mercury (mercury intrusion porosimetry, MIP). For the MIP, the nonwetting liquid volume of mercury is measured as a function of the applied impregnation pressure. This technique permits to discriminate also the size of the open pores: the smaller the open pores, the higher the required impregnation pressure. The intrusion medium can also be gas (generally helium), in which case the pressure increase in an unvarying gas volume is measured for a cell containing the sample or empty. Archimedean porosimetry permits to quantify the closed (nonconnected to the surface) voids content by measuring the "dry" weight and the "wet" weight (immersion into water) of the sample. Depending on the method and intrusion medium, these techniques permit to quantify one or more of the voids connected to the coating surface (i.e., open voids), closed voids (voids not connected to any surface), and the specific surface area of the open voids. The main drawbacks are the facts that measurement is not direct and that the impregnation of the whole open voids network is not guaranteed. Two main limitations impede the accuracy/reliability of such methodologies, such as for MIP, the high pressure applied to ensure percolating of the liquid through the microstructure, which can induce material failure on the one hand and the minimum sample volume needed for analysis on the other hand (common for all these methods). The stereological protocols facilitate the observation of the coating cross section (implementing mostly scanning electron microscopy, SEM) coupled with appropriate image treatments, and statistical models known as stereological models permit to quantify voids content as a function of void sizes and when considering some specific models of their shape and crack density and orientation. These methods permit to analyze the void content regardless of the connectivity of the network. An appropriate magnification and image resolution must be determined to reproduce the all-necessary details (i.e., microcracks). Experience indicates that image characteristic dimension should be between ten and fifteen times larger than the objects of interest (voids) to be analyzed to account with the representative elementary volume (REV) of the structure. Two main limitations impede the accuracy/reliability of such methodologies: the limited SEM resolution on the one hand, which makes difficult to take into consideration features smaller than 0.1 m and the artifacts (i.e., pullouts, scratches, etc.) on the other one resulting from cutting and polishing steps. Electrochemical methods (electrochemical impedance spectroscopy) involve the percolation of an electrolyte inside the interconnected voids network, allowing quantifying the void fraction connected to the substrate by analyzing the chemical reaction (passivation most of the time) at the substrate/electrolyte interface. The electrochemical impedance spectroscopy (EIS) technique consists, hence, in measuring the impedance of the electrochemical cell. The immersed substrate surface behaves as the working electrode. The connectivity of a voids network is related to the quantity of voids that connects the substrate to the surrounding atmosphere through the coating thickness. The simplistic test of deionized water droplet percolation through the coating permits to determine the smallest open pore diameter into which the water is able to percolate. For example, contact angle, θ, between zirconia and deionized water is measured to be about 59 degrees, and the surface energy, γ, of deionized water is 72.8 $mN\,m^{-1}$. At atmospheric pressure (about 105 Pa), pure water percolates into open voids of equivalent diameter equal or larger than 1.5 μm. The simple calculation proves that the classical coating connectivity measurement by EIS cannot be representative of the connected voids network since the electrolytic solution is diluted and has a surface energy of the same order than that of deionized water. So the ionic solution cannot percolate through the entire connected network. The EIS analysis does, hence, not seem to be a reliable protocol to quantify the connectivity of SPS coatings.

5.5 APPLICATIONS OF SPS COATING: AEROSPACE AND BIOMATERIALS

The development of SPS-TBCs are directed toward (i) the development of new classes of ceramic TBC materials, such as pyrochlore, $YSZ:Eu$, or $Ta_2O_5\,/\,Nb_2O_5$ doped Y_2O_3, among others; (ii) the decrease in the ceramic TBC material microstructural features (grains and voids size), from micrometer-sized to submicrometer-sized and even nanometer-sized, resulting in reduction of their apparent thermal conductivity together with an increase in their thermal reflectivity; and (iii) the concomitant development of new processes/the adaptation of existing processes to manufacture such microstructures, among which thermal spray processes are one of the major players.

The SPS-TBC development for aerospace applications is a major goal in the development of new TBCs for aeroengines and land-based turbines today and addresses the severe thermal gradient or thermal shock, which can (i) reduce manufacturing costs, (ii) increase reliability (and therefore reduction in maintenance frequency), and (iii) improve insulation performance. The first goal requires the implementation of processes with a high ratio of deposition rate to operating cost [64]. Second goal needs development leading to improved resistance to thermal ageing and thermal shocks together with more stable TGO, enhanced resistance to erosion [62]. The third goal requires a decrease in the apparent thermal conductivity of the TBC, permitting an increase in the system operating temperature to improve its thermodynamic efficiency. For instance, different from the frequent thermal cycling of aeroengines, a heavy-duty gas turbine continuously operates at high temperature over 8,000 h, and the thermal stress, which mainly originates from the start and stop, is almost released

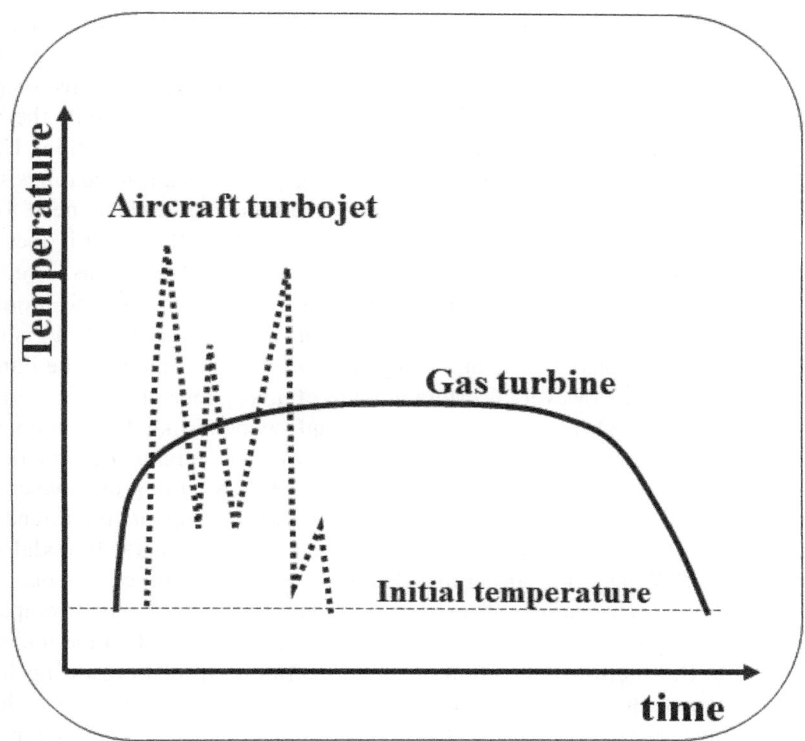

FIGURE 5.50 The typical service temperature versus time condition for different SPS-TBC applications.

by material creep in the long-term service. Moreover, the inlet air of a heavy-duty gas turbine is filtrated and clean. (See *Figure 5.50*.)

The SPS-TBC development for biocompatible applications occurs with the increasing public attention to oral health and how oral medicine has transformed from solely a recovery function to a balance between function and beauty. Although titanium and its alloys are widely used as an implant material, it has shown certain drawbacks, including poor chemical stability, low biocompatibility, and tissue toxicity. Therefore, the choice of dental implant material has shifted from metals to bioceramics [65]. Among the family of bioceramics, zirconia ceramics have shown good potential for dental implants, owing to the high fracture strength, good chemical stability, high corrosion resistance, and excellent toughening effect of this material. The zirconia coatings have presented satisfactory osseointegration and good tissue biocompatibility after a sufficient healing period. The suspension, dispersion, and stability of yttria-zirconia ($3Y-ZrO_2$) and $3Y-ZrO_2$/graphene oxide (GO) ceramic can be used for SPS coating application [66]. The other ceramic system includes hydroxyapatite (HAp), as a bioceramic material that has Ca/P ratio close to that of natural bone, i.e., < 1.67, owing to substitution of Ca ions by metabolic ions, such as Mg, Na, K, and others. The excellent osteoconductive nature of *HAp* makes it an ideal candidate as a coating material in biomedical applications, such as hip and dental implants. The use of plasma spraying for coating is the only thermal spray process which is approved by the FDA (Food and Drug Administration), USA. This combination of plasma spraying and ceramic suspension leads to coatings suitable for medical implants. Ideal *HAp* coatings for orthopedic implants need to be thin ($< 50\,\mu m$), with high adhesion and strong cohesive strength so that it does not delaminate or crack under surface shear forces, and high hardness so as to reduce the wear, and possess enough roughness as well as sufficient porosity to promote the ingrowth of bone tissues. Moreover, high coating crystallinity ($> 45\,\%$) is also needed as the amorphous phase can undergo in-vivo dissolution under the human body fluid conditions. However, a certain amount of amorphous phase is beneficial for osseointegration and for precipitation of secondary, bone-like *HAp* on contact with biofluid. The current trend in both research as well as the industrial practice is therefore to fabricate thin ($< 50\,\mu m$) *HAp* coatings while retaining the abovementioned characteristics. However, the challenge with thin coatings is that it tends to dissolve rapidly [67]. The proper suspension in these systems include bimodal size distribution $\phi : 30\%$, $D_{50} : 100\,\mu m$ in distilled water to create thin, continuous, and adherent $Ca_5(PO_4)_3OH$ coatings. The other common system includes hydroxyapatite (HAp)/tricalcium phosphate (TCP). The structure revolves around (i) the nodular-shaped due to the shallow deposition of molten ceramic droplets resulting from the well-known shadowing effect and its consequent result on producing nodules around the substrate surface imperfections. A lower solid load and lower mean solute particle size can result in lower suspension droplet size and hence lower droplet momentum, resulting in rougher coating surface. And (ii) the column-like or treelike, due to subsequent shallow deposition of low momentum droplets over the asperities because the ensuing passage of the plasma torch leads to the shadow effect. And then (iii) very fine spherical particles: bigger molten ceramic droplets formed due to the coarser initial powder size and coarser suspension droplet size, respectively, can travel through the core of the plasma before impacting the substrate for a significant duration of their in-flight journey. Moreover, due to their higher molten ceramic droplet mass, the droplets may not cool down as fast as the fine molten ceramic droplets do. Fine molten ceramic droplets may resolidify to form resolidified spherical particles in flight before impacting the substrate and arrive cold under resolidified condition at the substrate. Also, (iv) the fine splats creating typical pancake or disc shape, coarse feedstock (i.e., powder) particles with higher solid loading and higher mean solute particle size of suspension that makes atomization (breaking up the suspension stream into droplets) difficult, which form larger fully/partly molten ceramic droplets resulting in larger splats. Coarser ceramic molten droplets can arrive hot under fully/partly molten conditions at the substrate, which upon impact form proper pancake-shaped splats. The large portion of the plasma heat is consumed in solvent (water) evaporation and agglomeration/sintering of fine *HAp* solute particles prior to the phase degradation of a *HAp* solute particle in the suspension droplet:

(a) A particle is protected by the liquid shell of the suspension droplet for a significant amount of time during its in-flight journey.
(b) A significant amount of plasma heat is already consumed in evaporating the solvent.
(c) The spray distance used in ASPS is typically very short (70 mm in this work), reducing the residence time significantly. The amount of heat left for the phase degradation of the *HAp* solute particles in the suspension droplet depends on the solvent type and the atomized suspension droplet size. Since the solvent type was the same (i.e., water), the size of the atomized

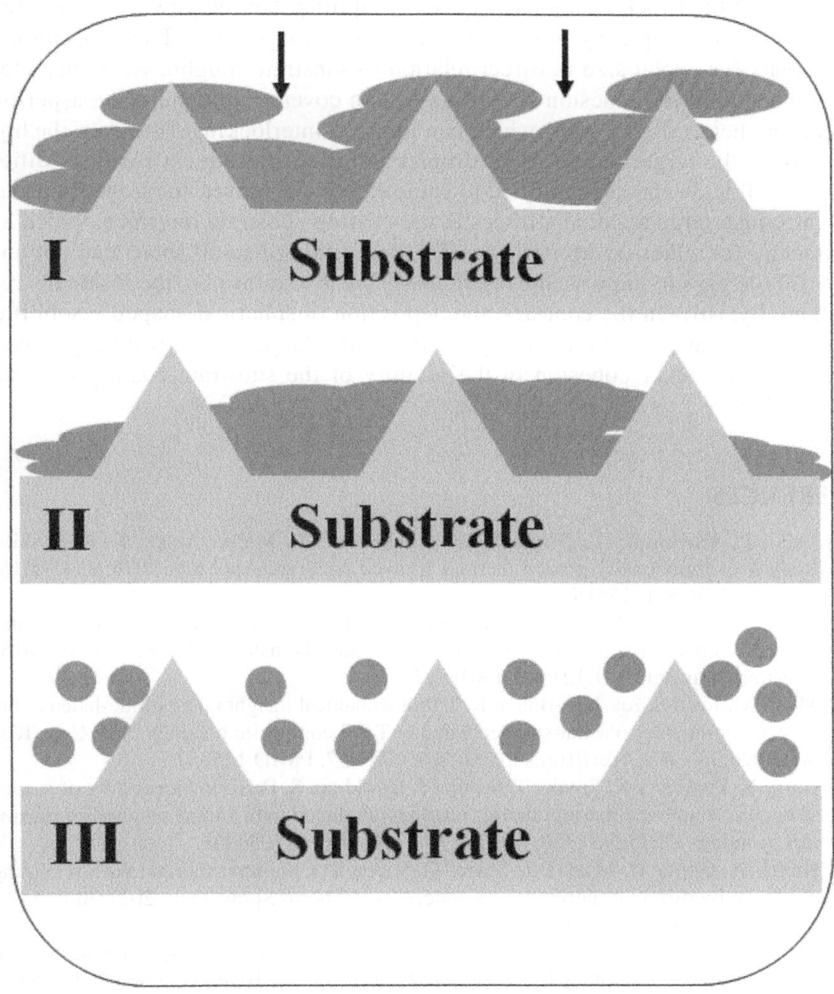

FIGURE 5.51 The SPS-HAp deposition from molten particles of micrometer-sized suspensions on substrates.

suspension droplet can be considered as the major factor deriving the phase degradation in different deposited *HAp* coatings. The larger the suspension droplet size, the lower the amount of heat exchange between the plasma and the HAp powder particles surrounded in a liquid shell. Moreover, larger droplet size gives higher droplet momentum, resulting in lower residence time in the plasma plume. The higher solid load content and higher mean solute particle size increase the suspension droplet size.

The SPS-HAp coatings occur with multilayer deposition. The three possible cases of initial first few layers of splats and substrate asperities interaction, depending on the initial feedstock characteristics: (i) Assuming that the powder particle/solute particle are completely molten during its in-flight journey. The deposition of large splats (i.e., splat size in direct relation to substrate roughness), as depicted, can be beneficial for adhesion strength as it can cover several substrate asperities at once and better fill the gaps, which can improve interlocking. However, the high viscosity of the larger molten HAp droplet and their subsequent rapid solidification can still leave the gaps unfilled to some extent. Moreover, the large splats can also introduce large residual stresses at the coating-substrate interface, which are detrimental for adhesion strength. (ii) The deposition of small splats can not only better fill the gaps to improve the interlocking but also minimize the residual stress substantially. (iii) On the contrary, the deposition of spherical-shaped resolidified particles can leave interasperity gaps unfilled to a large extent, resulting in poor adhesion and weaker cohesion in the vicinity of the substrate-coating interface [68] (see *Figure 5.51*).

REFERENCES

[1] Łatka L, Pawłowski L, Winnicki M, Sokołowski P, Małachowska A, Kozerski S. Review of functionally graded thermal sprayed coatings. Appl Sci. 2020;10(15):5153. doi: 10.3390/app10155153.

[2] Dalir E, Dolatabadi A, Mostaghimi J. Investigating the in-flight droplets' atomization in suspension plasma-sprayed coating. Int J Heat Mass Transfer. 2022;182. doi: 10.1016/j.ijheatmasstransfer.2021.121969, PMID 121969.

[3] Mistri M, Joshi S, Kar KK, Balani K. Tribomechanical insight into carbide-laden hybrid suspension-powder plasma-sprayed tribaloy T400 composite coatings. Surf Coat Technol. 2020;396. doi: 10.1016/j.surfcoat.2020.125957, PMID 125957.

[4] Curry N, VanEvery K, Snyder T, Susnjar J, Bjorklund S. Performance testing of suspension plasma sprayed thermal barrier coatings produced with varied suspension parameters. Coatings. 2015;5(3):338–56. doi: 10.3390/coatings5030338.

[5] Kumar N, Gupta M, Mack DE, Mauer G, Vaßen R. Columnar thermal barrier coatings produced by different thermal spray processes. J Therm Spray Tech. 2021;30(6):1437–52. doi: 10.1007/s11666-021-01228-5.

[6] Fan W, Bai Y. Review of suspension and solution precursor plasma sprayed thermal barrier coatings. Ceram Int. 2016;42(13):14299–312. doi: 10.1016/j.ceramint.2016.06.063.

[7] Lewis JA. Colloidal processing of ceramics. J Am Ceram Soc. 2000;83(10):2341–59. doi: 10.1111/j.1151-2916.2000.tb01560.x.

[8] Yaghtin M, Yaghtin A, Tang Z, Troczynski T. Improving the rheological and stability characteristics of highly concentrated aqueous yttria stabilized zirconia slurries. Ceram Int. 2020;46(17):26991–9. doi: 10.1016/j.ceramint.2020.07.176.

[9] Khatibnezhad H, Ambriz-Vargas F, Ettouil FB, Moreau C. An investigation on the photocatalytic activity of sub-stoichiometric TiO2-x coatings produced by suspension plasma spray. J Eur Ceram Soc. 2021;41(1):544–56. doi: 10.1016/j.jeurceramsoc.2020.08.017.
[10] Carnicer V, Orts MJ, Moreno R, Sánchez E. Engineering zirconia coating microstructures by using saccharides in aqueous suspension plasma spraying feedstocks. Ceram Int. 2020;46(15):23749–59. doi: 10.1016/j.ceramint.2020.06.149.
[11] Fauchais P, Etchart-Salas R, Rat V, Coudert JF, Caron N, Wittmann-Ténèze K. Parameters controlling liquid plasma spraying: solutions, sols, or suspensions. J Therm Spray Tech. 2008;17(1):31–59. doi: 10.1007/s11666-007-9152-2.
[12] Krishnasamy J, Ponnusami SA, Turteltaub S, van der Zwaag S. Modelling the fracture behaviour of thermal barrier coatings containing healing particles. Mater Des. 2018;157:75–86. doi: 10.1016/j.matdes.2018.07.026.
[13] Carpio P, Salvador MD, Borrell A, Sánchez E, Moreno R. Alumina–zirconia coatings obtained by suspension plasma spraying from highly concentrated aqueous suspensions. Surf Coat Technol. 2016;307:713–19. doi: 10.1016/j.surfcoat.2016.09.060.
[14] Mewis J, Wagner NJ. Thixotropy. Adv Colloid Interface Sci. 2009;147–148:214–27. doi: 10.1016/j.cis.2008.09.005, PMID 19012872.
[15] Zhang C, Jiang Z, Zhao L, Guo W, Gao X. Stability, rheological behaviors, and curing properties of 3Y—ZrO2 and 3Y—ZrO2/GO ceramic suspensions in stereolithography applied for dental implants. Ceram Int. 2021;47(10A):13344–50.
[16] Carnicer V, Orts MJ, Moreno R, Sánchez E. Influence of solids concentration on the microstructure of suspension plasma sprayed Y-TZP/Al2O3/SiC composite coatings. Surf Coat Technol. 2019;371:143–50. doi: 10.1016/j.surfcoat.2019.01.078.
[17] Jeon H, Lee I, Oh Y. Changes in thermal conductivity and thermal shock resistance of YSZ-SiO2 suspension plasma spray coatings according to various raw material particle sizes. Ceram Int. 2021;47(3):3671–9. doi: 10.1016/j.ceramint.2020.09.219.
[18] Xie S, Song C, Liu S, He P, Lapostolle F, Klein D, Deng C, Liu M, Liao H. Dense nanostructured YSZ coating prepared by low-pressure suspension plasma spraying: atmosphere control and deposition mechanism. Surf Coat Technol. 2021;416. doi: 10.1016/j.surfcoat.2021.127175, PMID 127175.
[19] Zhao Y, Wang Y, Peyraut F, Liao H, Montavon G, Planche M-P, Ilavsky J, Lasalle A, Allimant A. Evaluation of Nano/submicro pores in suspension plasma sprayed YSZ coatings. Surf Coat Technol. 2019;378. doi: 10.1016/j.surfcoat.2019.125001, PMID 125001.
[20] Ganvir A, Gupta M, Kumar N, Markocsan N. Effect of suspension characteristics on the performance of thermal barrier coatings deposited by suspension plasma spray. Ceram Int. 2021;47(1):272–83. doi: 10.1016/j.ceramint.2020.08.131.
[21] Ganvir A, Calinas RF, Markocsan N, Curry N, Joshi S. Experimental visualization of microstructure evolution during suspension plasma spraying of thermal barrier coatings. J Eur Ceram Soc. 2019;39(2–3):470–81. doi: 10.1016/j.jeurceramsoc.2018.09.023.
[22] Bernard B, Quet A, Bianchi L, Schick V, Joulia A, Malié A, Rémy B. Effect of suspension plasma-sprayed YSZ columnar microstructure and bond coat surface preparation on thermal barrier coating properties. J Therm Spray Tech. 2017;26(6):1025–37. doi: 10.1007/s11666-017-0584-z.
[23] Shahien M, Suzuki M, Tsutai Y. Controlling the coating microstructure on axial suspension plasma spray process. Surf Coat Technol. 2018;356:96–107. doi: 10.1016/j.surfcoat.2018.09.055.
[24] Bernard B, Bianchi L, Malié A, Joulia A, Rémy B. Columnar suspension plasma sprayed coating microstructural control for thermal barrier coating application. J Eur Ceram Soc. 2016;36(4):1081–9. doi: 10.1016/j.jeurceramsoc.2015.11.018.
[25] Łatka L. Thermal barrier coatings manufactured by suspension plasma spraying—a review. Adv Mater Sci. 2018;18(3):95–117. doi: 10.1515/adms-2017-0044.
[26] Tarasi F, Alebrahim E, Dolatabadi A, Moreau C. A comparative study of YSZ suspensions and coatings. Coatings. 2019;9(3):188. doi: 10.3390/coatings9030188.

[27] Xiao B, Huang X, Robertson T, Tang Z, Kearsey R. Sintering resistance of suspension plasma sprayed 7YSZ TBC under isothermal and cyclic oxidation. J Eur Ceram Soc. 2020;40(5):2030–41. doi: 10.1016/j.jeurceramsoc.2019.12.046.

[28] Tesar T, Musalek R, Medricky J, Cizek J. On growth of suspension plasma-sprayed coatings deposited by high-enthalpy plasma torch. Surf Coat Technol. 2019;371:333–43. doi: 10.1016/j.surfcoat.2019.01.084.

[29] VanEvery K, Krane MJM, Trice RW, Wang H, Porter W, Besser M, Sordelet D, Ilavsky J, Almer J. Column formation in suspension plasma-sprayed coatings and resultant thermal properties. J Therm Spray Tech. 2011;20(4):817–28. doi: 10.1007/s11666-011-9632-2.

[30] Aranke O, Gupta M, Markocsan N, Li X-H, Kjellman B. Microstructural evolution and sintering of suspension plasma-sprayed columnar thermal barrier coatings. J Therm Spray Tech. 2019;28(1–2):198–211. doi: 10.1007/s11666-018-0778-z.

[31] Zhao Y, Wen J, Peyraut F, Planche M-P, Misra S, Lenoir B, Ilavsky J, Liao H, Montavon G. Porous architecture and thermal properties of thermal barrier coatings deposited by suspension plasma spray. Surf Coat Technol. 2020;386. doi: 10.1016/j.surfcoat.2020.125462, PMID 125462.

[32] Kabir MR, Sirigiri AK, Naraparaju R, Schulz U. Flow kinetics of molten silicates through thermal barrier coating: a numerical study. Coatings. 2019;9(5):332. doi: 10.3390/coatings9050332.

[33] Vaßen R, Kaßner H, Mauer G, Stöver D. Suspension plasma spraying: process characteristics and applications. J Therm Spray Tech. 2010;19(1–2):219–25. doi: 10.1007/s11666-009-9451-x.

[34] Gupta M, Li X-H, Markocsan N, Kjellman B. Design of high lifetime suspension plasma sprayed thermal barrier coatings. J Eur Ceram Soc. 2020;40(3):768–79. doi: 10.1016/j.jeurceramsoc.2019.10.061.

[35] Dalir E, Dolatabadi A, Mostaghimi J. Modeling the effect of droplet shape and solid concentration on the suspension plasma spraying. Int J Heat Mass Transfer. 2020;161. doi: 10.1016/j.ijheatmasstransfer.2020.120317, PMID 120317.

[36] Gupta M, Markocsan N, Li X-H, Kjellman B. Development of bondcoats for high lifetime suspension plasma sprayed thermal barrier coatings. Surf Coat Technol. 2019;371:366–77. doi: 10.1016/j.surfcoat.2018.11.013.

[37] Karaoglanli AC, Doleker KM, Ozgurluk Y. State of the art thermal barrier coating (TBC) materials and TBC failure mechanisms. In: Öchsner, A., Altenbach, H. (eds). Properties and Characterization of Modern Materials. Adv Structured Mater. Singapore, Springer; 2017;33. https://doi.org/10.1007/978-981-10-1602-8_34.

[38] Gupta M, Musalek R, Tesar T. Microstructure and failure analysis of suspension plasma sprayed thermal barrier coatings. Surf Coat Technol. 2020;382. doi: 10.1016/j.surfcoat.2019.125218, PMID 125218.

[39] Hu Z-C, Liu B, Wang L, Cui Y-H, Wang Y-W, Ma Y-D, Sun W, Yang Y. Research progress of failure mechanism of thermal barrier coatings at high temperature via finite element method. Coatings. 2020;10(8):732. doi: 10.3390/coatings10080732.

[40] Zhou D, Malzbender J, Sohn YJ, Guillon O, Vaßen R. Sintering behavior of columnar thermal barrier coatings deposited by axial suspension plasma spraying (SPS). J Eur Ceram Soc. 2019;39(2–3):482–90. doi: 10.1016/j.jeurceramsoc.2018.09.020.

[41] Gu JJ, Joshi SS, Ho Y-S, Wei BW, Huang TY, Lee J, Berman D, Dahotre NB, Aouadi SM. Oxidation-induced healing in laser-processed thermal barrier coatings. Thin Solid Films. 2019;688. doi: 10.1016/j.tsf.2019.137481, PMID 137481.

[42] Huang J, Wang W, Li Y, Fang H, Ye D, Zhang X, Tu S. Improve durability of plasma-splayed thermal barrier coatings by decreasing sintering-induced stiffening in ceramic coatings. J Eur Ceram Soc. 2020;40(4):1433–42. doi: 10.1016/j.jeurceramsoc.2019.11.074.

[43] Sampath S, Schulz U, Jarligo MO, Kuroda S. Processing science of advanced thermal-barrier systems. MRS Bull. 2012;37(10):903–10. doi: 10.1557/mrs.2012.233.

[44] Loghman-Estarki MR, Edris H, Razavi RS, Jamali H, Ghasemi R, Pourbafrany M, Erfanmanesh M, Ramezani M. Spray drying of nanometric SYSZ powders to obtain plasma sprayable nanostructured granules. Ceram Int. 2013;39(8):9447–57. doi: 10.1016/j.ceramint.2013.05.062.
[45] Yamazaki Y, Matsuura S, Hamaguchi T, Nagai M, Habu Y. Improved thermal fatigue resistance in thermal barrier coatings via suspension plasma spray technique. Mater Lett. 2020;280. doi: 10.1016/j.matlet.2020.128608, PMID 128608.
[46] Zhou C, Wang N, Wang Z, Gong S, Xu H. Thermal cycling life and thermal diffusivity of a plasma-sprayed nanostructured thermal barrier coating. Scr Mater. 2004;51(10):945–8. doi: 10.1016/j.scriptamat.2004.07.024.
[47] Ganvir A, Joshi S, Markocsan N, Vassen R. Tailoring columnar microstructure of axial suspension plasma sprayed TBCs for superior thermal shock performance. Mater Des. 2018;144:192–208. doi: 10.1016/j.matdes.2018.02.011.
[48] Evans HE. Oxidation failure of TBC systems: an assessment of mechanisms. Surf Coat Technol. 2011;206(7):1512–21. doi: 10.1016/j.surfcoat.2011.05.053.
[49] Li C-J, Dong H, Ding H, Yang G-J, Li C-X. The correlation of the TBC lifetimes in burner cycling test with thermal gradient and furnace isothermal cycling test by TGO effects. J Therm Spray Tech. 2017;26(3):378–87. doi: 10.1007/s11666-017-0530-0.
[50] Góral M, Kubaszek T, Pytel M. Isothermal oxidation of thermal barrier coatings deposited using LPPS, CVD, and PS-PVD methods on MAR M247 nickel superalloy. Adv Manuf Sci Technol. 2020;44(1):9–14. doi: 10.2478/amst-2019-0001.
[51] Xie F, Li D, Zhang W. Long-term failure mechanisms of thermal barrier coatings in heavy-duty gas turbines. Coatings. 2020;10(11):1022. doi: 10.3390/coatings10111022.
[52] Ganvir A, Vaidhyanathan V, Markocsan N, Gupta M, Pala Z, Lukac F. Failure analysis of thermally cycled columnar thermal barrier coatings produced by high-velocity-air fuel and axial-suspension-plasma spraying: a design perspective. Ceram Int. 2018; 44(3):3161–72. doi: 10.1016/j.ceramint.2017.11.084.
[53] Giamarchi T. Umklapp process and resistivity in one-dimensional fermion systems. Phys Rev B Condens Matter. 1991;44(7):2905–13. doi: 10.1103/physrevb.44.2905, PMID 9999878.
[54] Bäker M, Rösler J, Heinze G. A parametric study of the stress state of thermal barrier coatings part II: cooling stresses. Acta Mater. 2005;53(2):469–76. doi: 10.1016/j.actamat.2004.10.004.
[55] Zhou D, Mack DE, Gerald P, Guillon O, Vaßen R. Architecture designs for extending thermal cycling lifetime of suspension plasma sprayed thermal barrier coatings. Ceram Int. 2019;45(15):18471–9. doi: 10.1016/j.ceramint.2019.06.065.
[56] Mohammadkhani S, Jalilvand V, Davis B, Ettouil FB, Dolatabadi A, Roué L, Moreau C, Guay D. Suspension plasma spray deposition of CoxNi1-xO coatings. Surf Coat Technol. 2020;399. doi: 10.1016/j.surfcoat.2020.126168, PMID 126168.
[57] Mahade S, Jahagirdar A, Li X-H, Kjellman B, Björklund S, Markocsan N. Tailoring microstructure of double-layered thermal barrier coatings deposited by suspension plasma spray for enhanced durability. Surf Coat Technol. 2021;425. doi: 10.1016/j.surfcoat.2021.127704, PMID 127704.
[58] Mahade S, Zhou D, Curry N, Markocsan N, Nylén P, Vaßen R. Tailored microstructures of gadolinium zirconate/YSZ multi-layered thermal barrier coatings produced by suspension plasma spray: durability and erosion testing. J Mater Process Technol. 2019;264:283–94. doi: 10.1016/j.jmatprotec.2018.09.016.
[59] Mahade S, Curry N, Jonnalagadda KP, Peng RL, Markocsan N, Nylén P. Influence of YSZ layer thickness on the durability of gadolinium zirconate/YSZ double-layered thermal barrier coatings produced by suspension plasma spray. Surf Coat Technol. 2019;357:456–65. doi: 10.1016/j.surfcoat.2018.10.046.
[60] Shen Z, He L, Xu Z, Mu R, Huang G. Rare earth oxides stabilized La2Zr2O7 TBCs: EB-PVD, thermal conductivity and thermal cycling life. Surf Coat Technol. 2019;357: 427–32. doi: 10.1016/j.surfcoat.2018.10.045.

[61] Yaghtin M, Yaghtin A, Najafisayar P, Tang Z, Troczynski T. Deposition of columnar-morphology lanthanum zirconate thermal barrier coatings by solution precursor plasma spraying. J Therm Spray Tech. 2021;30(7):1850–61. doi: 10.1007/s11666-021-01258-z.
[62] Algenaid W, Ganvir A, Calinas RF, Varghese J, Rajulapati KV, Joshi S. Influence of microstructure on the erosion behaviour of suspension plasma sprayed thermal barrier coatings. Surf Coat Technol. 2019;375:86–99. doi: 10.1016/j.surfcoat.2019.06.075.
[63] Kumar N, Mahade S, Ganvir A, Joshi S. Understanding the influence of microstructure on hot corrosion and erosion behavior of suspension plasma sprayed thermal barrier coatings. Surf Coat Technol. 2021;419. doi: 10.1016/j.surfcoat.2021.127306, PMID 127306.
[64] Xie S, Song C, Yu Z, Liu S, Lapostolle F, Klein D, Deng C, Liu M, Liao H. Effect of environmental pressure on the microstructure of YSZ thermal barrier coating via suspension plasma spraying. J Eur Ceram Soc. 2021;41(1):535–43. doi: 10.1016/j.jeurceramsoc.2020.08.022.
[65] Zhang W, Gu J, Zhang C, Xie Y, Zheng X. Preparation of titania coating by induction suspension plasma spraying for biomedical application. Surf Coat Technol. 2019;358:511–20. doi: 10.1016/j.surfcoat.2018.11.047.
[66] Mahade S, Mulone A, Björklund S, Klement U, Joshi S. Incorporation of graphene Nano platelets in suspension plasma sprayed alumina coatings for improved tribological properties. Appl Surf Sci. 2021;570. doi: 10.1016/j.apsusc.2021.151227, PMID 151227.
[67] Ganvir A, Nagar S, Markocsan N, Balani K. Deposition of hydroxyapatite coatings by axial plasma spraying: influence of feedstock characteristics on coating microstructure, phase content and mechanical properties. J Eur Ceram Soc. 2021;41(8):4637–49. doi: 10.1016/j.jeurceramsoc.2021.02.050.
[68] Abir MMM, Otsuka Y, Ohnuma K, Miyashita Y. Effects of composition of hydroxyapatite/gray titania coating fabricated by suspension plasma spraying on mechanical and antibacterial properties. J Mech Behav Biomed Mater. 2022;125:104888. doi: 10.1016/j.jmbbm.2021.104888.

Index

T

V

Y

Z